Endorsements

"Henrik Schoenefeldt has made fascinating discoveries of many hitherto unknown features of this complex system. Dr Reid produced one of the earliest forms of 'stack ventilation' which has until now been an underappreciated aspect of this remarkable building.

We learn how the architecture of the palace was influenced by the ventilation system and how the architect and the 'Ventilator' as Reid was known, came to disagree to the point at which the system itself had to be divided between architect and engineer.

This is a vital study of the inner workings of a building which is being analysed overall in preparation for a major programme of refitting, repair and conservation. It will be interesting to see which – if any – of Reid's original principles, either of a mechanical or natural type, explained so clearly here by Dr Schoenefeldt, could be adopted in the forthcoming programme of works."

Adam Watrobski, Principal Architect of the Houses of Parliament

"Having been shown by Henrik Schoenefeldt at first hand several of the original flues and passageways and how they were used, I have eagerly awaited this book. It succeeds on all levels. He has investigated dozens of ventilation ducts, how they work, how they were operated and who operated them. One of the intriguing discoveries was the surprisingly modern concept of seeking 'feedback' from MPs and Peers post-occupancy to inform how the system was functioning.

The ventilation system has always been hidden because the intention was to conceal every vestige of the thousands of flues, the steam engines, fans and even the air inlets and outlets by using the glorious decorative details covering every surface. By employing the latest BIM modelling, the original features of the system, even those parts since lost – including those for the original Commons Chamber destroyed during World War II – may now be fully appreciated."

Dr Mark Collins, Estates Archivist at the Houses of Parliament

Henrik Schoenefeldt's research provides a unique insight into the development, construction, and maintenance of the Palace of Westminster. The personal, political, and engineering challenges that emerged from the early nineteenth century until today are expertly detailed and expounded upon. This research will be a vital asset as MPs and Peers evaluate how best to preserve and restore the Palace for generations yet to come."

Rt Hon Sir Edward Leigh, MP for Gainsborough

"A fascinating account of how David Boswell Reid influenced the design of the Palace of Westminster. Previous studies have tended to dismiss Reid as a crank whose grand ventilation scheme did not work, but Professor Schoenefeldt dispels that view through archival research and study of the hidden structure of the Palace: many of the "Gothic" towers and turrets of the Palace we see today were built to disguise Reid's ventilation stacks; his pioneering experiments with the temporary parliamentary chambers were highly influential on later Victorian buildings; and elements of his system remained in use for over 90 years.

Schoenefeldt also sees implications for the long-overdue restoration of the Palace. Reid understood that ventilation is the key to health and well-being in a heavily used public building and tried to maximise natural as opposed to mechanical ventilation. Two hundred years later, Reid's ideas point the way towards sustainability and low-energy design."

Dr Richard Ware, Adviser on Parliamentary Relations, Restoration and Renewal Sponsor Body

Ware was former director of the Palace of Westminster Restoration and Renewal Programme from 2013 to 2016.

"The Palace of Westminster, built in the middle decades of the 19th century, was an engineering *tour de force*, designed and completed at the height of Britain's industrial revolution. The work of Charles Barry and A.W.N. Pugin, and of their craftsmen collaborators, has formed a cornerstone in our understanding of design history since the time of its completion. This focus on its aesthetic qualities has sometimes been to the detriment of appreciation of the Palace as a pioneering modern, fully serviced building and an outstanding technical achievement of its time.

As our understanding of cultural heritage has expanded beyond architectural history in recent decades to include wider environmental, social and economic values, so the role of historic engineering in the context of built heritage has been re-evaluated and given new prominence. Henrik Schoenefeldt's outstanding study illuminates the history, function and performance of the heating, cooling and ventilation of the Palace of Westminster. In this book the work of the enterprising physician, Dr Boswell Reid, who designed the Palace's ventilation systems, which were largely realised in the completed building, emerges vividly as the pioneer of modern engineered building services. Schoenefeldt's work provides a new narrative on the architectural heritage of Britain's most famous building, and is particularly timely in the age of climate emergency, when the need for zero carbon solutions calls once again for new approaches to the design of conditioned environments."

Patrick Duerden, Conservation Architect and Practice Director, Donald Insall Associates

Rebuilding the Houses of Parliament

Rebuilding the Houses of Parliament explores the history of the UK Houses of Parliament in Westminster from an environmental design perspective, and the role David Boswell Reid played in the development of the original ventilation and climate control system in Parliament.

This book retraces and critically examines the evolution of the environmental principles underlying the design of the Houses of Parliament, engaging with fundamental questions about air quality, energy efficiency and thermal comfort. This yields insights into the historic methods of environmental design that were characterised by physical experimentation and post-occupancy evaluation. *Rebuilding the Houses of Parliament* examines the history of the buildings' operation, studying the practical reality of its performance in use and offers the opportunity to reflect on current challenges faced by architects and engineers adapting to the realities of climate change.

This book is an ideal read for academics, politicians and practitioners with an interest in architectural history and heritage, theory, engineering and conservation.

Henrik Schoenefeldt is Professor of Sustainability in Architectural Heritage at the University of Kent in England, National Teaching Fellow and AHRC Leadership Fellow. He trained as an architect and specialised in environmental design with an MPhil and PhD from the University of Cambridge. His main research interest is in historic principles of environmental design, both as a field of scholar research and architectural practice.

Routledge Research in Architecture

The *Routledge Research in Architecture* series provides the reader with the latest scholarship in the field of architecture. The series publishes research from across the globe and covers areas as diverse as architectural history and theory, technology, digital architecture, structures, materials, details, design, monographs of architects, interior design and much more. By making these studies available to the worldwide academic community, the series aims to promote quality architectural research.

The Complexities of John Hejduk's Work
Exorcising Outlines, Apparitions and Angels
J. Kevin Story

Writing Architecture in Modern Italy
Narratives, Historiography and Myths
Daria Ricchi

The Resistant Object of Architecture
A Lacanian Perspective
Petra Čeferin

Rebuilding the Houses of Parliament
David Boswell Reid and Disruptive Environmentalism
Henrik Schoenefeldt

Le Corbusier in the Antipodes
Art, Architecture and Urbanism
Antony Moulis

Kenosis Creativity Architecture
Appearance through Emptying
Randall S. Lindstrom

For more information about this series, please visit: https://www.routledge.com/Routledge-Research-in-Architecture/book-series/RRARCH

Rebuilding the Houses of Parliament

David Boswell Reid and Disruptive Environmentalism

Henrik Schoenefeldt

Routledge
Taylor & Francis Group
LONDON AND NEW YORK

First published 2021
by Routledge
2 Park Square, Milton Park, Abingdon, Oxon OX14 4RN

and by Routledge
52 Vanderbilt Avenue, New York, NY 10017

Routledge is an imprint of the Taylor & Francis Group, an Informa business

British Library Cataloguing-in-Publication Data
A catalogue record for this book is available from the British Library

Library of Congress Cataloging-in-Publication Data
A catalog record has been requested for this book

ISBN: 9781138741522 (hbk)
ISBN: 9781315182803 (ebk)

Typeset in Sabon
by Deanta Global Publishing Services, Chennai, India

This book is dedicated to my wife, whose love, care and companionship has sustained me, and Professor Alan Short, University of Cambridge, who has strongly influenced my work.

Contents

Figures

Acknowledgements

The research for this book would have been impossible without the support of friends, family colleagues and organisations. First I would like to express my gratitude for the continued support, financial and in kind, that I have received from Parliament and the University of Kent, England. A generous six-month sabbatical, granted by the university, enabled me to write this book, consolidating the findings of research that I have undertaken over the past eight years. The research for this book was supported by the award of a research fellowship by the Arts and Humanities Research Council as well as through grants from the House of Lords and House of Commons.

For the last four years I have been seconded to Parliament to lead a large research project within the Palace of Westminster Restoration and Renewal Programme (R&R).[1] This secondment was important as it enabled me to undertake original research requiring access to the building and security clearance, but also to allow the new research to directly inform the work of the programme. I am particularly grateful to Andrew Piper, design director of R&R, and two former directors of the R&R programme, Richard Ware and Tom Healy, for their continued support during the project. In the execution of the project I was assisted by Andrei Iulian, Elizabeth Venning and Regina Avancini. I also received invaluable support from departments, in particular the Sergeant-at-Arms, Black Rod and the Speaker of the House of Commons and the Parliamentary Maintenance and Operations.

The work was based on extensive archival research undertaken in different repositories and library access in England. This included the British Library, Cambridge University Library, the University of Kent's Templeman Library, the National Archives (Kew) and the architectural archives of Historic England in Swindon. I am especially grateful for the help I received from Mari Takyanagi and her colleagues at Parliamentary Archives.

Throughout this period I had the support of academic friends and colleagues. Dean Hawkes, whom I have known since my student years at the University of Cambridge, for writing the foreword for this book and also many inspiring conversations about the history of environmental design. At the University of Kent it was Gordana Fontana-Giusti and Marialena Nikolopoulou, Gerry Adler, Michael Richards and Fran Beaton who supported my development as a researcher and educator. The research also received inspiration from many conversations with Richard Lorch, Chief editor of Building & Cities, Sebastian Macmillan and members of the CIBSE Heritage Group, in particular Neil Sturrock, with whom I share a mutual interest in the work of David Boswell Reid, and Andrew More, current chair of the heritage group.

The research also underpinned the development of the new Conservation Management Plan and Policies for the Palace of Westminster. The research allowed the significance of the technological and environmental heritage to be fully recognised and also to develop new strategies for their preservation or potential reuse. I collaborated with Patrick Duerden and Hannah Parham of the conservation architects Donald Insall Associates and with Mark Collins, Tessa Blundy, Adam Watrobski from the architectural conservation department at the Palace of Westminster.

Moreover, I collaborated with the surveyors Plowman Craven, BDP and R&R to create a virtual model of the historic system using Building Information Modelling (BIM). I worked with Thomas Leroy, former lead surveyor at Plowman Craven, in a series of pilot surveys between 2016 and 2017, and between April 2019 and February 2020 I also collaborated on the BIM project with a team of surveyors and specialist BIM modellers at Plowman Craven. This team was composed of project managers Jim Gregory and Kate Anstey, production manager Charles Luckhurst, and seven modellers: Agnieszka Ciesla (Lead Modeller), Alexandru Macovei, Olga Lesiak, Francesco Izzo, Suraj Athwal, Gus Macedo and Ray Richfield.

Note

1 The R&R programme has now been superseded by two new bodies: a delivery authority, which is an arms-length organisation charged with managing the delivery of the restoration, and the Sponsor Board, which acts as the client body.

Foreword by Dean Hawkes

> 'architectural history as it has been written up till the present time has seen no reason to apologise or explain away a division that makes no sense in terms of the way buildings are used and paid for by the human race, a division into structure, which is held to be valuable and discussable, and mechanical servicing, which has been almost entirely excluded from historical discussions to date.'
>
> Reyner Banham, *The Architecture of the Well-tempered Environment, 1969*

With these words, written in 1969, Reyner Banham brought the environmental function of buildings into the realm of architectural history. In *The Architecture of the Well-tempered Environment*[1] Banham laid the ground for the establishment of a significant new branch of architectural history that has, in the last half century, become firmly established. The key to this history lies in events at the beginning of the 19th century, when the new sciences and technologies of the Industrial Revolution were brought to bear on conditioning the environments within buildings. For the first time, it became possible to conceive of buildings that were to some degree liberated from the constraints of the external climate. They could be made warm in winter, ventilated when the air was still and light after dark.

As cities grew in size throughout the industrialised world, buildings of many types – theatres, law courts, banks, museums, libraries, town halls, schools, prisons and hospitals – often designed by the leading architects of the time, embraced the new environmental technologies fundamentally to transform the very nature of architecture. Banham's pioneering account offered just a sketch of all of this that has now been greatly fleshed out by the work of a growing band of scholars from around the world.

The most significant building project in Britain in the 19th century was the rebuilding of the Palace of Westminster, following the destruction of the earlier building by fire in October 1934. In view of the national importance of this building, indeed its international significance, it is curious that it received just two minor references in *The Architecture of the Well-tempered Environment*, neither regarding the design itself. This omission is comprehensively addressed in Henrik Schoenefeldt's study that is the subject of the present book. The outcome of eight years of research, with access to the most comprehensive archival sources and, crucially, to the innermost recesses of the building itself, Schoenefeldt has pieced together an account of the process by which this remarkable building was conceived, developed and inhabited in the years following its completion in 1852.

There are two parallel strands to the narrative. First is the meticulous reconstruction of the design of the environmental systems of the building, as these evolved in a complex interaction with the overall architectural conception. Second is the account of the relationship between the architect Sir Charles Barry and Dr David Boswell Reid, who was appointed as a consultant to advise on the design of a ventilation system for the building. These themes are equal in complexity and the book succeeds in bringing clarity to both. The personal drama of Barry and Reid might almost be read as a Victorian novel – the battle of Jarndyce v. Jarndyce in Dickens' *Bleak House* is brought to mind – but this is in fact a fascinating and objective study of how the profession of architecture came to collaborate with the new force of environmental consultant – or 'ventilator' as we learn Reid was sometimes referred to. The building itself is, however, the central character in this wonderful story and this is where Schoenefeldt's deep scholarship brings the greatest rewards. By providing key archival drawings from all stages of the design's long development and presenting his own beautifully drawn reconstructions of major spaces and systems, he illustrates and illuminates the building as never before.

This project has become a central contributor to the ongoing programme of restoration of the building, validating historical research as an element of contemporary practice. But the overriding value of this book is in bringing into view the hidden wonders of the environmental systems of this great building and, thereby, completing a significant gap in environmental history in architecture.

Dean Hawkes
Cambridge.

Note

1 Reyner Banham, *The Architecture of the Well-tempered environment*, London, 1969, The Architectural Press.

Preface

> *The mechanical and electrical engineering systems that keep the place lit, heated, cooled, drained and dry are already well past their use-by date. The risk of a catastrophic failure such as a fire or a flood rises exponentially every five years that we delay. We should be in absolutely no doubt: there will be a fire. There was a fire a fortnight ago and there are regularly fires. People patrol the building 24 hours a day to ensure that we catch those fires.*
>
> Chris Bryant, MP for Rhondda[1]

PREFACE

This book is the outcome of academic research that I have undertaken at the University of Kent over the past eight years. My research began with more traditional archival research, but evolved into a much larger project that allowed the research to directly inform the inquiry into the forthcoming restoration of the Palace of Westminster. Over the last four years I have been on secondment to the *Palace of Westminster Restoration and Renewal Programme* leading a large research project called 'Between Heritage and Sustainability – Restoring the Palace of Westminster's Nineteenth-century Ventilation System'. The first two years of this secondment were funded through an Arts and Humanities Research Council (AHRC) research fellowship, whilst the last two years were funded through grants from the House of Commons and the House of Lords.

In this book, I examine the architectural history of the Palace of Westminster from an environmental perspective. It is the first study to systematically investigate the design, history and performance of the original 19th-century ventilation and climate control system, combining archival research with on-site investigations, technical analysis and virtual modelling. I focus on the contribution of David Boswell Reid (1805–63) – a Scottish physician and chemist – who with the architect Charles Barry (1795–1860) developed the original ventilation and climate control system in the Palace of Westminster. To begin, I retrace the evolution of this original system from 1835 to 1852 and then consider its wider operational history, which covered a period of 100 years. I show that its design was not static but constantly evolving as a result of trial and error. The system was subject to numerous performance evaluations and inquiries into alterations to enhance and adapt the original system throughout its lifetime.

Until now, the post-occupancy history of historical buildings has received scant attention from architectural historians, even amongst those who have focused on the environmental design history of architecture. For the last 30 years the architectural profession has come under increasing pressure to study the performance of their buildings in use, but even today only a small proportion of architectural firms regularly undertake building performance evaluations. This critique equally applies to the teaching of and research into architectural history, theory and criticism. A substantive cultural change in architectural education and practice will be necessary if they are to meet our climate change obligations. I believe that the tendency of more traditional architectural history to present architectural works as pristine and static objects is problematic as it has prevented us from looking at the practical reality of historical structures, as revealed, for example, by the experience of those who used or operated them. I argue in this book that the post-occupancy history of the Houses of Parliament illuminates not only how the environmental systems in the mid-19th century were designed but also how they were evaluated and improved through a process of trial and error. As will be seen, this history shows that these systems relied on a wide network of collaboration, in which a wide-variety of stakeholders – designers, clients, users, facilities management staff and specialist consultants – were key actors.

In addition to contributing to the field of academic scholarship, I believe that this research project has illuminated the practical value that research about the history of environmental design can provide to the field of architectural conservation and building services engineering. One of the objectives of the forthcoming restoration is to explore whether the disused 19th-century infrastructure can be revitalised and whether it could provide the basis for a more sustainable and energy-efficient approach to ventilation and climate control. I believe that this research is a first important step towards addressing these issues. These issues are motivated by a conservation-led approach to the design of building services. By harnessing the practical value of our environmental heritage, we can prevent it from being preserved merely as redundant archaeological remains. The two perspectives of architectural conservation and sustainable design are often perceived as incompatible with one another. I believe that the study of environmental principles that have been used in the past can bridge the gap between the two. Finally, I hope to demonstrate that research in the arts and humanities can help us to address the urgent questions of how we can live sustainability and abate climate change.

Professor Henrik Schoenefeldt
Canterbury, 2020

Note

1 Palace of Westminster: Restoration and Renewal, *Hansard's Parliamentary Debates: The Official Report* (25 January 2017, Vol 620, cols. 97WH–98WH).

Introduction

> *A striking example is the deliberate choice of the Gothic style for the nineteenth century rebuilding of the British Parliament, and equally deliberate decision after World War II to rebuild the parliamentary chamber on exactly the same basic plan as before. The historic past into which the new tradition is inserted need not be lengthy, stretching back into the assumed mists of time.*[1]
>
> Eric Hobsbawm, *Inventing Traditions*, 2012.

Disruptive environmentalism

In 1834, the majority of the original medieval buildings that comprised the Houses of Parliament, also known as 'the Palace of Westminster', were destroyed in a fire. Parliament decided that the Houses should be rebuilt in Gothic style and, subsequently, in 1835 held an architectural design competition for it. The architects Charles Barry and his partner Augustus Welby Pugin won the competition. In the quotation above that starts this book, the social historian Eric Hobsbawm is primarily writing about the motivation of Parliament to specify the Gothic style as appropriate for the new Palace and is describing the design of the Houses as an illusion of historical continuity. Any sense of continuity, however, is factitious, not only in a historical sense but in its Gothic design: whilst reminiscent of Elizabethan England, it disguises the impact of its 19th-century environmental technology. The famous picturesque silhouette of gothic towers, spires and turrets was not intended in the original architectural plans by Barry and Pugin. With the exception of the Elizabeth (or 'Clock') Tower and Victoria Tower, the remaining towers, spires and turrets were added later not solely as architectural embellishments but as functional features of a 19th-century ventilation system.

A first preliminary inquiry into the necessary building services for a building of the scale of the Houses of Parliament was undertaken by a Select Committee in 1835 whose aim was to inform the technological brief for the design of the new Palace. During this inquiry, they considered many of the existing environmental technologies and whether they could provide the debating chambers with a more tightly regulated atmosphere. The transformation of the new Houses of Parliament from architectural plan to functioning building begun in 1839 with the appointment of David Boswell Reid, a Scottish physician, as a specialist consultant. Reid was to develop a ventilation system for the new Palace and collaborate with Barry to integrate it into the extant architectural plans. Adapting these plans to accommodate Reid's proposed ventilation system was a lengthy, complex and, at times, highly-disruptive process. This collaboration resulted in a significant deviation from the original plans and a highly-invasive physical transformation of the existing structures as it affected the overall architectural form and layout of the building as well as the details of the architectural interior and furniture, construction and fire safety.

The transformation of the new Houses of Parliament from conception to construction can be understood as an example of 'disruptive innovation' in 19th-century architecture. In modern business studies, this term is used to describe transformations of extant technologies or practices driven by emerging technologies. The Palace of Westminster was designed early in the development of the mechanically-serviced building. This period, which began in the late 18th century with the invention of multiple central heating technologies, would culminate in the creation of mechanically air-conditioned buildings in the 20th century. Writing about this period in his book *The Environmental Imagination*, Dean Hawkes writes: 'A profound change occurred in the nature of architecture. For the first time the environmental function of the form and construction of a building came to be assisted by mechanical devices'.[2]

Work by Robert Brucemann, Robert Bruegmann, Vidar Lerum and Todd Willmert, amongst others, on the history of building services has shown that central heating technologies were widely used by architects in the 1830s.[3] Two prominent examples of this in the early 19th century are the Dulwich Picture Gallery (completed in 1817) and Westminster Law Courts (constructed between 1822 and 1827), both designed by the architect John Soane.[4] As such, the specific technologies used in the new Palace of Westminster were not revolutionary by any means. In contrast to the new Palace of Westminster, the early 19th-century buildings were relatively small and simple, with climate control limited to the use of central heating technologies. Reid's plans for the Palace, however, were much more ambitious and he was to attempt to integrate multiple technologies into a comprehensive climate control system in, as Brucemann and Prowler put it, 'what was probably the most ambitious and costly mechanical system of the century'.[5] The final ventilation system occupied more than 25% of the entire internal space of the Houses of Parliament and adopting it was a main contributor to the higher costs of the project. The final cost of the system would equate to no less than 25% of the final cost of £2,000,000 or 60% of the original estimate set in 1837.[6]

The ventilation systems inside the two debating chambers – the House of Commons and the House of Lords – incorporated cooling, heating, air filtration and humidity control methods, and its operation was governed using sophisticated monitoring and control regimes. These regimes were introduced to make the system responsive to change, reacting to variations in the weather, user activities or the level of air pollution. In contrast to modern intelligent buildings, which are controlled by computers, Reid's regimes relied entirely on people in order to work. Such operational aspects, which engage with questions of facilities management, were a central concern for Reid. In *Illustrations of the Theory and Practice of Ventilation*, he writes:

> *Architectural arrangements, either for ventilation or other purposes, where once finished, may be regarded as being completed for ever, but the management of ventilation in a crowded and varying assembly, is like the management of a musical instrument, which must not only be tune to a proper key, but sustained, continuously with an impulse apportioned to the strengths and tone with which it ought to breath.*[7]

The design of his system was characterised by three distinct, yet interdependent, elements:

(1) *the permanent architectural infrastructure*, comprising the towers and an extensive web of horizontal and vertical shafts;
(2) the mechanical *plant*, such as fans and central heating; and
(3) the ephemeral *operational architecture*, which, in addition to physical features such air valves and monitoring equipment, included monitoring and control practices.

The principles of natural ventilation, enabled through the permanent architectural infrastructure, was enhanced or complemented by the mechanical plant as well as through intelligent operational architecture. The operational architecture consisted of sophisticated socio-technical practices in which staff and users were important actors.

Re-appraising Reid's contribution

Previous studies on the history of the Palace of Westminster have focused largely on the difficulties in Reid's relationship with Barry. The design of Reid's ventilation system and its lasting impact on the architecture of the Palace has remained largely unexamined.[8] Studies by Robert Cooke (1987), Michael Port (1976), James Pope-Hennessy (1975), Barnett Cocks (1977), Christine Ridings & Jacqueline Riding (2000), and more recently by Moritz Gleich (2012), Rosemary Hill (2007) and Caroline Shenton (2016) have focused on the personal tensions between Reid and Barry, and all consider Reid's plan for the ventilation system as unrealistic.[9] For example, Shenton writes that 'there is no doubt that personally he [Reid] was vain and bombastic, and wholly unwilling to work cooperatively with other experts', dismissing Reid's scheme as a 'crackpot plan'.[10] Hill writes that there was

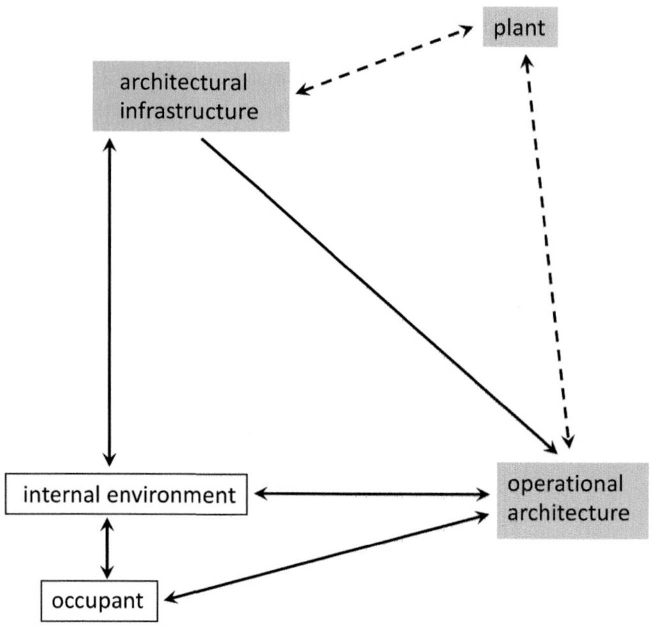

Figure 0.1 Scope of Reid's environmental systems (author's own diagram).

a 'state of war' between Barry and 'the heating engineer, Dr David Reid, whose enormous ventilating apparatus threaten to overwhelm the building'.[11] The architectural historian James Stevens Curl argues that Reid's scheme was based on 'megalomaniac ideas'.[12] In a similar vein, Denis Smith claims that the development of the ventilation system was the 'worst of the architect's many problems' and that it was accentuated by Reid's 'inability to make his mind, his megalomaniac demands and his readiness to take offence'.[13] Finally, Pope-Hennessy has described Reid's scheme as a 'bogus heating apparatus which at one moment threatened to take up a third of the cubic content of Westminster Palace'.[14]

It is argued in this book that past studies have not adequately examined the nature of the collaborative relationship that existed between Reid and Barry and have failed to recognise how much of Reid's scheme for the Palace of Westminster was in fact realised. Although the interpersonal tensions between Reid and Barry were in part personal, they were largely the result of three significant practical challenges as a result of designing a ventilation system for a large and complex public building of the scale of the Palace of Westminster.

The first major practical challenge was to overcome the substantive and complex design problems and to manage the disruption caused by the late adoption of the ventilation system. Construction on the new Palace had already begun by the time Reid was commissioned to design a ventilation system. Thus, the system had to be designed within the constraints of an existing architectural plan and physical space.

The second major practical challenge was to navigate the institutional structures and power relations arising from both the involvement of Select Committees of the Houses acting independently from one another and the poor management structure of the Office of the Woods. The Office was initially responsible for overseeing its procurement.

The third major practical challenge was to establish a framework for the cross-disciplinary collaboration between a scientist and an architect (and his practice), which had never before been attempted at this scale. Although Reid was an expert on ventilation, his experience of working with civil engineers or architects was limited. Consequently, Barry and his staff had to adapt their well-established working practices to accommodate Reid's empirically-driven ideas.

Although the construction of the new Palace was beset by substantial delays and complications, some of which were a direct result of the dysfunctional relationship that developed between Reid and Barry, recent research by the author of this book has shown that, despite this, almost two-thirds of the structural carcass of the Palace of Westminster – which contained the ventilation network set out in Reid's original plans – was completed. In other areas of the Palace, such as the House of Lords, this network was completed under the direction of Barry and involved some deviations from Reid's original proposals. There are only a few areas, such as the front-facing Old Palace Yard, in which the ventilation network was designed without Reid's involvement.

The memory of Reid's original plans has not been erased. Although some changes were subsequently made to the ventilation system, these only had a minor impact on the design of Reid's underlying physical infrastructure. The ventilation system provided the primary infrastructure on which the ventilation of the Houses of Parliament depended for over 90 years, and only fell into disuse in the second half of the 20th century with the introduction of mechanical ventilation and steel and copper ducts. The surviving infrastructure – recently the subject of new surveys undertaken by the author himself between 2016 and 2019 – represents an architectural palimpsest that represents multiple generations of alterations made over the past 180 years.

Physical reconstruction of a historic(al) system

The research for this book, which was undertaken over a period of eight years, combines the study of archival records and on-site investigations of the surviving features. The archival research presented here involved the study of several thousand historic documents, comprising drawings, sketches, reports, letters and Parliamentary papers. The largest number of documents involved in this research are held by Historic England in Swindon and at the National Archives in Kew. Other important documents are held by the Victoria and Albert Museum, the Royal Institute of British Architects, the Parliamentary Archives, the Parliamentary Art Collection and within the estate archives of the Palace of Westminster.

The on-site investigations show that much of the historic infrastructure has survived, including the extensive network of voids hidden within the depths of the walls, floors, roofs, ceilings and buttresses. Some of the original thermometers, valves and control mechanisms used by the attendants, who monitored and controlled the temperature and ventilation, have also survived.

The author also collaborated with Plowman Craven and the Palace of Westminster Restoration and Renewal Programme between April 2019 and February 2020 to create a virtual model of the entire ventilation system using Building Information Modelling (BIM). In this model, information relating to the existing and lost features of the original system are represented.[15] In addition to enabling a detailed reconstruction of the physical infrastructure, the surveys also allowed important questions of historical interest to be addressed, which could not be answered through the study of archival evidence alone. The surveys have also allowed the scale of Reid's lasting contribution to the design and construction of the Palace of Westminster to be verified, revealing how much of his original proposal was actually built and how much of it was not.

Focusing on the period from 1835 to 1855, this book retraces and critically examines the evolution of Reid's scheme and the underlying environmental discourse and scientific investigations. Reid's involvement in the rebuilding of the Palace of Westminster lasted 17 years and can be divided into four main phases around which this book will be structured. Part One of this book, **'The experimental phase 1835–39'**, examines Reid's initial involvement with the new Palace of Westminster which was limited to the development of hypothetical proposals and experimental demonstrations of his principles of ventilation.

Chapter 1 introduces the background to the issues of ventilating the Houses of Parliament. Reid had first presented a proposal for the ventilation of the debating chambers of the Houses of Parliament to an 1835 Select Committee, which had been appointed to investigate the technical requirements of the Palace of Westminster. Impressed by his ideas, the Office of Woods subsequently commissioned Reid to test his proposals in the two debating chambers of the temporary Houses of Parliament. Although Reid had not received an official commission to work with the architect Barry on the design of the Palace of Westminster, Barry was instrumental in the promotion of Reid's ideas amongst both the MPs and Lords, and the Commissioners of Woods, Forests, Land Revenues, Works and Buildings, all of who were responsible for overseeing its rebuilding.

Chapter 2 reconstructs Reid's ventilation system in the temporary Houses of Parliament and shows how he implemented monitoring and control social feedback processes to maintain and improve its operation. The performance evaluations of Reid's experimental system have been documented in historical records, such as logbooks with monitoring data, letters, diaries and technical reports of scientific experiments, as well as in the Parliamentary papers themselves. Indeed, Reid, who had a team of attendants that manually undertook the monitoring and control procedures, directly involved the primary users of the Houses, the MPs and Lords, who would provide regular feedback about the indoor conditions during sittings. These procedures enabled Reid both to gain an intimate knowledge of how the system performed under real-life conditions, and to test and refine his concepts before they might be adopted in the new Palace of Westminster.

Chapter 3 focuses on the important role of MPs and Lords – especially through the mechanism of the Select Committee – in determining how Reid's system was received and consequently developed. The Parliamentary papers associated with these committees (e.g. official reports and debates as reported in Hansard) offer an intimate insight into the methods used to empirically evaluate building performance during the mid-19th century. As will be seen, issues such as thermal comfort, which became a central focus of the Select Committee inquiries, were the direct result of the feedback provided by the MPs and Lords themselves.

Part Two of this book, '**Attempted application, 1840–46**', is composed of two chapters and retraces the development of Reid's first scheme for applying his principles of ventilation to the Palace of Westminster. Although this scheme was not fully realised, the historical records allow the original proposal to be reconstructed. The intimate involvement of the various Select Committees in issues relating to the ventilation of the Houses continued throughout this second phase of Reid's involvement (see Figure 0.2). Ad hoc Select Committees were appointed to scrutinise and approve the cost and design concept of Reid's original scheme (1841–42), and to review and resolve issues regarding the collaboration between Barry and Reid (1843–46).

Reid proposed to ventilate the entire Palace of Westminster, including the two debating chambers, using one central system. Reid proposed adapting the Elizabeth and Victoria Towers (which were part of Barry's original plans) to function as fresh air shafts, and adding a third tower above the central lobby to function as the central outlet for the air and smoke produced inside the new Palace.

Chapter 4 covers the period from 1839 to 1842 and retraces the development and formal approval of Reid's general design concept. The costs of the third tower began to spiral and Reid's scheme became the subject of formal inquiries led by Select Committees from both the House of Lords and House of Commons between 1841 and 1842, leading to delays in implementing the ventilation system.

Chapter 5 explores the development of Reid's proposed ventilation system from 1842 to 1846, and examines the practical difficulties that Barry

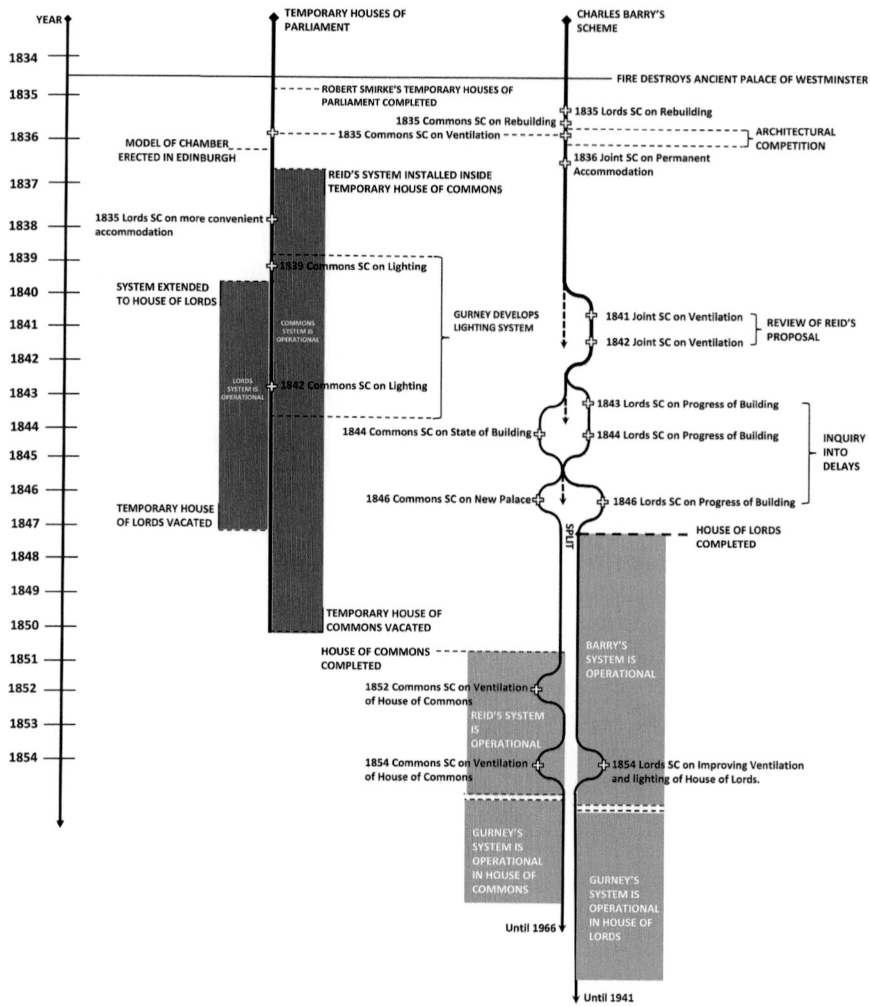

Figure 0.2 Timeline of inquiries by the select committee through which MPs and Lords interfered in the design process and performance evaluations (author's own diagram).

and Reid faced during this period. These difficulties ultimately led to a breakdown in their collaboration and the resulting delays led the House of Lords to limit Reid's involvement in the delivery of their own part of the system in 1846. Following several Select Committee inquiries, Reid's involvement was eventually limited to the delivery of the ventilation system for the House of Commons. Responsibility for completing the ventilation system in other parts of the Palace, including the House of Lords, was transferred to Barry's team. By this time, a large part of the physical infrastructure required for Reid's system, including that of the House of Lords and the River Front, had already been completed. It was subsequently reutilised and adapted for the final ventilation scheme developed between 1847 and 1852.

Part Three of this book, **'Realisation within new borders, 1847–52'**, explores how the original ventilation system was eventually reorganised into a series of local systems, replacing Reid's original concept of a centralised system. In this context, the Central Tower, once the core of Reid's original scheme, was disconnected from the House of Commons and became an independent system.

Chapter 6 investigates how Reid's original plan for the House of Lords was completed and adapted by Barry in 1847. This is followed by an examination of experimental investigations that Meeson and the Office of the Woods had undertaken in order to evaluate and improve its performance in response to criticism from users. This illuminates the practical challenges that ultimately led to the decommissioning of Barry's original system after seven years.

Chapter 7 provides a reconstruction of Reid's final design for the House of Commons, which was completed between 1847 and 1852. Its development was significantly delayed by lengthy and difficult negotiations between Reid and Barry about the design of the new arrangement that was necessary for the ventilation of the House of Commons to function as an independent system.

The fourth and final part of this book, **'Critical evaluations and adaptation, 1852–1950'**, examines the post-occupancy history of the Houses of Parliament. Chapters 8 focuses on the performance evaluations of Reid's system from February 1852 to March 1854. Although Reid's ventilation system was only operational for just over two years, the archival records show how operational staff acquired, recorded and utilised knowledge of the building's performance gained through its day-to-day operation and through separate and more in-depth studies. These records also show how this knowledge was utilised to improve existing arrangements by informing the design of physical and operational modifications. Reid's system was decommissioned after unfavourable reviews of its performance – in particular, of thermal discomfort – by two Select Committees in 1852 and 1854. It was eventually replaced (mirroring the fate of Barry's system in the House of Lords, discussed in Chapter 6) with a new system developed by the physician Goldsworthy Gurney.

The last chapter, Chapter 9, examines selected periods in the post-occupancy history of the two debating chambers after Gurney's interventions were completed in 1855. During these periods, performance evaluations of, and physical alterations to, the ventilation system continued. These historical inquiries illustrate how – over a period of nearly 100 years – Parliament (i.e. the organisation itself) continued to acquire, record and utilise knowledge about the performance of their own building.

This book does not only intend to be a study of technical design, but seeks to engage with wider questions about architectural practice and facilities management. These questions are as pertinent today as they were in the 19th century. The systems in the Houses of Parliament were empirically evaluated using a combination of different methods, including air flow tests, user consultations, environmental monitoring as well as fine-tuning operational procedures. As such, they should be interpreted as historic examples of post-occupancy evaluation.

Unfortunately, the performance or operation of a building is rarely covered in the history of architecture. However, the way a building accommodates its users and whether the building works for its users are essential to gain a full understanding of the practical reality of gaining environmental control in a 19th-century building, including the challenges from a facilities management perspective. This is also a concern today in the procurement of modern buildings, as it was in the 19th century.

Historical principles and sustainability

This book examines the environmental design history of the Palace of Westminster and yields insights that could help to address issues of sustainability in the context of architectural conservation. The renewal of the ageing buildings services in the Palace of Westminster will be a major focus of the forthcoming restoration programme. The 'Pre-feasibility study and preliminary strategic business case' report, published in October 2012,[16] and the final report of the Joint Select Committee on the restoration and renewal, published in September 2016,[17] both highlight the urgency of the current situation by pointing out that current building services are prone to suffering technical failures and pose a serious fire risk. Whilst the House of Commons was lost – destroyed in the war and rebuilt using modern technology in the 1940s – most of the original 19th-century infrastructure has survived in other parts of the Palace and currently lies dormant. The restoration programme has afforded the opportunity to systematically re-examine the original ventilation system and to explore how it was revitalised and integrated into the design of the future building services.

Conceptually, the Palace of Westminster is a 19th-century example of a mechanically-serviced building. Similar to its 20th- and 21st-century counterparts, it was intended to immerse its occupants in a hermetically-sealed and climate-controlled environment. Reid was aware that his technological approach would have an environmental impact. As will be shown in Chapters 4, 5 and 7, Reid wanted to to minimise energy use of his ventilation system by incorporating methods of natural ventilation, cooling and heat recovery. Although not an example of a low-energy design, one of the main aims of Reid's system was to minimise both the energy consumption associated with the use of fossil fuel for heat and power and the air pollution produced as a result by using architectural features, such as large towers, to harness the natural stack effect (thermal buoyancy) and wind pressure to permit the ventilation of the Houses.

Fifty years ago, the Palace of Westminster was mainly discussed in the context of the history of mechanical building services. However, the recent 'sustainability' agenda has motivated renewed interest in 19th-century methods of 'natural' ventilation. Vidar Lerum (2016) and Alan Short (2017) have argued that 19th-century buildings have the potential to inform contemporary sustainable practice.[18] Although modern concepts of sustainable design were not available to Reid and his contemporaries, the principles underlying the ventilation system of the Palace of Westminster closely resemble those deployed in 'advanced naturally-ventilated buildings'. [19]Notable examples of these buildings include the Everyman Theatre in Liverpool,[20] the Building Research Establishment's Environmental Building in Garston, the University College London's School of Slavonic and Eastern European Studies,[21] the Centre for Mathematical Sciences in Cambridge, Senedd in Cardiff and 'The Shed' at the National Theatre in London. By integrating natural and mechanical methods of ventilation, the systems in the Palace of Westminster followed what modern engineers would call a 'hybrid' or 'mixed mode' approach.

However, despite the conceptual similarities between the 19th- and 21st-century approaches to stack ventilation, there are fundamental technological differences. Nineteenth-century systems were designed to be assisted by the waste heat produced by the lighting and heating technologies of the day (e.g. gaslights, fireplaces and boiler furnaces) and were also manually controlled. Compared to modern stack-ventilation systems, this waste heat allowed much stronger convection currents to be produced within the ventilation towers and air to be pulled over considerably longer distances. Over the summer, when the available waste heat was limited, the air current inside the towers was also enhanced artificially through the use of coke fires, steam jets and heating pipes. By contrast, modern stack-ventilation systems use comparatively small quantities of waste heat produced by electric lighting, people and electronic devices, which produce weaker currents, and low velocity fans are sometimes used to enhance the flow artificially.[22] Understanding these differences is an essential first step in addressing the question of whether reusing historical environmental principles can provide a new approach to achieving a more sustainable approach to ventilation and climate control in historical buildings.

Notes

1 Eric Hobsbawm, 'Introduction: Inventing Traditions', in *The Invention of Tradition*, ed. by Eric J. Hobsbawn and Terence O. Ranger (Cambridge: Cambridge University Press, 2012), pp. 1–2.
2 Hawkes, Dean, *The Environmental Imagination*, 2nd Edition (Abington: Routledge, 2020) p. 257.
3 Todd Willmert, Heating Methods and their Impact on Soane's Work, Lincoln's Inn Fields and Dulwich Picture Gallery, *Journal of the Society of Architectural*

Historians, 52 1993, pp. 26–58, Robert Bruegmann, Central Heating and Forced Ventilation: Origins and Effects on Architectural Design, *Journal of the Society of Architectural Historians*, 37 (3) 1978, pp. 143–160.

4 Dean Hawkes, *Architecture and Climate*, (Abington: Routledge, 2012).

5 Robert Brucemann and Donald Prowler, 19th Century Mechanical System Designs Journal of Architectural Education, 30 (3) *Energy and Architecture* (Feb., 1977), pp. 11—15.

6 Michael Harry Port, *The Houses of Parliament* (Newhaven: YUP, 1976).

7 David Reid, *Illustrations of the Theory and Practice of Centilation,* (London, 1844), p. xiii–xiv.

8 Robert Brucemann, '19th Century Mechanical System Designs', *Journal of Architectural Education*, 1977–30, pp. 11–15; Robert Bruegmann, 'Central Heating and Forced Ventilation: Origins and Effects on Architectural Design', JSAH, 37 143–160; Vidar Lerum, *Sustainable Building Design: Learning from Nineteenth-Century Innovations* (London: Routledge, 2016); Peter Collins, *Changing Ideals in Modern Architecture, 1750–1950* (Quebec: McGill-Queens University Press, 1998).

9 Barnett Cocks, 1977 *Mid-Victorian Masterpiece* (London: Book Club Associates); Robert Cooke, *The Palace of Westminster* (London: Burton Skira, 1987), pp. 107–25; Moritz Gleich, 'Architect and Service Architect: The Quarrel between Charles Barry and David Boswell Reid', *Interdisciplinary Science Reviews* 37–2012, pp. 332–44; Rosemary Hill, *God's Architect: Pugin and the Building of Romantic Britain* (London: Allen Lane, 2007); Pope-Hennessy, James, The Palace of Westminster (London: Michael Joseph, 1975), pp. 24–25, 14; Michael Harry Port, *The Houses of Parliament* (Newhaven: YUP, 1976); Alexandra Wedgwood, 'The New Palace of Westminster', in Christine Ridings and Jacqueline Riding (eds), *The Houses of Parliament – History, Art, Architecture* (London: Merrel, 2000) pp. 113–35; Caroline Shenton, *Mr Barry's War, Rebuilding the Houses of Parliament after the Great Fire of 1834* (Oxford: OUP, 2016).

10 Caroline Shenton, *Mr Barry's War – Rebuilding the Houses of Parliament after the Great Fire of 1834.* (Oxford: OUP, 2016), p.166 & 142.

11 Rosemary Hill, *God's Architect*, pp. 316–17.

12 James Stephen Curl, *Victorian Architecture Diversity and Invention* (Reading: Spire, 2007), pp. 61–3.

13 Denis Smith, Building Services, in M. H. Port, *The Houses of Parliament* (Newhaven: YUP, 1976), pp. 218–45.

14 James Pope-Hennessy, The Palace of Westminster (London: Michael Joseph, 1975), pp. 24–25, 14.

15 Henrik Schoenefeldt, Historic Research as an Applied Science – A Collaborative Project at the Palace of Westminster Shows How Academic Research Can Provide a Better Understanding of Environmental Technology in Historic Buildings, *Contrast: Journal of the Institute of Historic Building Conservation*, July 2018, pp. 15–17.

16 Restoration and Renewal of the Palace of Westminster: Pre-feasibility study and preliminary strategic business case, October 2012

17 Joint Committee on the Restoration and Renewal of the Palace of Westminster, First Report of Session 2016–17 (HL Paper 41 / HC 659)

18 C. Alan Short, *The Recovery of Natural Environments in Architecture: Air, Comfort and Climate* (New York: Routledge, 2017); Vidar Lerum, *Sustainable Building Design: Learning from Nineteenth-Century Innovations* (London: Routledge, 2016).

19 The concept of advanced naturally ventilated building is discussed in: Kevin Lomas, Malcolm Cook and C. Alan Short, 'Commissioning Hybrid Advanced

Naturally Ventilated Buildings: a US Case Study Research', *Building Research and Information*, 37 (2009), p. 397.

20 Case Study: Everyman Theatre Liverpool, *CIBSE Journal*, February 2015, pp. 54–62 http://portfolio.cpl.co.uk/CIBSE/201502/case-study-everyman/

21 C. Alan Short, Malcolm Cook and Kevin Lomas, Delivery and Performance of a Low-Energy Ventilation and Cooling Strategy, *Building Research and Information*, 37–1 (2009), pp. 1–30.

22 David Etheridge, *Natural Ventilation of Buildings: Theory, Measurement and Design* (Chichester: Wiley, 2011); Ulrike Passe and Francine Battaglia, *Designing Spaces for Natural Ventilation: An Architect's Guide* (Abington: Routledge, 2015).

PART I

The experimental phase: 1835–39

The naysayers wanted to put us in a glass pod—a temperature-controlled pod to ensure that we were kept at the perfect temperature, boiled to the right level in Fahrenheit or centigrade, whichever you prefer. This is a building that has survived for 800 years, not a hot air heating system.

It seems to me that we could have sat there in our overcoats, as that would have solved the problem in the winter. And in the summer, some hon. Members more racy than I am might have felt it possible to take off their jackets. It seems to me that there is an easy, affordable solution whereby we maintain a Chamber in our historic residence. That is what we should do and that is what we should vote for.

Jacob Rees-Mogg, MP for North-East Somerset, January 2018[1]

Note

1 Restoration and Renewal, House of Commons, *Hansard's Parliamentary Debates: The Official Report* (31 January 2018, Hansard, vol 635, col. 919).

1 The origins of an empirical approach to design: 1832–36

The louvres are ingeniously contrived to open or shut, as the wind may play against- that is, when the wind is beating against one side, or at the angle-for though quadrangular at the base, it assumes the form of an octagon, just at the commencement of the louvres, thus be fanning any one side, or angle, the opposite apertures remain open; thus the foul air readily finds egress.[1]

Illustrated London News, 16 December 1854.

Illusions of unity

In October 1834, a large fire seriously damaged the old Palace of Westminster, leaving behind only the empty masonry shells of many an important space. Christopher Wren's original 17th-century interior of the House of Commons debating chamber was lost in its entirety. Without a place to hold sittings, MPs and Lords had to be squeezed into two temporary chambers. The chambers, designed by the architect Robert Smirke, were built within the ruins of the old Palace, reusing the surviving masonry walls, and were simple timber structures. The temporary House of Commons was constructed within the shell of the old House of Lords chamber, whilst the Lords were accommodated with a much smaller debating chamber inside the old painted chamber. These structures were occupied by February 1835, and in the following months, Parliament began its inquiry into the rebuilding of the Palace of Westminster.

Although the 1834 fire did not entirely destroy the ancient Palace of Westminster, only a small part of the original structure was retained and restored. Historical illustrations of the ruined Palace, housed at the Parliamentary Art Collection, show that much of the original masonry survived the fire. As with the recent fires at the Notre Dame cathedral in Paris in 2019 or the Glasgow School of Arts in 2018, it was primarily the timber roofs and floors that were lost. These ruins were subsequently cleared to make space for a new, purpose-built parliament building. In addition to Westminster Hall, only the remains of the chapel of St Mary Undercroft, which dates from the 13th century, and the adjacent 16th-century cloister were retained, restored and integrated into the design for a new Palace of Westminster. This new palace was designed to meet the functional requirements of Parliament in the early 19th century, including a sophisticated system of ventilation and climate control. The climate control system was adopted to overcome longstanding issues dating back to the 17th century with the ventilation inside the debating chambers.[2] The system was revolutionary, not only in terms of the extent to which the environmental functions of buildings, such as lighting, ventilation and climate control, were addressed technologically, but also in terms of the relationship between technology and architectural form. According to Peter Collins (1998), it marked 'the first occasion when mechanical services had a real influence on architectural design',[3] whilst Denis Smith (1976) has argued of building services that the 'unprecedented way in which they dictated to the architect, make it our most important building in this aspect of nineteenth-century technology'.

The exterior architecture gives the impression of a harmonious relationship between environmental technology and architectural form. Many of its Gothic towers and turrets, which add to the picturesque qualities of its silhouette, were not introduced as architectural embellishments, but were the functional features of a 19th-century ventilation system. The quote at the beginning of this chapter describes the function of the intricate system of louvres in one of these towers.

Its external appearance, however, is deceptive. This history of the rebuilding of the House of Parliament illustrates that the technology for ventilation and climate control was not developed as an integral part of the architecture from the beginning. The original architectural design that Charles Barry and Augustus Pugin had produced for the architectural competition in 1835 was conventional from an environmental perspective. From 1836 to 1840 inquiries into the design of building services, led by David Boswell Reid, were confined to speculative design studies and experiments and conducted separately from the development of Barry's plans. Reid did not meet Barry to discuss the application of his system within the actual Palace until the autumn of 1839.

This chapter, which focuses on the period from 1835 to 1837, retraces the development of the architectural scheme and its relationship to Reid's environmental inquiries in the early stages of the project. The evolution of this relationship is crucial to understand the causes of later difficulties with integrating Reid's technical scheme into Barry's plans, which will be explored in Chapters 4–6.

The roots of separation between architectural and environmental strands

The architectural design for the new Palace of Westminster was procured through a public competition. This procurement process, which lasted from 1835 to 1837, was led by parliamentarians (rather than professional engineers or architects): Select Committees of the Houses of Lords and Commons in collaboration with the Treasury and the Office of Woods, Forests, Land Revenues, Works and Buildings (hereafter referred to as Office of Woods). The Office of Woods was a government department responsible for managing the delivery, maintenance and operation of buildings of the state, including the parliamentary estate in Westminster. It was headed by a board of three commissioners: from 1835 to 1841 the role of First Commissioner was held by Viscount Duncannon, a member of the House of Lords; and two civil servants, Sir Benjamin Stephenson and Alexander Milne, acted as Second and Third Commissioners respectively. Viscount Duncannon had served under Viscount Melbourne's Whig government.

The future design of the new Palace of Westminster also became the subject of formal debates, and motions and important decisions that had to be approved by vote. The fact that it was led by MPs and Lords, the Palace's future users, is significant as the development of its design was strongly influenced by their specific concerns, views and expectations.

In March 1835, the House of Lords and the House of Commons appointed two separate select committees to develop the design brief and schedule of accommodation for the rebuilding of the Palace. It specified the architectural style, which was to be 'either Gothic or Elizabethan'[4], and defined the space requirements, which were based on evidence presented by senior staff from different parliamentary offices. The design of the new debating chamber of the House of Commons was also specified, and included seating for 420–460 MPs on the main floor. In addition, two division lobbies (voting lobbies), which was a new feature introduced as part of a reform of voting procedures (an outcome of another inquiry in March 1835), were to be constructed.[5] Technical or environmental requirements of the debating chambers or other spaces, however, were not given much consideration. In the written specification for the new building, published in the final reports of the Select Committees on 3 June 1836, references to environmental requirements were confined to a brief note saying that 'special provision be made for the due ventilation of all the rooms in the new Palace'[6]. By requesting all drawings to be drawn at a scale of 1:1500,[7] the competition rules also limited the degree of detail that architects could show within their proposals.

Although the MPs and Lords considered technical requirements outside the scope of the architectural competition, it became subject of another, more specialist parliamentary inquiry. In July 1835, several weeks after the competition had been formally launched, Benjamin Hawes, MP for Lambeth, made a successful motion for the appointment of an ad hoc Select Committee to lead an investigation into the principles of ventilation, climate control, artificial lighting and acoustics that could be applied to the design of the Palace of Westminster once an architectural design had been procured. It held six sittings between July and September and interviewed several eminent British scientists and one engineer. These included the physicist and chemist Michael Faraday, civil engineer John Sylvester, chemist William Thomas Brande and two physicians, George Birkbeck and David Boswell

Reid.[8] The fact that none of these witnesses, except for Sylvester, had backgrounds in civil or mechanical engineering is significant as it resulted in questions of environmental design being approached from a largely scientific perspective: their focus was on the climatic, chemical and physiological rather than the mechanical engineering aspects of ventilation.

The transcripts of the testimonies illustrate that the scientists provided advice on fundamental issues of thermal comfort, indoor air quality and external pollution. This was not unusual in the 1830s as 'building services engineering' was not formally recognised as a distinct profession within the UK until the foundation of the Institution of Heating and Ventilating Engineers in 1897. Instead, its roles were fulfilled by individuals with backgrounds in civil engineering or the natural sciences, with chemists, physicists and medical doctors having particular prominent roles.[9] It also reflects the development of the 'science' and 'mechanics' of ventilation as two closely interrelated, yet distinct strands. Both strands were critical in the design of the debating chambers as the problems of its environmental control could not be resolved using the skills of engineers alone. It also relied on the specialist knowledge of other fields to engage with environmental and human factors, such as health, pollution and thermal comfort.

In terms of their knowledge and skills, Reid and the other scientists are more accurately understood as the predecessors of modern building scientists, rather than building services consultants. Reid had no prior experience of working on the design of public buildings: his reputation was based entirely on his academic research and on the success of his design for a purpose-built teaching laboratory and lecture hall in Edinburgh. His work was known to some MPs and Lords, including Lord Henry Brougham and former Prime Minister Earl Grey, who had visited his laboratories in 1834 as part of a delegation of the *British Association for the Advancement of Science*. There they witnessed demonstrations of its ventilation, fume extract system and acoustics. Reid had demonstrated similar principles of acoustic design in a temporary pavilion on the grounds for Edinburgh High School: a large hall with a floor area of 113 ft × 101 ft (33 m × 31 m), used to host 2,800 guests attending a formal banquet for Earl Grey. According to the *Edinburgh Philosophical Journal* 'each individual speaker could be distinctly heard'.[10] Lord Sudeley, Charles Henry-Tracy, who also knew of Reid's work, invited him to give evidence to the Select Committee on the basis of his purpose-built laboratory.[11] Although it was a different building type, the laboratory provided a potential precedent for the design of a debating chamber. Both were spaces that required good acoustics and ventilation to satisfy their function as spaces for the performance of speeches and the gathering of large numbers of people.

The skills and knowledge offered by scientists and how they relate to those offered by architects and engineers were discussed during the inquiry. In his interview on 4 August 1835, Birkbeck admitted that he could only provide theoretical guidance and would have to rely on the skills of architects or engineers to translate them into more concrete, practical solutions.[12] Brande was highly critical of the architectural profession, subsequently arguing on 12 August that it 'commonly overlooked' the requirements of ventilation and 'at all events' view them as secondary concerns.[13] The chair of the Committee, Hawes, had on 7 August already alluded to the possibility of resolving this issue by 'combining the architect with some scientific man'[14] and he also advocated the idea of testing and improving designs with the aid of physical experiments. In the last two sittings of the inquiry, the Committee explored the possibility of using the temporary House of Commons as a site for testing Reid's principles under real-life conditions. The architect Smirke, who was also consulted, was equally critical of his own profession. During discussions about his involvement in remodelling the temporary debating chamber for the proposed trials, he argued that Reid, not the architect, should be in charge of coordinating its implementation. Smirke thought this was essential 'to ensure the complete and efficient execution of his plan in all its details'[15] and offered to take a more subordinate role. At this stage, although discussions about the roles of different disciplines were entirely hypothetical, it was revisited in 1840 when the conditions for a collaboration between Reid and Barry were first discussed. This led to the adoption of a model of collaborative practice in which Reid and the architect Barry had almost equal authority (see Chapter 4).

These discussions illustrate an awareness of the need for cross-disciplinary collaborations, and in this, Reid's specific skills as a medical doctor afforded new ways of addressing the fundamental problems of ventilation and climate control in the debating chambers of the House of Parliament.

Drawing on the skills of a medical doctor

Reid had studied medicine at the Royal College of Surgeons of Edinburgh and for 15 years pursued a career as an educator and researcher within the field of chemistry. In 1827, whilst working as a lecturer under Professor Thomas Charles Hope, chair for chemistry at the College, Reid began developing a curriculum for courses in practical chemistry.[16] Its focus was on teaching the practical skills of experimental research, and its applications within the context of industry and the design of the built environment. The study of ventilation in buildings was a prominent feature in his curriculum. Reid later noted that 'they were students of every department of practical chemistry, but the air question was the one to which I gave pre-eminent attention'.[17] Some of these methods are illustrated in the three educational textbooks for students of practical chemistry that Reid published in the 1830s: *Elements of Practical Chemistry* (1830), *Text-Book for Students of Chemistry* (1836) and *The Rudiments of Chemistry* (1837).[18] These textbooks contained instructions for conducting certain experiments and explained how these could be used to examine different aspects of ventilation, providing methods for the study of fluid dynamics or the assessment of indoor air quality. Reid was not successful in establishing a separate chair for practical chemistry at the University of Edinburgh and left the institution.

In the early 1830s Reid established a private chemistry laboratory with new purpose-built facilities for teaching and conducting experimental research. The laboratory facilities were located at Roxburgh Place on a site immediately behind the College's Surgeons' Hall.[19] The laboratory facilities, completed in 1833, were based on Reid's own plans: it was his first, and at the time only, construction project, enabling him to gain experience with the design and operation of stack ventilation systems. Although it was designed to meet the specific ventilation requirements of a chemistry laboratory, Reid saw it as a potential precedent for the legislative chambers. He argued that the type of ventilating shaft inside his laboratory provided the high level of control that would be required to deal with the sudden and large variations in the number of MPs commonly using the House of Commons.[20] In his 1837 paper *Brief outlines illustrative of the alterations in the House of Commons*, Reid wrote that the issue of managing noxious fumes and heat, produced by numerous experimental furnaces, was significantly more severe than the ventilation of legislative chambers.[21] In the debating chamber itself, the main problem was mitigating the impact of artificial lighting and large crowds on the internal environment. The opportunity to design a purpose-built chemistry laboratory allowed him to address these problems from first principles.

The facility comprised a large open space with a teaching laboratory and lecture theatre for chemistry demonstrations. The ventilation was driven by a large central stack, which was assisted by 14 smaller shafts. The smaller shafts were activated at times when the central stack was not sufficient to sustain the required ventilation rate. The stack system had two different functions: first, to extract the noxious fumes produced by the chemical experiments; and second, to discharge used air from the main laboratory space. As shown in *Ventilation in American Dwellings*, the central stack was connected to a system of underfloor air ducts, which joined the numerous local outlets in the chemistry benches;[22] the pull produced by the hot air inside the shaft was used to extract fumes close to their source. The ventilation of the room was facilitated through an array of apertures within the wall of the shafts, which was regulated manually with the aid of adjustable dampers.[23] These allowed the system to respond to changes in the quantity of noxious fumes as well as large fluctuations in the number of students. The thermal convection was sustained artificially using a wood furnace. Reid argued that unassisted natural stack ventilation systems were unreliable as their performance was dependent on wind conditions and external temperature.[24] The fire provided a more reliable source of power.[25] Reid noted that without constantly maintaining a high rate of ventilation, it would only take a few minutes until the concentration of fumes in the internal atmosphere would reach such high levels that students would need to be evacuated. The fire was to ensure that the noxious fumes were directly extracted before contaminating the general atmosphere and getting into the lungs of students and instructors.[26]

A laboratory for the practice and teaching of building science

In conjunction with his role as an educator, Reid also conducted scientific research into indoor air quality and thermal comfort, and erected physical mock-ups of rooms to trial different methods of controlled ventilation. In this context he developed experimental methods to empirically evaluate the performance of ventilation arrangements that took into account physical, chemical as well as physiological aspects of indoor environments.

His earlier inquiries into ventilation focused on the physiological aspects of ventilation, addressing issue of thermal comfort and air quality.[27] The objective of this research was to define new standard ventilation rates, which addressed problems of air quality as well as thermal comfort. He criticised typical levels of air supply in public building as being too low, resulting in atmospheric conditions that were unpleasant and made occupants feel unwell.[28] Reid was not the first scientist to set ventilation standards: engineer Thomas Tredgold had recommended a ventilation rate of 4 CFM, whilst Neil Arnott had recommended 2–3 CFM. Reid argued that these only defined the minimum ventilation required to satisfy a person's metabolic needs and did not account for maintaining thermal comfort or a pleasant air quality.[29] These low rates, he wrote, 'would not give the comfort and maintain the constitution in such good condition as a larger allowance'.[30] His objective was to achieve an 'agreeable and refreshing atmosphere' and to maintain the human body in a 'firm and healthy tone'. To determine the air supply required to achieve this higher standard, Reid worked with various types of experimental chambers. These were 'wood boxes' and were equipped with ventilation systems that allowed one to control the air supply.[31] In these studies Reid relied entirely on the subjective feedback of participants about the quality of the air to which they had been exposed. The experiments involved groups of up to 250 participants, many of which were Reid's students who 'volunteered their services when he was anxious to ascertain the effect of a given supply of air, at a regulated temperature, renewed in the manner he had proposed'.[32]

Reid concluded from these experiments that 10 CFM was an 'ample allowance for an adult',[33] but in warm weather the ventilation rate would need increasing to 40–60 CFM if a comfortable temperature was to be maintained without artificial cooling.[34] He also observed that this rate could only be raised to a certain level before strong internal air currents became a major cause of thermal discomfort.[35]

In the early 1830s, Reid had constructed various special rooms inside his new laboratory to test different methods of introducing and extracting air in sealed rooms.[36] He presumed that sealed rooms with controlled stack ventilation would enable a tighter level of control over the environmental conditions than naturally-ventilated rooms with operable windows.[37] Full-scale models of rooms – simple utilitarian structures that could be quickly erected, tested and modified – were erected in which air was admitted and discharged exclusively through perforated ceilings, walls and floors, or through traditional interior architectural features (such as skirting boards, cornices and dado rails).[38] Air was introduced and discharged through a network of ducts connected to the laboratory's ventilation system. Human participants reported on the quality of the air and the physical sensations produced by air currents due to the varying velocities, degree of diffusion and temperature. Through a system of supply and discharge ducts 'air could be made to enter and be withdrawn in any required proportion'.[39]

In his 1855 *The Revision of Architecture in Connection with the Useful Arts* and 1856 *Progress of Architecture* lectures, Reid described three experimental rooms, constructed to investigate ways of distributing air inlets over the largest possible area.[40] In one mock-up, all the fresh air was introduced through a perforated ceiling; in a second, the porous surface was extended along the walls, allowing the incoming air currents to be distributed over an even larger area; in a third model, the air was admitted through a perforated floor and discharged via a perforated ceiling. These experiments show that Reid's inquiries crossed the boundaries between experimental research and the experimental validation of design solutions: on the one hand, he created architectural spaces to function as the technical equipment of scientific research; on the other hand, he used experiments to test and refine potential technical solutions inside of real physical models. The latter, which might be described as an experimental approach to design, is of particular significance as it strongly influenced Reid's first proposal to the Select Committee in August 1835 for the debating chambers and provided the foundation for further experimental design studies undertaken to develop his proposal between 1836 and 1839.

Reid's hypothetical scheme for a debating chamber

In contrast to the other witnesses, whose contributions were limited to briefings on general principles and critical reviews of ventilation arrangements used in other buildings, Reid presented a detailed proposal for the debating chamber: in three sittings he gave oral accounts of his proposal and underlying concepts and presented a set of five drawings, which included an interior perspective, floor plans and a diagrammatic cross-section.[41]

The architecture of its interior was utilitarian in character as the form and materiality of the chamber had been determined following exclusively functional criteria, focusing on ventilation, climate control and acoustics. Reid did not concern himself with formal architectural criteria of style, proportion or composition, nor did he conform to the historical form of the lost chamber. He proposed a radically different form that he considered ideal from an acoustic perspective. The original space was tall, narrow and

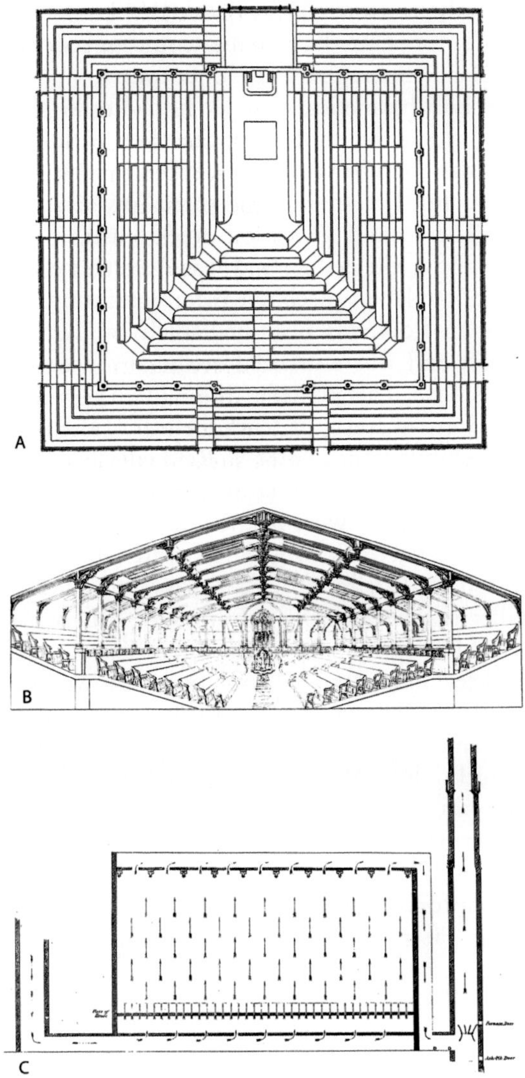

Figure 1.1 Reid's sketches outlining his proposal for the debating chamber (Select Committee on Ventilation, 1835). Key: A. Floor plan of the square chamber, B. Interior perspective, C. Diagrammatic cross-section showing stack ventilation system.

rectangular in plan. By contrast, his chamber had a square floor plan (90 ft × 90 ft [~27 m^2]) and the entire seating was accommodated on raked floors without projecting galleries; similar to his laboratory, the space was covered with a low and pitched ceiling, which provided a surface for the reflection of sound. Although this was much larger than the current debating chamber of the House of Commons – which only measures 45 ft × 64 ft (13.7 m × 19.6 m) – if the attendance was low, the size of Reid's chamber could be reduced to a core area of 70 ft × 70 ft (~21 m^2) through the use of curtains.

Reid envisaged the chamber to be a hermetically-sealed and climatically-controlled environment. In his sketches he showed a chamber with two towers. One tower, connected to the chamber through a subterranean network of air passages, acted as a central fresh-air inlet and allowed air to be sourced from a high altitude. The second tower, referred to as the 'up-cast shaft', was linked to a space above the ceiling and was used to extract the vitiated air. The system did not rely on mechanical fans: utilising the principle of stack ventilation, the air flow was sustained solely by thermal convection inside the up-cast shaft and was enhanced artificially with the aid of a coke furnace at the shaft's base. This provided the motive power to both remove the vitiated air and draw in fresh air from the top of the inlet shaft. Although his drawings only showed one chamber, Reid envisaged the ventilation towers to be part of a much larger ventilation network, which – in addition to the two debating chambers – were to include 30 committee rooms. He likened the system to the network of gas and water mains, in which tubes and dampers are used to distribute fluids across cities.[42] This principle was trialled at a smaller scale within his laboratory. A network connected to the central stack could be expanded or reduced in size to serve a larger or smaller number of spaces; dampers inside the ducts allowed the experimental rooms and chemistry benches to be temporarily connected or disconnected from the central chimney at any time.

In addition to presenting a technological solution, Reid believed that physiological factors would need to be taken into account in order to have both adequate ventilation and thermal comfort. Balancing ventilation and thermal comfort was challenging: even if the air was kept fresh, the ventilation could make people feel uncomfortable when the air was too cold or when currents became too strong. He proposed closely-controlled indoor climate using facilities for cooling, heating and humidity control, but recognised that to achieve comfort MPs would need protection from internal currents. The internal currents of the chambers could get intense and uncomfortable, in particular during hot weather or when the chamber was fully occupied.

To overcome this problem Reid proposed several ways of introducing fresh air into the chamber. For example, one proposal was to reduce the intensity of the air currents by covering the entire floor of the chamber with several thousand small apertures: he claimed that if the fresh air was kept at the right temperature and sufficiently diffused, it would be possible to have high ventilation rates without causing uncomfortable currents. The velocity and temperature were to be constantly monitored, and adjustment was to be made in response to changes in attendance.[43] Reid argued that there would not be 'the slightest danger of any feelings of cold'.[44] If the temperature could not be kept at the right temperature, e.g. during very cold weather, Reid proposed to switch the supply to the ceiling where the air could enter without exposing MPs to the cold currents.[45] In another statement he proposed an arrangement where the air was introduced at 'some height' rather than through the floor surface and noted that even with 10,000 apertures 'there is some danger of the members feeling the effect of the direct introduction of the cold air in very cold weather'.[46] Reid's sketches show an arrangement in which air is introduced partly through iron gratings in the floor and partly through the back of the seats.[47]

The idea of testing Reid's concept

Reid's proposal made a strong impression on the Committee and during the last two sittings, held on 26 and 28 August 1835, it engaged Reid himself and the architect Smirke in discussions about testing it inside the temporary chambers. Since their opening in February 1835, both chambers had had problems with ventilation, which were particularly severe in the House of Lords (the issue had been raised in debates and became the subject of a formal investigation by a select committee in July 1835).[48] During a sitting on 29 June 1835 Lord Brougham claimed that 'its temperature was bad, for though it might require healthy ventilation it was full of draughts, so that he was in absolute horror when he heard of opening apertures for air'.

Viscount Duncannon also asked 'whether any apertures could be made to admit the air so as to ventilate the House in a better manner than at present'.[49] Smirke did not deny that there were problems and reported that the fresh air for the House of Commons was supplied through 78 operable valves in the floor but 'from the difficulty of regulating its temperature and velocity, Members often complain of the inconvenience from it, and close all apertures near them'.[50]

The Select Committee asked Reid to examine Smirke's ventilation system in the House of Commons and to show how it could be improved using the system outlined in his earlier proposal. On 26 August 1835 Reid presented the first set of sketches.[51] Although Smirke again did not deny that there were problems, he objected to the use of the actual debating chamber to trial Reid's system.[52] He warned that the cost and scale of physical alterations required to adapt the existing space would be substantial and feared that 'if the experiments fail, or succeed imperfectly', it could cause serious disruptions to the business of Parliament.[53] Smirke was doubtful that Reid could succeed in achieving optimal climate conditions within a hermetically-sealed interior. To avoid these risks he recommended adapting another, less important space for the trials. In response to Smirke's criticisms, Reid proposed to erect a life-size model of the debating chamber for the trials, arguing that the cost of the model – estimated at £1,000 to £2,000 – would be negligible in the design of a public building of this scale.[54] This proposal directly built on methodologies that Reid had used during earlier studies in Edinburgh, conducting experiments with full-scale mock-ups, which according to Reid, allowed to 'confine ourselves to principles alone, and not trouble ourselves with ornament'.

Despite Smirke's objections, the Committee insisted on testing Reid's principles within the temporary House of Commons, but proposed to confine it to periods when it was not occupied. In its final report, dated September 1835, the Committee wrote that Reid's principles should 'be submitted to the test of actual experiment" with the objective of "ascertaining with accuracy the soundness of the principles' and also to determine the feasibility of its possible application within the new Palace of Westminster after the competition. Plans for undertaking the trials during the autumn recess were discussed by the Office of Woods after the report and Reid produced set of more detailed drawings for its application in the temporary House of Commons.[55] However, these plans were postponed after Smirke restated his objections in another letter to the Commissioners in November 1835.[56] Although Reid's trials faced an uncertain future, Reid did not give up in his efforts to provide evidence against these reservations.

A first demonstration of Reid's principles within a physical model

Although Reid's trials faced an uncertain future, Reid did not give up in his efforts to provide evidence against these reservations. Using his own initiative and at his own expense, Reid constructed a physical mock-up of his proposal to prove its feasibility through physical demonstrations. Reid wrote that it was constructed with the 'view of imitating the various circumstances that present themselves during the debates'.[57] This model, completed in Spring 1836, was erected on the site of his laboratory in Edinburgh.[58] It comprised a hermetically-sealed room, which was equipped with arrangements for ventilation, climate control and artificial lighting: fresh air was supplied through a network of underfloor channels and entered the model through 50,000 holes within the floor[59]; and the vitiated air was extracted at the ceiling, which was linked to the main ventilation shaft of the laboratory. The model was also equipped with a system of flues and dampers, which allowed the direction and intensity of the incoming air currents to be altered.[60]

Reid demonstrated two different approaches to integrating gas lighting inside the chamber without exposing MPs to exhaust fumes or heat. In one arrangement the gas flames were concealed behind a glazed cornice, which ran along the edge of the ceiling. (**Figure 1.2, 'C,' 'D' & 'E'**). In the other arrangement the burners were placed inside four glass pendants in the centre of the ceiling[61] (**Figure 1.2, 'A,' 'B' & 'C'**). In both cases the flames were isolated from the atmosphere of the chamber and the exhaust fumes were extracted through separate flues that terminated inside the ventilation shaft.[62] With a length of 30 ft (9 m) and width of 20 ft (6 m),[63] the model

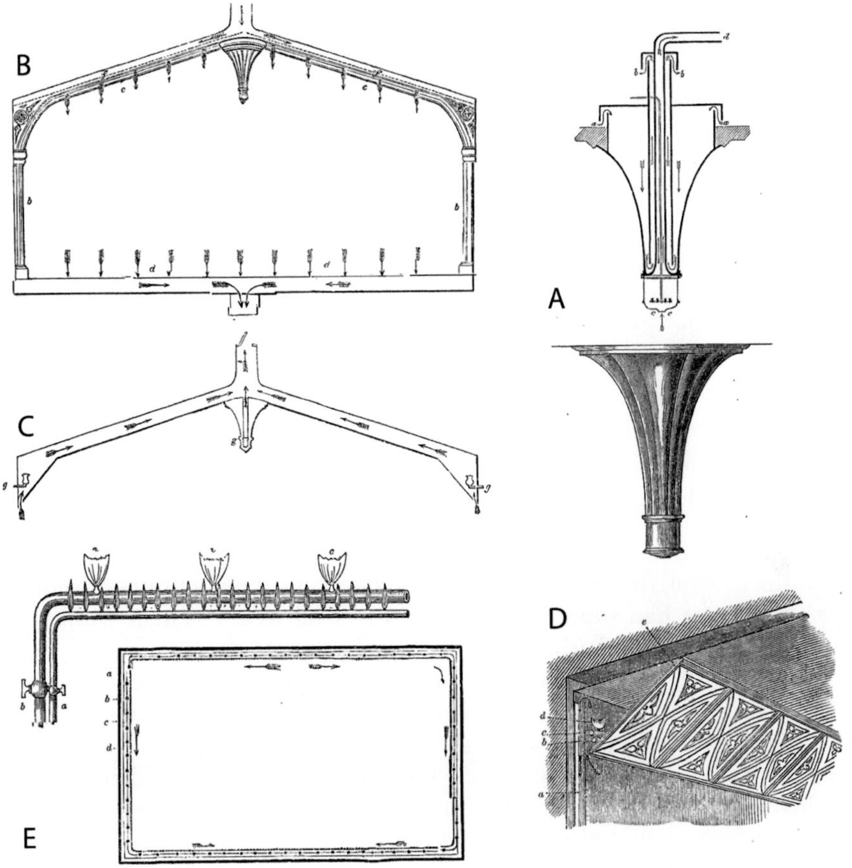

Figure 1.2 Reid's model of the debating chamber, 1836. Key: A. Detail of pendant gas light (*Illustrations*, p. 31), B. Cross section showing ventilation operated in downward mode (*Illustrations*, p. 303), C. Cross section showing gas light above glazed cornices, with fumes being extracted through void above ceiling (*Illustrations*, p. 303), D. Detail of ornamental glass cornice with lights above (*Illustrations*, p. 304), E. Plans showing gas pipes with rows of burners inside glass cornice (*Illustrations*, p. 305).

was not full-size but was large enough to undertake trials involving human participants.

Groups of 12–234 participants were involved in these trials in order to simulate the impact of people on the state of the internal environment, and also to report their experience.[64] The first trials, involving groups of over 100 participants, were undertaken in the summer of 1836: Reid exposed participants to different climatic conditions by adjusting the temperature, relative humidity and intensity of internal air currents.[65] In July 1836 the

Caledonian Mercury reported that Reid tested the ability of his system to dissipate odours by releasing 'ether' and nitrous oxide into the atmosphere and also conducted trials of introducing air at different temperatures.[66] It wrote that the 'air was completely renewed by a slow and insensible current every five minutes, and the various changes so gradually induced, that it was impossible to tell when they commenced'.

Reid also tested his proposal for switching the supply to the ceiling of the chamber, which was to be deployed over periods when the rising currents around the feet and legs caused discomfort[67] (**Figure 1.2, 'B'**). In July 1836, Reid took members of the Philosophical Society to his debating chamber for a live demonstration of the ventilation,[68] including Sir Thomas Brisbane, president of the Royal Society in Edinburgh.[69] In 1839 the Royal Society even held a dinner inside the model itself, during which Reid entertained his scientific guests by demonstrating the various artificial environmental conditions that could be produced with his system.[70]

Architectural responses to the requirements of light and air

The inquiries undertaken by the Select Committee on ventilation had little influence on the outcome of the architectural competition. Although the Committee had supported proposals to trial Reid's system within the temporary House of Parliament, it was reluctant to recommend any specific system for the new Palace of Westminster. The Committee asked the architects to reserve space above and below the debating chambers and to revisit the question after the competition. In its final report of 1835, the Committee noted that the architects should be made aware of their responsibility for ensuring the successful integration of the building services, and that drawings should also be completed prior to the start of construction. This was necessary not only to 'ensure its success' but also 'to prevent needless expense and inconvenience' of retrospective changes.[71]

However, these recommendations were not followed. The review of the competition entries was undertaken by the Commission for Selection of plans for the rebuilding of the House of Parliament. This panel was composed of five MPs and chaired by Viscount Duncannon, First Commissioner of the Woods. Between October 1835 and January 1836, the panel had reviewed 97 submissions – with nearly all proposals including brief notes on heating, lighting and ventilation[72] – and produced a shortlist of four architectural designs. They had also collaborated with the architects of these designs in further improving them through revisions to their original proposals.

The architectural scheme of Barry and Pugin was formally announced as the winner of the competition in February 1836.[73] During the selection or revision process, the potential implications of adopting Reid's proposals and systems were not discussed. With the exception of the addition of the space the debating chamber, the findings of the Select Committee were not used to inform the revisions to the architectural scheme. The original drawings show that Barry introduced spaces for the accommodation of 'warming and ventilating apparatus' on the ground floor below the two debating chambers.[74] Although the selection panel included one MP who, due to his previous membership in the ventilation committee, was familiar with Reid's ideas, did not consider himself sufficiently qualified to assess these technical aspects. In its final report, dated 9 May 1836, the panel wrote 'that with the imperfect knowledge we possess on the theories of sound and artificial ventilation, we deemed ourselves scarcely authorized to allow those subjects to have weight in determining our preference'.[75]

Despite its reluctance to engage with Reid's scheme, the panel had given some attention to the requirements of natural ventilation and lighting, reviewing the schemes in terms of their ability to provide 'equal distribution of light and air through every part of the structure'.[76]

In the spring of 1836, another select committee was appointed by the House of Commons to undertake a critical appraisal of the procurement, involving interviews with Barry and individual MPs from the selection panel. According to their testimonies, the panel had focused on addressing the fundamental architectural requirements of the traditional principles of natural ventilation and lighting. These principles were distinct from the architectural requirements of a mechanically-serviced building of a type that Reid had outlined in his proposal. In the final phase of the selection process, the panel shared and discussed with Barry the findings of their reviews, highlighting their concerns, amongst others, about insufficient access to natural light and air. Barry substantially amended his drawings to address these issues. In his oral statement, given on 22 April 1836, Barry noted that these changes were largely undertaken for the 'purposes of increased light and ventilation'.[77] The original drawings and sketches show that both the number and size of the courtyards were significantly increased whilst the depth of interior spaces was reduced. These changes optimised the spaces for the use of daylight and traditional methods of cross-ventilation.[78] These changes show that Barry and the Commissioners were following a more conventional approach to ventilation – one which relies on the use of operable windows within courtyards and external elevations – and did not consider the architectural implications of Reid's proposal for a building that was hermetically sealed and equipped with a system of 'artificial ventilation'. Reid's system was deferred for three years.

In May 1836, the Select Committee approved Barry's plan in principle, but made its adoption contingent on the review of its cost. The final decision was postponed for six months to give Barry's team time to produce more accurate estimates. After the Select Committee of 1836 had completed its inquiry, full responsibility for managing the delivery of the project was transferred to the Office of Woods. During the six-month hiatus, Barry's practice produced more detailed drawings and specifications, and, with the assistance of Messrs consulting engineers James Walker and Alfred Burges, provided a new estimate. These documents were presented to the Office of Woods for an internal review in January 1837. However, although they offered a greater level of detail, they still did not cover the cost of the mechanical ventilation system. In April 1837 the Office of Woods submitted a final report.[79] The total estimate was £707,104, which included the architect's fee of 10%. The report also highlighted the fact that this figure excluded the cost of a number of important and expensive features that could not be accurately estimated until the design was in a more advanced stage. Amongst these omissions were the costs for 'internal lighting, warming and ventilating the several parts of the building'.[80] The records show that it took a further four years until a general design and the costs for a ventilation scheme were developed and approved by Parliament.

From theory towards the experimental phase

The approval of the general budget marked the end of the first phase of the development of the new Palace of Westminster. It resulted in the successful procurement of an architectural scheme, and also involved an initial inquiry into how environmental technology might be used (although the latter was confined to entirely theoretical investigations). Reid's scheme for the debating chambers was hypothetical as its potential application within the actual Palace of Westminster was not explored. Although it is not unusual for technical requirements to receive scant attention in the early stages of an architectural design, it had serious implications for the later development of Barry's scheme (see Chapters 4 and 5).

In retrospect it is evident that the cost, time and physical implications of adopting Reid's principles were seriously underestimated. As the adoption of Reid's principles in the new Palace of Westminster had not yet been agreed, the cost of doing it, the time it would take and its physical implications were seriously underestimated. This would not come for three years as it was contingent on the success of further trials of his principles. In the Autumn 1836 Reid was commissioned by the Office of Woods to coordinate independent trials of his principles within the temporary Houses of Parliament. As a result the environmental strand, initiated in 1835 through the appointment of a select committee, continued to evolve in isolation from the development of Barry's architectural plans. This period, which will be examined in Chapters 2 and 3, may be described as Reid's 'experimental phase' during his involvement in Westminster. This phase shows how Reid adapted his scientific working methods to test and refine technical solutions under real-life conditions. It was a critical step in the transition from scientific research, undertaken within the artificial setting of his laboratory, towards the field test of systems inside an actual building. It also enabled MPs and Lords to gain direct experience with the type of controlled environments that Reid was hoping to realise within the new Palace of Westminster.

Notes

1 Description of a ventilation shaft on the roof of the Palace of Westminster, *Illustrated London News*, 16 December 1854.
2 Elizabeth Hallam Smith, 'Ventilating the Commons, Heating the Lords, 1701–1834', *Parliamentary History*, 38 (2019), pp. 74–102.
3 Peter Collins, *Changing Ideals in Modern Architecture, 1750–1950* (Quebec: McGill-Queens University Press, 1998), p. 238.
4 House of Commons (HC) 262 1835, *Report from the Select Committee on Rebuilding Houses of Parliament*, p.5.
5 HC 66 1835, *Report of the Select Committee Appointed to take into Consideration Mode of Taking Lists of Divisions of the House*.
6 HC 262 1835, Select Committee on Rebuilding, p.3.
7 The competition brief requested drawings on a scale of 20 ft to 1 in (600 cm to 2.5 cm).
8 HC 583 1835, Report from Select Committee on the Ventilation of the Houses of Parliament.
9 Robert Bruegmann, 'Central Heating and Forced Ventilation: Origins and Effects on Architectural Design', *Journal of the Society of Architectural Historians*, 37(3) (1978), pp. 143–60.
10 'Proceedings of Societies', *The Edinburgh Philosophical Journal*, 64 (1863), pp. 301–3.
11 David Boswell Reid, *Illustrations of the Theory and Practice of Ventilation* (Longman, 1844), p. xiv; Reid, *Brief outlines illustrative of the alterations in the House of Commons* (Edinburgh: Neill & Company, 1837), p. 1; Reid, *Ventilation in American dwellings: with a series of diagrams, presenting examples in different classes of habitations* (New York: Wiley & Halsted, 1858), p. xxv.
12 Oral statement by G. Birkbeck, 4 August 1835 in HC 583 1835, *Select Committee on Ventilation*, Q33.
13 Oral statement by W. Brande, 12 August 1835, in HC 583 1835, *Select Committee on Ventilation*, Q368.
14 Oral statement M. Faraday, 7 August 1835, in HC 583 1835, *Select Committee on Ventilation*, Q306.
15 Oral statement by R. Smirke, 28 August 1835, in HC 583 1835, *Select Committee on Ventilation*, Q542.
16 'Proceedings of Societies', *The Edinburgh Philosophical Journal*, 64 (1863), pp. 301–3.
17 Oral statement, 16 March 1846, HC 1846, Q853.
18 Reid, *Elements of Practical Chemistry* (Edinburgh: Maclachlan & Stuart, 1830); *Text-Book for Students of Chemistry* (Edinburgh: Maclachlan & Stuart, 1836); *Rudiments of Chemistry with Illustrations of the Chemical Phenomena of Daily Life* (Edinburgh: William and Robert Chambers, 1836).
19 Town Plan of Edinburgh (south east part), Scale: 1:1056, Ordinance Map Office, Southampton, 1852.
20 Oral statement by D. B. Reid, 21 August 1835, in HC 583 1835, *Select Committee on Ventilation*, Q412–420.
21 *Brief outlines*, p. 13.
22 *American Dwelling*, p. xx.
23 Ibid.
24 HC 583 1835, *Select Committee on Ventilation*, pp. 34–40.
25 Ibid.
26 Oral statement by D. B. Reid, 21 August 1835, in HC 583 1835, *Select Committee on Ventilation*, Q387.

27 Reid, 'Progress of Architecture', in *Smithsonian Annual Reports,* 1856 (Washington: Smithsonian Institution), pp. 152f.
28 HC 277 1837–38, Ventilation of the House: Reid, D.B. Letter to Viscount Duncannon, in reply to observations addressed to His Lordship by Sir Frederick Trench.
29 Ventilation of the House: Reid, D.B. Letter to Viscount Duncannon, in reply to observations addressed to His Lordship by Sir Frederick Trench (HC 277 1837–38)
30 'Progress', p. 153.
31 *Elements of Chemistry,* p. 668.
32 Reid, Letter to the Council of the University College, London, in 'Testimonials regarding Dr D.B, Reid's qualification as a lecturer on chemistry and teacher of practical chemistry, April 1837', Hume Tracts, UCL Library.
33 Reid, 'Progress', p. 153.
34 HC 277 1837–38, Ventilation of the House: Reid, D.B. Letter to Viscount Duncannon, in reply to observations addressed to His Lordship by Sir Frederick Trench.
35 Reid, 'Progress', pp. 153f.
36 Reid, 'The Revision of Architecture in Connection with the Useful Arts', in *The Builder,* 5 May 1855, pp. 208f.
37 'Progress', pp. 163–7; 'Revision of Architecture', pp. 208f.
38 'Revision of Architecture', pp. 208f; *Brief outlines*, pp. 23f; See also 'Ground Plan' (see note 36).
39 Reid, 'Progress', p. 153.
40 Reid, 'Revision of Architecture', pp. 208f.
41 HC 583 1835, *Select Committee on Ventilation*, Plates 1–5.
42 Oral statement by D. B. Reid, 21 August 1835, in HC 583 1835, *Select Committee on Ventilation*, Q409.
43 Ibid, Q426–7.
44 Ibid., Q511.
45 Ibid, Q508.
46 Ibid, Q439.
47 HC 583 1835, *Select Committee on Ventilation*, Plate 1.
48 *Report of Select Committee appointed to consider a plan for a more convenient temporary accommodation for the House of Lords, 3 July 1835* (Parliamentary Archives: Arc/Pro/Work 11/10/2, no. 450); *House of Lords, Minutes of Proceedings, 12 June to 24 August 1835* (Parliamentary Archives, Arc/Pro/Work 11/10/2).
49 House of Lords, *Hansard's Parliamentary Debates: The Official Report (29 June 1835, vol. 29, cols. 3–5) (London: Hansard).*
50 Oral statement by R. Smirke, 28 August 1835, in HC 583 1835, *Select Committee on Ventilation*, Q542, Q552–55.
51 Hawes, B. Letter to the Commissioners of Woods, 26 August 1835 (Parliamentary Archives, Work 11/12, no. 2).
52 Smirke, R. Letter to the Commissioners of Woods, 11 November 1835 (National Archives, Work 11/12, no. 6).
53 Oral statement on 26 August, HC 583 1835, Q527–41, Q542.
54 Oral statement on 28 August, Q581.
55 Hawes, B. Letter to Viscount Duncannon, 8 September 1835 (National Archives: Work 11/12, no. 3); Humes, W. Letter to J. Philipps, 27 September 1835 (Work 11/12, no. 5).
56 Smirke, R. Letter to the Commissioners of Woods, 11 November 1835 (National Archives: Work 11/12, no. 6).

57 D.B. Letter to Viscount Duncannon, 28 March 1838 (HC 277 1838).
58 'Ground Plan of Dr Reid's premises' (Parliamentary Archives, OOW/23).
59 'Philosophical Society', *Caledonian Mercury*, 28 July 1836.
60 Drawings of the damper system are shown in HC 277 1837–38, Reid. D.B. Letter to Viscount Duncannon, 28 March 1838, p. 9: figures 6 and 7.
61 Reid, *Illustrations*, pp. 302–4.
62 Ibid., pp. 30–31 (detail) and p. 303 (in context).
63 'Ground Plan of Dr Reid's premises', Roxburgh Place, Edinburgh (Parliamentary Archives, OOW/23).
64 HC 277 1837–38, Ventilation of the House: Reid, D.B. Letter to Viscount Duncannon, in reply to observations addressed to His Lordship by Sir Frederick Trench. Reid, D.B. Letter to Viscount Duncannon, 28 March 1838 (HC 1838 277), pp. 6–7.
65 Reid, *American Dwelling*, pp. xv-xxxvii; *Illustrations*, pp. 177, 86, ; 'Progress', p. 164–77.
66 'Philosophical Society', *Caledonian Mercury*, 28 July 1836.
67 Reid, *Illustrations*, p. 273; See also HC 277 1837-38, Ventilation of the House: Reid, D.B. Letter to Viscount Duncannon, 28 March 1838, in reply to observations addressed to His Lordship by Sir Frederick Trench.
68 'Philosophical Society', *Caledonian Mercury*, 28 July 1836.
69 Reid, D.B. Letter to Viscount Duncannon, 28 March 1838.
70 Reid, D.B. Letter to the Commissioners of Woods and Forests, 1 December 1839 (National Archives, Work 11/12 no. 24).
71 HC 583 1835, p. iii.
72 *Catalogue of the designs offered for the new Houses of Parliament, now exhibiting in the National Gallery* (London: W. Clowns and sons, 1836) (National Art Library, V&A: 33.C.30).
73 HC 66 1836, *Report of the commissioners appointed to consider the plans for building the new Houses of Parliament*; HC 245 1836, *Report from Select Committee on Houses of Parliament.*
74 Drawings of basement floor, undated (Parliamentary Archives: Work 29/87).
75 HC 66 1836, *Report of Commissioners*, p.2.
76 Ibid.
77 Oral statement by C. Barry on 22 April, HC 245 1836, *Select Committee on Plans*, Q131.
78 Principal floor plan, undated, (Parliamentary Archives, Work 29/81).
79 HC 216 1837, *Report of the Commissioners of Woods, Forests, &c. on Mr Barry's plans and estimates for building the proposed new Houses of Parliament*, pp. 1–2.
80 Ibid., p. 2.

2 Reconstructing the temporary Houses of Parliament: 1835–51

Introduction

On 8 September 1843, Lord Sudeley testified that:

> *The pestilential atmosphere of the House of Commons was notorious; its baneful effects on the health and energies of its members were painfully felt and admitted: means from time to time were resorted to, to correct the evil, till scarcely a hope if any, remained even that it could be lessened, and the most sanguine never dreamt that it could be cured, much less that the ventilation of the House could be brought to such a degree of perfection as to defy the chills of winter and heat of summer, or the effect of the numbers, however great, congregated within the walls of lessening its effects.*[1]

By this time, six years had passed since David Boswell Reid had been authorised to test his system inside the temporary House of Commons. Lord Sudeley's letter represented a piece of qualitative evidence that testified to the success of Reid's system in the temporary House, countering doubts about its feasibility, which had been initially raised by the Commissioners of the Woods in Autumn 1835. The 1835 Select Committee on Ventilation saw this as a means to overcoming the defects of the existing arrangements as well as an opportunity to demonstrate the technical feasibility of his novel idea. Reid's idea was to transform the temporary debating chambers into a hermetically-sealed and climatically-controlled environment. The trial was initially postponed after concerns about it were raised by Robert Smirke, architect of the temporary Houses of Parliament. In response to growing pressure from MPs, the Commissioners of the Woods revisited the plans in Spring 1836. Although the debating chamber were only intended to be temporary, MPs were not prepared to tolerate the bad quality of the ventilation and acoustics within them. Indeed, in May 1836 the *Preston Chronicle* reported that 'legislators in both Houses complain bitterly of being killed by bad air',[2] and in the same month Reid was re-engaged by the Commissioners of Woods to examine the existing ventilation arrangements and to produce drawings.[3] During a budget debate on 10 August 1836,

Benjamin Hawes, former chairman of the Select Committee on Ventilation, asked the Chancellor of the Exchequer to authorise funds for the trial.[4] Three radical Members – Joseph Hume, MP for Middlesex; Henry Warburton, MP for Bridport; and Thomas Wakley, MP for Finsbury – supported this motion with the view that Reid's plans should be implemented over the winter recess. John Bowring, MP for Kilmarnock Burghs, also stressed that it offered the opportunity to test Reid's principles before they were applied at the Palace of Westminster, noting that 'they would have the experience and benefit of that experiment'.[5]

In late August, the Treasury approved Reid's plans. This agreement was significant as it finally enabled Reid to test his concepts under real-life conditions, and also to directly involve MPs in the process of evaluating and improving its performance in use. The experimental system, which remained operational for 14 years, was extended to cover the House of Lords after the first two years.

This chapter provides a reconstruction of Reid's experimental system and its operation, and will also examine how feedback about its day-to-day performance was gathered and utilised. The empirical evidence that Reid collected over the first three years of the trial was instrumental in giving the Commissioners of the Woods confidence in his work and also in their appointment of him as the service architect for the Palace of Westminster.

Reid's experimental system in the temporary House of Commons

Reid was granted a budget of £3,657 for the construction of his experimental system, with an additional £390 to cover the cost of his time and personal expenses.[6] In Autumn 1836 he came to London for several weeks to coordinate the construction.[7] He was assisted by former members of the Select Committee on Ventilation, including Benjamin Smith and Benjamin Hawes, and by Lord Sudeley.[8] The design was discussed in various journals, newspapers and in the more specialist technical literature.[9] Reid published a first full account of his system in his article *Brief outlines illustrative of the alterations in the House of Commons*, with further explanations provided in his books *Illustrations of the Theory and Practice of Ventilation* and *Ventilation in American Dwellings*.[10] The underlying design objectives of Reid's system were also explained in various letters and select committee reports.[11] These records show that Reid's original 1835 plans for an ideal debating chamber had to be significantly adapted to apply them to the existing chamber.[12] In addition to introducing the technical infrastructure required for ventilation and climate control, Reid also remodelled the original interior, assisted by Lord Sudeley, who was in charge of architectural treatment, into a more functional space. Similar to his hypothetical 1835 proposal, its shape and materiality were dictated by the requirements of acoustics and ventilation. For example, acoustic considerations required the introduction of sound-absorbing carpets and the lowering of the galleries and ceiling, underneath which were installed sloping surfaces that acted as sound reflectors. It is evident from historical illustrations that Sudeley had given some attention to the architectural detailing of the ceiling panels and glazing, despite the temporary nature of the building.[13] A false ceiling – composed of a central panel and two sloping sides – was introduced only a few feet below Smirke's original ceiling,[14] primarily to improve the acoustics,[15] although Reid admitted that it contradicted the spatial proportions traditionally considered ideal for ventilation.[16]

Traditionally, a tall space was considered ideal as it was thought to provide a larger reservoir of air, thereby slowing down the speed at which air was vitiated by human respiration or artificial lighting. Moreover, air only needed renewing occasionally by opening windows. A low ceiling, which was a feature of the temporary Houses, was therefore thought to significantly reduce the volume of breathable air.

To overcome this problem, Reid adopted an alternative ventilation strategy: instead of providing occasional ventilation through the opening of windows, a slow but constant current of air rising from the floor to the ceiling was maintained.[17] The ceiling of Reid's model chamber had a similar height too but was wider than the implemented scheme. The floor and ceiling area

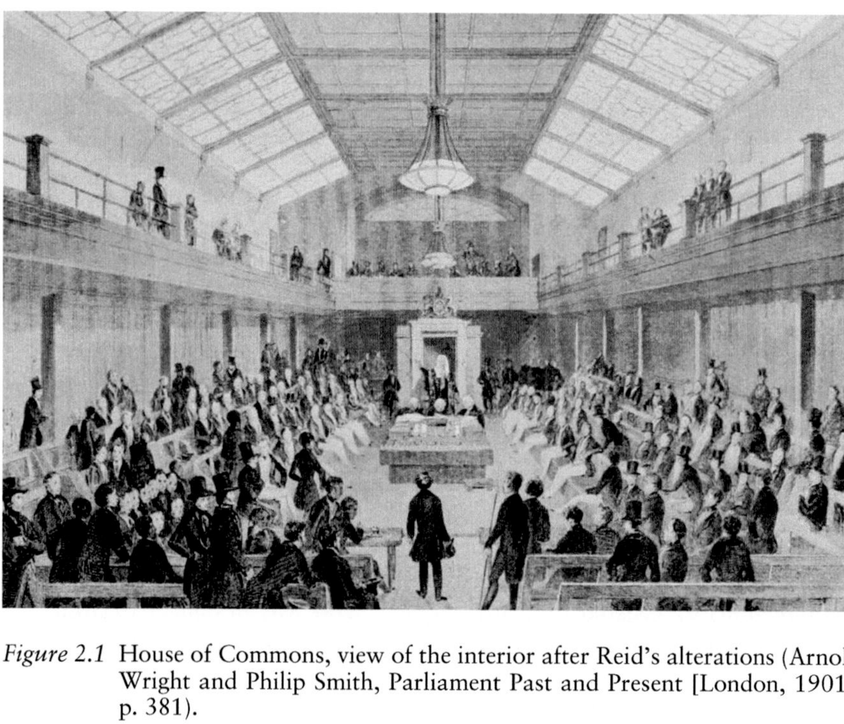

Figure 2.1 House of Commons, view of the interior after Reid's alterations (Arnold Wright and Philip Smith, Parliament Past and Present [London, 1901], p. 381).

available for the admission and discharge of air was significantly smaller than in the original 1835 proposal. The air openings within the ceiling were also confined to the central portion, which was composed of wooden panels. Gaps between the panels and the supporting beams allowed hot and vitiated air to escape into the discharge channel above. The slanting sides were filled with ornamental glazing, which admitted the natural light entering through the pre-existing clerestory windows behind them. The chamber was now completely sealed with no windows that opened directly into the chamber. Air was supplied exclusively through the floor and discharged through the ceiling. The ventilation was driven by a 210-ft-tall (64 m) ventilating chimney, located inside Cotton's Garden on the east side.[18] According to *Berrow's Worcester Journal*, the chimney was a freestanding shaft of utilitarian character, giving the building the 'outwards appearance of manufactury or gasometer'.[19] The base of the chimney was connected to the ceiling of the House through a vertical, underground tunnel.[20] Despite its significant height, the chimney is shown in very few historical images. Amongst these rare images is a watercolour painting by Henry Nibbs from May 1839.[21]

The intake shaft shown in Reid's original plans from 1835 were not implemented. Instead, fresh air was admitted through an intake at ground level. The intake comprised a perforated wall, which faced Old Palace Yard, and was covered internally with removable screens of moist fabric to catch the larger dust and soot particles contained in the external air. These screens were renewed daily. According to Reid, '200,000 visible portions of soot are sometimes excluded on a single evening' at periods when the atmosphere was extremely polluted.[22] Behind the screen was a 'receiving chamber', where the air was sprayed with water using a system of sprinklers to remove finer impurities.

The air passed from this chamber through two air passages into a 'lower air chamber'.[23] Situated at the north end of this chamber was a cold-air passage, whilst a second air passage with a heating apparatus was situated at the south end. During summers, the heating apparatus was also used for cooling air by circulating cold, rather than hot, water through the radiator cases. Both passages had doors, which acted as dampers, enabling the path the air would take to be controlled before entering the lower air chamber. A 'communication chamber', linking the two passages, was introduced to allow hot and cold air to be mixed. This allowed the temperature of the supply air to be adjusted before it was admitted into the debating chamber. The attendants who operated the doors had to work under uncomfortable conditions. Reid, speaking of one of the attendants, John Urquhart, expressed his sympathy for him as he, and other attendants, had to experience the 'effects of severe exposure in the various hot, cold, and mixed air-channels into which it was necessary for him to pass and repass continually'.[24]

Although Reid's alternative ventilation strategy followed the principle of a warm-air central heating system, which was a well-established technology in the 1830s,[25] he also incorporated facilities for air filtration, cooling,

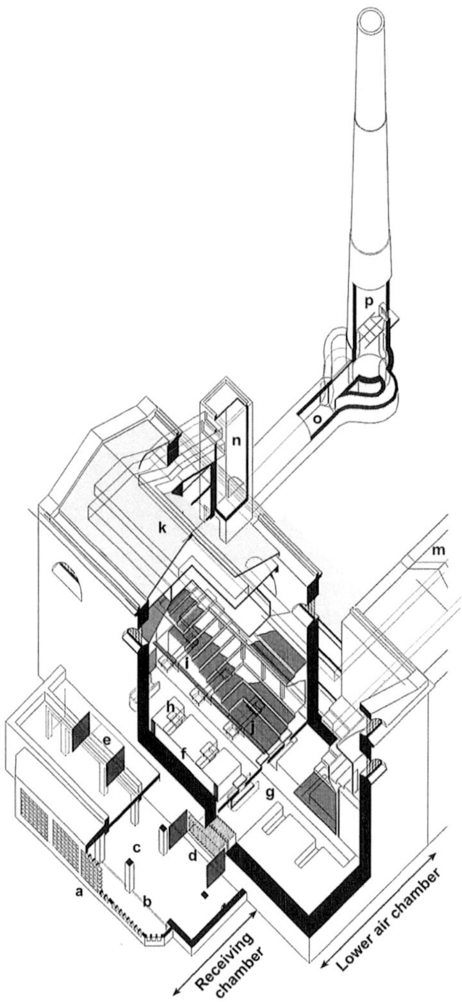

Figure 2.2 Axonometric projection of Reid's system in temperature at the House of Commons (author's own drawing). **Key:** A. Air intake facing the old Palace Yard with canvas screens for the filtering of soot particles, B. Nets containing lumps of ice for cooling during hot weather, C. Water sprinklers for air purification, cooling and humidification, D. Warm air passage, E. Cold air passage, F. Mixing channel linking passages 'D' and 'E', G. Central channel of lower air chamber, H. Square opening conveying air into equalising chamber, I. Equalising chamber, J. Perforated floor covered with horse-hair carpets, K. Vitiated air channel (Commons only) above ceiling, M. Second vitiated air channel linking shaft to House of Lords chamber, N. Descending shaft leading to air tunnel, O. Air tunnel leading to base of discharge shaft, P. Discharge shaft with coke fire.

heating and humidification.[26] Sturrock and Lawson-Smith have argued that the system was an early form of air-conditioning,[27] but it should be distinguished from modern 'mechanical air-conditioning' as it involved the use of largely 'passive' approaches to cooling or humidity control. A combination of three cooling methods were deployed: first, air was cooled by circulating cold water through the central heating pipes; second, through evaporation using the water sprinklers of the air purification system; third, air was drawn through nets filled with ice. However, this latter method was stopped early due to the prohibitive cost of ice.[28] The ability of these methods to physically lower the air temperature was limited, and the system was thus largely dependent on ventilation to improve comfort using 'ventilative cooling'.[29] At night-time, ventilation was used to cool the structural fabric (e.g. the masonry), whilst during sittings it was deployed to both counteract heat gain from people, sunlight and lighting, and enhance comfort by exposing MPs to currents, which cooled the body and increased evaporation of perspiration from the skin. Reid argued that even at temperatures as high as 75°F, people would feel comfortable if air movement was to be increased.[30]

Once inside the central passage of the lower air chamber, the fully-tempered air rose through an array of square openings, which were covered by dispersers, into the equalising chamber, which was located below the perforated floor of the debating chamber itself. The incoming air currents were uniformly distributed over the entire floor surface on both levels to reduce the strength of incoming currents to which Members were directly exposed.[31] The whole floor was perforated by several hundred-thousand conical holes.[32] MPs were even reported by one newspaper to be placed on this porous floor 'like spinach on a colander'.[33] The air was further diffused by hair-cloth carpet, the fibres of which provided the sound-absorbing surface required to reduce the reverberance to a level suitable for speech.[34] In addition to extracting air from the ceiling, the chimney put the whole chamber under a negative pressure, which forced fresh air into the House through the holes in the floor. During the winter, according to Reid, the thermal buoyancy produced by the heating system, lights and bodies was sufficient to propel the supply air into the building with only minimal assistance from the ventilating fire.[35] This fire was only kept at a low heat to provide the pressure required to force the hot air from the ceiling duct down the descending shaft into the underground tunnel.

Operation and maintenance under the stewardship of a scientist, 1837–51

In November 1836, after completing the installation of his system, Reid returned to Edinburgh. However, his involvement in Westminster and the development of the system did not cease. Reid was employed by the Office of Woods as a superintendent in charge of overseeing the system's maintenance and operation throughout the 14-year duration of its operational life from 1837–51. For the financial year of 1838–39 he received a significant wage of £200.[36] In addition, he was also responsible for overseeing the ventilation of the House of Lords chamber, from 1840–47.

Due to his commitments in Edinburgh, Reid was unable to constantly supervise operational procedures in person. Instead, the Office of Woods supplied him with weekly, written updates about its performance and any issues that had been encountered, including extracts of climate data and reports on technical issues.[37] During the first four years away from Westminster, Reid only attended in person during State Openings, other important debates or for further technical trials. In the summer of 1837, for instance, he returned to Westminster to coordinate the testing of cooling and humidification arrangements. This absence led to some difficulties staying up to date with any issues with the system, and he was at times also summoned to London by the Office of Works to discuss problems that MPs had raised.[38]

This situation changed in 1840 when Reid moved permanently to London in order to begin his collaboration with Barry on the design of the Palace of Westminster. Reid got more directly involved in the day-to-day operations, reporting in 1843 that he was 'rarely absent from the time the Houses meet till they adjourn'.[39] His presence in London enabled him to gain first-hand knowledge of its performance, engage with MPs and Lords in conversations about their experiences and expectations, and involve the operational staff in studies of external air pollution and technical experiments (see Chapter 3).

Intelligent control through 19th-century technology

As part of his role as superintendent, Reid constantly studied and improved the performance of his system. He also developed a sophisticated monitoring and control regime, which was critical as the effectiveness of the system was as dependent on the quality of its management as it was on physical infrastructure. In a letter to the First Commissioner, dated 28 March 1838, Reid emphasised that it was necessary to 'draw up those definite roles for the use of those who superintend daily the working of the ventilating apparatus without which it is impossible for them to work it with that precision of which it is susceptible'.[40]

This regime involved the systematic monitoring of the indoor climate, user activity and technical operations. It also incorporated procedures for the collection and processing of user feedback, which offered insights into how MPs and other users were affected by the climate conditions inside the chamber. From an operational perspective, the temporary House of Commons had all the qualities of an intelligent building. However, in contrast to its 21st-century counterpart, all monitoring and control functions were performed manually (without the aid of electronic sensors and computerised controls). Achieving a responsive system manually was a significant managerial challenge: Staff – who were trained in their new roles by Reid himself – had to constantly observe the state of the external weather and number of people inside the chamber, both of which were highly variable and directly affected the indoor climate and air quality. Reid noted that such a responsive system would not be feasible 'unless they are perpetually directing their attention to the passing changes in the same manner as a sailor steering a ship'.[41] Large quantities of data had to be gathered, interpreted and communicated before any adjustments were made to ventilation dampers and the heating, cooling and air filtration systems. Its performance relied heavily on the discipline of the staff and the effective coordination of the recording and adjustment procedures. Indeed, 'unless trained parties managed the ventilation, wherever plan might be introduced, it might be thrown away if it had not right people to conduct it'.[42]

When implementing the monitoring and control procedures, Reid had to collaborate with staff from both the Office of Woods and the Department of the Sergeant-at-Arms. The primary responsibility of the sergeant-at-arms was the security within the chamber, but during sittings he was also responsible for ensuring the internal environment was well-managed. The sergeant-at-arms from 1835 to 1848, Sir William Gosset, employed an 'attendant of ventilation',[43] who was in charge of overseeing the daily work of the technical staff. Reid reported that from 1843–46 between three to four technical staff – who were responsible for operating the shaft, boilers and adjusting the valves – were stationed below the Houses.[44] The Sergeant's messengers also assisted the attendant with collecting hourly temperature readings inside the debating chamber. This attendant of ventilation role was held by Benjamin Riches.[45] In interviews given in June 1839, Riches reported that the performance and operation of the system was systematically monitored and recorded in a central logbook; abstracts of these monitoring reports, covering the period 1837–39, show that the logbook contained columns for hourly readings of indoor and outdoor temperatures, the number of persons inside the chamber and information on the position of the valve used to regulate the air flow.[46] The interior of the debating chamber was equipped with three thermometers: one situated on the main floor behind the Speaker's chair, and one on each side, east and west, on the east side of the Members' gallery.[47]

The extent to which the system was monitored increased gradually over time to allow a better understanding of its performance. In 1838, following complaints about an excessively dry and dusty atmosphere, Reid charged his deputy in London with keeping a register of the relative humidity, which had previously not been monitored.[48] In 1843 he introduced additional thermometers above the ceiling and within the fresh air inlet.[49] New columns were added to the original logsheets for the inclusion of data on air velocity, use of water and ice for cooling, and fuel for heating and the ventilating furnace. In addition to the records of measured data, management of the system relied on direct observations as well as MPs and other users' feedback to gain an understanding of the full range of environmental conditions encountered within the chamber. Important factors, which were known to affect thermal comfort, such as internal air movement, radiant heat or air quality, were not routinely monitored using quantitative methods. Air purity was not measured except for a few ad hoc measurements of the carbon dioxide level, which was used as an indicator for this.[50]

Formal processes for the collection of user feedback, which the Chancellor of the Exchequer referred to as a 'system of complaint', were introduced and fully integrated into the environmental control regime.[51] These enabled staff to stay informed about how MPs were affected by the thermal environment, and to, if necessary, make adjustments to it. As our experience of thermal comfort is entirely subjective, Reid stressed that the 'thermometer is no indicator whatsoever of the probably influence of the state of the atmosphere upon the feelings'.[52] Under normal conditions attendants followed a series of instructions: one for the winter and one for the summer.

During the winter, attendants were instructed to maintain a temperature between 60°F (15°C) and 63°F (17°C).[53] The House was heated to 62°F (16°C) before the House sat, whilst during a sitting, a temperature between 63°F (17°C) and 70°F (21°C) was to be maintained.[54] The temperature inside the chamber between 5 and 11 February 1839 ranged from 59°F (15°C) to 66°F (19°C). Although these temperatures might seem very low, temperatures between 62°F (16°C) and 70°F (21°C) were considered acceptable in the Victorian era, largely due to the comparatively heavy clothing worn by Victorian men.[55] During the summer months, attendants were ordered to keep the interior temperature below 67°F (19°C).[56] When the weather was warm, the temperature was largely regulated through adjusting the rate of ventilation, which the attendants achieved by lowering or raising a larger shutter at the shaft and by increasing the heat in the furnace at the base of the shaft.[57] Examination of the log sheets show that the shutter was adjusted hourly in response to changes in the interior temperature of the chamber and the number of people in it. In June and July 1839 the shutters were fully opened most of the time.[58]

In addition to observing routine instructions, staff also made ad hoc adjustments to the internal climate based on qualitative feedback from its users.[59] This practice was grounded on the presumption that a more optimal climate could be achieved by both monitoring the physical environment and collecting self-reports of its effect on users themselves.[60] If users complained, Reid's staff would have 'Some means of knowing the precise conditions at the moment of the external atmosphere, which we can control, and their own feeling, which we cannot control'.[61]

Nevertheless, Reid was aware that it was not possible to achieve a standard climate that would satisfy every individual. Indeed, in his original 1837 briefing paper, published only a few days after the temporary House was first occupied, Reid informed MPs that 'no rule can be given so as to regulate the system of ventilation in such a manner as to be equally

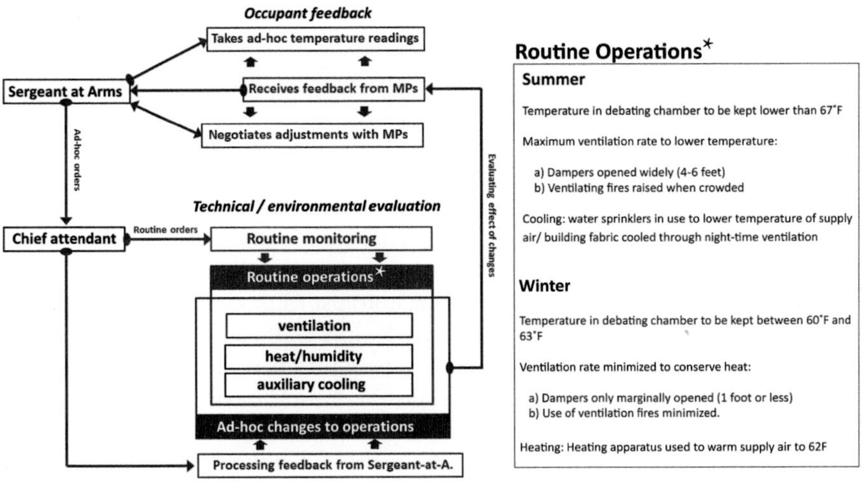

Figure 2.3 Diagram of control regime (author's own drawing).

agreeable to all'.[62] This was largely due to the fact that (the perception of) thermal comfort is affected by both personal factors and environmental factors. For example, thermal comfort could be affected by such personal factors as levels of physical exercise, body weight, type of clothing that is worn and an increased metabolism after a meal. Another important factor was the length of time that individual MPs were sitting in the debating chamber. This was of particular relevance to the speaker and sergeant-at-arms, who both regularly spent several hours within the chamber. In the early 1840s Reid experimented with making local adjustments to the ventilation around their chairs in order to meet their needs.

As a direct result of the variable experience of users, the House was confronted with conflicting feedback from individual MPs. According to Reid, opinions 'constantly clash'[63] and if they were to attempt to respond to every individual complaint, 'nothing but confusion would ensue'.[64] This problem was addressed by introducing a moderation process, which was convened by the sergeant-at-arms and, at times, by the speaker of the House of Commons. The sergeant acted as the primary liaison within the chamber, mediating between MPs and the attendants, and reviewing the range of feedback received before instructing the attendants to make any ad hoc adjustments to the climate or ventilation. In 1839, the Sergeant reported that there were often conflicting views about the ideal temperature, which made it difficult to satisfy each individual MP's request:

> *People come in very hot, and say, 'How cold the House strikes;' and another man says 'I have been sitting here half an hour, and I am in*

> *fever' and if I see the thermometers are too high or too low, I give directions accordingly.*[65]

The sergeant or speaker therefore had to make adjustments that would satisfy the majority of MPs. Through their engagement with MPs, they acquired a good 'knowledge of the general expression of opinion', which was 'always a safer guide than that of individual Members'.[66]

The procedures in Reid's system highlight the central role of human factors in its design. Reid evaluated the internal environment from (what today would be called) a 'psychophysical' perspective. Psychophysics was established by the German experimental psychologist Gustav Fechner in 1860.[67] According to the 20th-century physicist Ralph Galbraith Hopkinson, psychophysics is a scientific field that aims to gain an understanding of how humans respond to or perceive their thermal, luminous or acoustic environments. It requires that scientists 'learn how to use people as meters to register for us their experience in the environment'.[68] Unlike these later methods of psychophysical research, Reid's approach to studying the perception of climates was not systematic: the parameters for the climate inside the debating chamber were developed based on trial and error, using measurements and user-feedback collected inside it. Changes to the 'settings' for the heating were made to accommodate MPs' subjective feedback: although attendants had been instructed to keep the temperature of the supply air between 62°F and 65°F, MPs felt too cold, in particular under crowded conditions when the ventilation rate was at its highest. The supply air was thought to be too low to mitigate the cooling effect of the strong currents entering through the floor. Thus, the temperature had to be temporarily raised to 70°F or higher.[69]

It is important to note that national standards for indoor thermal comfort, comparable to those introduced by the Chartered Institution of Building Services Engineers (CIBSE UK) or the American Society of Heating, Refrigerating and Air-Conditioning Engineers (ASHRAE), did not exist in the 19th century. Some basic guidance was given in the technical literature of the time,[70] but the parameters for the indoor climate inside the temporary Houses were determined by staff through their own field observations. In the context of the day-to-day operations, attendants acquired an intimate knowledge of how the indoor climate affected users, covering factors including the thermal sensations of internal air currents and air temperature, humidity, seasonal differences in the perception of temperatures and the experience of thermal transitions between the exterior and interior. In 1844 Reid wrote, 'As far as I have been able to observe, a temperature of 65°F, with an atmosphere moving in a very gentle stream, so as not to be perceptible, is the most agreeable in rooms that are not overcrowded'.[71]

Referring to relative humidity, he reported that 'when there is a difference of 5°F between the dry thermometer and wet-bulb thermometer next to it, I have the least number of complaints'.[72]

The Temporary House of Lords and the creation of an extended network, 1839–47

Over the course of several years, Reid's experimental ventilation system was gradually extended to form part of a much larger network. Three years after Reid had installed his system in the temporary House of Commons, he was commissioned to make alterations to the ventilation system in the temporary House of Lords. The Lords had been complaining about the air quality and climate conditions for some time and in June 1835 appointed the Select Committee for a more convenient temporary accommodation for the House of Lords to undertake an inquiry.[73] Robert Smirke produced a design for a new temporary Lords chamber, providing for more space and better ventilation than the existing structure. The chamber, constructed within the shell of the medieval painted chamber, was small and narrow with 50% of the floor area of the House of Commons.[74] Due to the high costs, the proposal was abandoned. Instead, Smirke was instructed to improve the existing ventilation arrangements (although there is no evidence that he actually undertook any physical alterations to them).[75] In the autumn of 1839 the Commissioners of Woods commissioned Reid to remodel the ventilation system in the House of Lords.[76] It was originally anticipated that the House of Lords would be occupied for only a short number of years, and the Treasury therefore only granted Reid a small budget of £750, which limited the scale of alterations.[77]

Reid provided a detailed written description in *Illustrations of the Theory and Practice of Ventilation* and between 1841 and 1852 also gave several oral accounts of how the system was operated.[78] The House of Lords system was different from that of the House of Commons, which relied entirely on the ventilation stack to provide the power for both the supply and discharge of air. For the House of Lords ventilation system, the ventilating stack was only used for the extraction of air, with a separate plenum system introduced to deal with the supply.[79]

The House of Lords became an integrated part of an extended ventilation network: first, by providing a second air supply for the House of Commons; and second by sharing the same discharge shaft as the House of Commons. The latter feature allowed Reid to test the feasibility of his idea to ventilate multiple rooms using a central ventilation shaft, an idea that he had presented during the preliminary inquiry in 1835. Although it was evidently much smaller than in the original 1835 proposal, the idea of a central tower became a core feature of his subsequent plans for the Palace of Westminster.[80] A diagram in his book *Illustrations of the Theory and Practice of Ventilation* shows that, in addition to the two debating chambers, the stack became connected to the sewer and also to several rooms, including several offices and Reid's own experimental room. The hot air from the House of Lords was extracted through circular apertures in the ceiling, leading into a separate air channel and then the chimney.[81] To protect the atmosphere surrounding the temporary Houses from the smoke pollution emitted by its own chimneys, Reid also connected the ventilation shaft to various furnaces and fire places.[82] As the central stack was not sufficient to ventilate all spaces simultaneously, the connection had to be switched between the Houses.[83] Whilst one House was temporarily disconnected from the tower, the ventilation in the other could be boosted for a certain time before the stack was reconnected with the other House.[84]

The fresh air, introduced through a new set of inlets facing Cotton's Garden, was purified, humidified, heated or cooled in an air chamber below the floor of the House of Lords.[85] This chamber was also connected to another air chamber below the House of Commons to form a shared supply network with access to two separate inlets. This allowed attendants to switch the air supply if one of the inlets was exposed to too much air pollution, which was dependent on the wind conditions.[86]

In contrast to the House of Commons, the floor in the House of Lords was not completely perforated: air was injected through openings under the tables and benches, and through porous wall panels and grills in the risers of the raked floor.[87] This allowed fresh air to be introduced in horizontal currents.[88] However, sections of the porous surfaces were closed after only four months of operation as the Lords felt uncomfortable with having their backs exposed to currents. In May 1840 following their complaints, the Office of Woods instructed Reid to close the area from the top of the benches to a 'height above the Peers heads',[89] with Reid later reporting that the 'openings near the head were stopped up, for the draft was too strong, but it came over the head, and then passed down into the House'.[90]

An extended operational regime, 1839–40

After completing the installation of the ventilation system inside the House of Lords, Reid became superintendent inside both Houses, and the environmental monitoring and control regime established in the House of Commons was extended. A sample of the standard log pages adopted after 1839 show that eight new columns were added for the temporary House of Lords. Inside the chamber, readings were taken from thermometers located near the throne, table and bar. The log also covered data on the number of people present, the position of the valve at the shaft, and also had three columns for attendants to qualitatively rate – as '*neutral*', '*hot*' or '*cold*' – the thermal state of the incoming air.[91] Readings of velocity were taken at the bottom of the shaft.[92] A separate attendant, playing a similar role to the 'attendant in ventilation' in the House of Commons, was employed to be inside the House of Lords throughout sittings.[93] This person managed the technical staff who operated the plant, received feedback from the Lords and also supervised the monitoring and control procedures.

Reid also used the House of Lords to test a different approach to climate control than that used in the House of Commons. His objective was

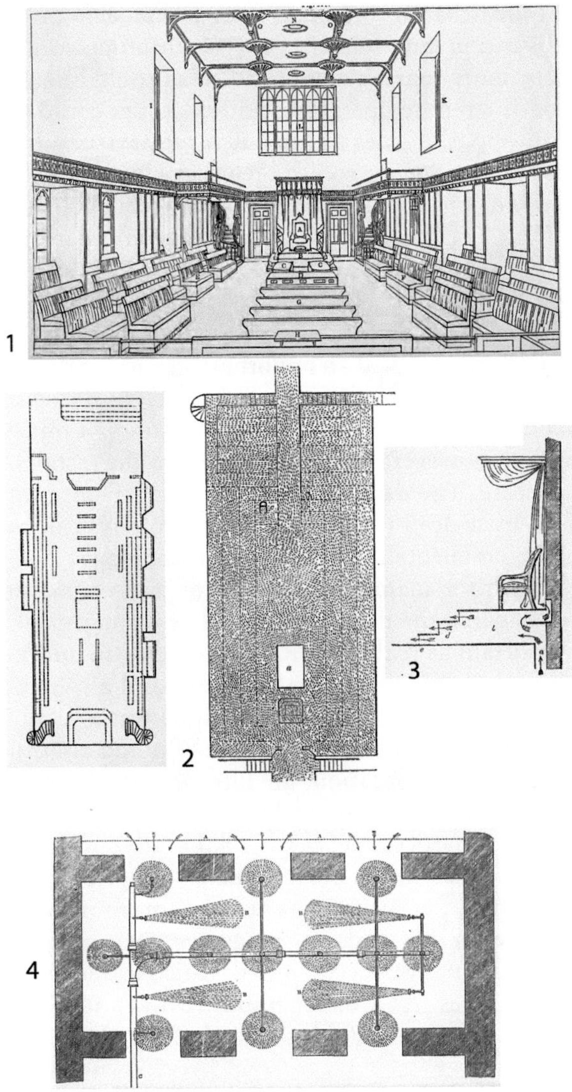

Figure 2.4 Interior of the House of Lords after Reid's interventions (*Illustrations*, 1844, pp. 286, 288–91). Key: 1. Interior perspective showing vitiated air outlets in ceiling, 2. Floor plans of Lords (left) and Commons chambers (right), showing extent of opening in floor, 3. Cross-section showing fresh air supply arrangement for throne. 4. Experimental air purification system in vaults below the House of Lords. Key for drawings '1': A. Throne, E. and F. Benches, N. Outlets for vitiated air in ceiling, B. Woolsack, C. Judges Seat, D. Table, E. Ministerial seats, F. Opposition seats, G. Cross-benches, H. The Bar. Key for drawings '4': A. Wire gauze screen for removal large soot particles, B. water jets for removal for finer impurities (and cooling and humidification), C. pipe for supply to water jets.

to explore how far user satisfaction could be increased by delivering local climates to different parts of the debating chamber.[94] For instance, in more crowded areas where there was a greater tendency for overheating, cooler air could be introduced, whilst more sparsely populated areas could be supplied with warmer air. Air of different temperatures and velocities could be introduced at the benches on opposite sides as well as in areas at the bar and throne end of the chamber, producing cooler or warmer zones in the chamber. The air introduced into one section could be as low as 52°F and as high as 75°F in another. [95]

Conclusion: from theory to practice

This extension to the House of Lords marked an important step in the development of Reid's experimental ventilation system, as it enabled him to test the idea of an integrated ventilation and climate control system on a much larger scale. Reid's contribution, however, was not confined to the installation of the physical infrastructure. The design of the temporary Houses of Parliament was characterised by a deep understanding of the physical, as well as the performative and experimental, dimensions of technology.

Reid engaged extensively with the managerial aspects of his system, an area that today would be considered the domain of facilities management. Entering this domain was important as it allowed Reid to acquire an understanding of its performance in use under real-life conditions and also enabled him to test and refine his principles in situ. The outcome was a highly sophisticated system, which incorporated complex monitoring and control procedures in addition to combining ventilation, air filtration and climate control. The procedures were developed over several years through trial and error, and were the outcome of a participatory and collaborative process. Reid collaborated with the staff of the Houses of Parliament to establish and sustain practices of user engagement, ensuring that feedback from its MPs and Lords was used to improve its day-to-day management. Social feedback processes alongside technical ones were established, which formed two distinct, yet complementary activities. As a direct result its operation was dependent on a close collaboration between Reid, the staff of the Houses of Parliament, and its users/occupants.

These practices might be fruitfully understood through the lens of the actor–network theory, the ventilation system could be interpreted as a network of human and technical actors. According to this theory the term 'actor' can refer to people or technologies (e.g. thermometers and valves) that causally affect the production of a shared knowledge or understanding. The term 'network' can refer to an assemblage of actors (human or non-human) and their relationships (as defined by their specific function and place within the network).[96] In the ventilation system network, human actors have two primary functions: first, to perform the control and monitoring procedures; and second, to facilitate the social interactions between staff and users. As the primary users of the Houses of Parliament, the MPs and Lords were directly involved in reporting their own experience of the quality of the environment. These operational practices, which included the more ephemeral processes of social interaction and data collection, are important in understanding Reid's later contribution to the design of the Palace of Westminster. They strongly influenced the discourse underlying the development of his scheme between 1840 and 1852 (see Chapters 4 and 5), and also directly underpinned the design of more sophisticated operational practices introduced by Reid in the new House of Commons in 1852 (see Chapter 7). These feedback processes could be interpreted as an early example of an 'occupant survey', which is commonly used in modern post-occupancy evaluations. However, the approach to assessing user perception in Reid's time was less systematic than modern occupant surveys (e.g. Building Use Studies [BUS]) and it did not include formal methods of rating perceived thermal comfort (e.g. Predicted Mean Vote). That said, the qualitative feedback procedures that Reid implemented were instrumental in providing designers and operators of the day with a deeper understanding of the environmental factors that can affect user satisfaction. In addition to informing the day-to-day operational adjustments, they yielded data that was used to underpin the development of standards for indoor thermal comfort within the debating chambers. These operational procedures manifest most clearly the influence of the physiological perspective that underlay Reid's early scientific investigations in Edinburgh.

Notes

1 Lord Sudeley. Letter to D.B. Reid, 8 September 1843, in Reid (1846) 'Extract from official documents, reports and papers referring to the progress of Dr Reid's plans for ventilation', pp. 8-9, Hume Tracts A5/1-26, UCL Library.
2 'Ventilating the New Houses of Parliament', *Preston Chronicle*, 21 May 1836.
3 Treasury, Letter to the Commissioners of Woods, 26 August 1836 (National Archives: Work 11/12, Nr. 7). The drawings are mentioned in Reid, D.B. Letter to Lord Canning, 11 May 1836, 'shewing the temporary houses of parliament &c. manner in which ventilation is affected by the position of the continguous department' (National Archives: work29/3000)
4 House of Commons, *Hansard's Parliamentary Debates: The Official Report (10 August 1836, vol. 35, cols. 1057-87) (London: Hansard).*
5 House of Commons, *Hansard's Parliamentary Debates: The Official Report (10 August 1836, vol. 35, cols. 1057-87) (London: Hansard).*
6 HC 277 1838 'Estimates', 31 March 1838, p. 10.
7 Treasury, Letter to the Office of Woods, 26 August 1836 (National Archives, work11/12, nr. 7); Office of the Woods, Letter to D.B. Reid, 13 September 1836 (National Archives, work11/12, nr. 8); 'Miscellanea', *Champion*, 3 October 1836, p. 1; 'London', *Caledonian Mercury*, 17 October 1836, p. 2.
8 Reid, 'Brief outlines'.
9 Walter Bernan (pseud. [i.e. Robert Stuart Meikleham.]), *On the History and Art of Warming and Ventilating Rooms and Buildings ... with notices of the the progress of personal and fireside comfort, and of the management of fuel. Illustrated, etc.* (London: George Bell, 1845), pp. 97-104; Reid, *Illustrations*,

pp. 273-328; Reid, *Brief outlines*, pp. 9-28; Reid, *American dwellings*, pp. xv-xxxvii; Robert Richie, *A Treatise on Ventilation: natural and artificial* (London: Lockwood & Co., 1862), pp. 136–49.

10 'Brief outlines', pp. 9-28; *American dwellings*, pp. xv-xxxvii.

11 HC 501 1839, Report from the Select Committee on Lighting the House; HC 536 1842, Report from the Select Committee on Ventilation of the New Houses of Parliament; HC 21 1837, 'Ventilation of the House, Letter from Dr Reid to Lord Duncannon, 4 February 1837, relative to the acoustic and ventilating arrangements lately made in the House of Commons'.

12 Reid, 'Brief outlines', p. 13.

13 Reid, *Illustrations*, p. xiv, pp. 270.; *American dwellings*, p. xxv.

14 'Alterations in the House of Commons', *The Champion*, 13 November 1836, pp. 66.

15 Reid, 'Brief Outlines', p. 13.

16 Oral statement, HC 583 1835, Q506-26.

17 'House of Commons', *Sheffield Independent,* 12 November 1836, p. 1.

18 Charles Tomlinson, *A rudimentary treatise on warming and ventilation; being a concise exposition of the general principles of the art of warming and ventilating domestic and public buildings, minds, lighthouses, ships, &c* (London: John Weale, 1858), p. 217.

19 'Latest News', *Berrow's Worcester Journal*, 29 December 1836, p. 4.

20 'The Alterations in the House of Commons From a Correspondent,' *The Times* 5 November 1836, p. 3.

21 Richard Henry Nibbs, *After the fire Westminster Hall – Abbey*, May 1839, watercolour, Museum of London, London (27/18.1).

22 *Illustrations*, p.276.

23 Reid, 'Brief Outlines', p.14.

24 *Illustrations*, p. 80.

25 Bruegmann, 'Central Heating'.

26 For a detailed study of the earlier air-conditioning methods used in the temporary Houses of Parliament see Henrik Schoenefeldt, 'The Temporary Houses of Parliament and David Boswell Reid's architecture of experimentation', *Architectural History* 57(1) (2014), pp. 175-215.

27 Neil Sturrock, & Peter Lawson-Smith, 'The Grandfather of Air-Conditioning: The Work and Influence of David-Boswell Reid, Physician, Chemist, Engineer', *Proceedings of the Second International Congress on Construction History*, pp. 2981–2998.

28 HC 277 1839 Lighting Committee, Q610.

29 For a detailed study on the use of ventilative cooling in the Houses of Parliament see Henrik Schoenefeldt, 'Historic practices of ventilative cooling - A case study on the House of Lords, 1836–1966', in *Innovations in Ventilative Cooling at building and urban scale*, ed. by Maria Kolokotroni, Giacomo Chiesa, & Per Heiselberg (Basel: Springer, 2020).

30 *Illustrations*, p. 294.

31 Reid, *Illustrations*, p. 274.

32 'House of Commons', *Sheffield Independent,* 12 November, 1836, p. 1; HC 21 1837, 'Ventilation of the House: Letter from Reid to Duncannon', 4 February 1837.

33 'Alterations in the House of Commons', *The Champion*, 13 November 1836, p. 66

34 HC 21 1837, 'Ventilation of the House: Letter from Reid to Duncannon', 4 February 1837; Reid, Brief Outlines, p. 13.

35 Oral statement by D.B. Reid on 26 July, HC 536 1842, Q82.

36 HC 277 1838 'Estimates', 31 March 1838.
37 Reid, D.B. 'Letter to the editor of *The Courier*, 26 December 1837', in *The Morning Post*, 29 December 1837, p. 1.
38 Office of Works. 'Letter to D.B. Reid', 17 June 1837, in *The Morning Post*, 29 December 1837, p. 1.
39 Oral statement on 10 March 1843, HL 202 1843, Q203.
40 Reid, D.B. Letter to Viscount Duncannon, 28 March 1838, in HC 277 1838, p. 8.
41 *Illustrations*, p. 296.
42 Oral statement by D.B. Reid, 24 March 1846, HC 574 1846, Q1072.
43 The attendant of ventilation, employed by the Sergeant-at-Arms, was paid £205 per annum. HM Treasury, Estimates Miscellaneous services for the year ending 31 March 1838, (HC 1838 (145)), pp. 10, 4.
44 Oral statement, 17 March 1846, HC 574 1846, Q785-87.
45 Reid, D.B. Letter to Viscount Duncannon, 4 February 1837, HC 21 1837–38.
46 *Illustrations*, p. 327; State of the three thermometers in February and June 1839, in HC 501 1839, pp. 75–6.
47 Oral statement by B. Riches, 20 June 1839, HC 1839, Q618-20; Reid, *Illustrations*, p. 327f; 1839 Lighting Committee, pp. 75f.
48 Trench, F. Letter to Viscount Duncannon, 22 April 1838.
49 Reid, D.B. 1838 Letter to Viscount Ducannon, 28 March 1838.
50 Reid, *Illustrations*, pp. 65-69; Reid, 'Progress', pp. 152-3. For more on Reid's inquiries into air quality, see Colin Porteous, 'Sensing a Historic Low-CO_2 Future', in *Chemistry, Emission Control, Radioactive Pollution and Indoor Air Quality*, ed. by Nicolas Mazzeo (London: IntechOpen, 2011), pp. 213-46.
51 Morrill Wyman, *A Practical Treatise on Ventilation* (Boston: James Monroe & Co., 1846), p. 224.
52 Oral statement by D.B. Reid on 17 June, in HC SC 1844, Q322.
53 Oral statement by B. Riches on 20 June, HC 501 1839, Q615.
54 Reid, *Illustrations*, p. 294.
55 Reid, 'Progress', p. 152.
56 Oral statement by B. Riches on 20 June, HC 501 1839, Q612-15.
57 Ibid, Q626-660; Reid, *Illustrations*, p. 326.
58 Ibid, Q660,; Oral statement by W. Gosett on 24 June, HC 501 1839, Q748, Q771.
59 For more on the user-feedback system, see Henrik Schoenefeldt, 'Powers of Politics, Scientific Measurement, and Perception: Evaluating the Performance of the Houses of Commons' First Environmental System, 1852–4', in *Experiencing Architecture in the Nineteenth Century*, ed. by Horatio Joyce and Edward Gillin (London: Bloomsbury, 2018), pp. 115-129.
60 Oral statement by D.B. Reid on 17 June, HC 448 1844, *Report of Select Committee on Houses of Parliament*, Q321.
61 Ibid., Q387.
62 Reid. D.B. Letter to Viscount Duncannon, 4 February 1837, HC 21 1837-38.
63 Oral statement, HL 202 1843, Q205.
64 Oral statement on 17 June 1844, HC 448 1844, Q387.
65 Oral statement by W. Gosett on 24 June, HC 501 1839, Q758.
66 Reid, *Illustrations*, p. 295.
67 Gustav Theodor Fechner, Elemente der Psychophysik [Elements of Psychophysics] (Leipzig: Druck und Verlag von Breitkopf und Hartel, 1860).
68 R.G. Hopkinson, *Architectural Physics: Lighting* (London: HMSO, 1963).
69 Oral statement by D.B. Reid on 6 August, HL 629 1844, Q557.
70 Thomas Tredgold, *Principles of warming and ventilating public buildings* (London: Architectural Library, 1824), pp. 3-17; Board of Health (1857), p. 24.

71 *Illustrations*, p. 185.
72 Parallel readings from dry- and wet-bulbs thermometers were used to measure relative humidity, with a lower humidity indicated by a greater difference between the two readings. See oral statement by D.B. Reid, 26 March, HC 402 1852, Second report from the Select Committee on Ventilation and Lighting of the House, Q361.
73 'The House of Parliament', *The Morning Post*, 26 December 1834, p. 4.
74 *Illustrations*, p. 290, shows plans of the House of Lords vs. House of Commons.
75 House of Lords, *Report of Select Committee appointed to consider a plan for a more convenient temporary accommodation for the House of Lords, 3 July 1835* (Parliamentary Archives: Arc/Pro/Work 11/10/2 nr. 450); House of Lords, *House of Lords, Minutes of Proceedings, 12 June to 24 August 1835* (Parliamentary Archives, Arc/Pro/Work 11/10/2)
76 Reid, *Illustrations*, pp. 271.
77 Letter from Lord Campbell, 11 September 1843, House of Lords, 27 June 1846, in Memoranda by Dr Reid in extracts from official documents reports (Hume Tracts, UCL Library).
78 Reid, *Illustrations*, pp. 284–93.
79 HC 402 1852, *Second report from the Select Committee on Ventilation and Lighting of the House*, Q511–542.
80 Oral statement by D.B. Reid on 21 August, HC 583 1835, Q409.
81 Reid, *Illustrations*, p. 284; Edward Wedlake Brayley & John Britton, *The History of the Ancient Palace and late Houses of Parliament at Westminster* (London: John Weale, Architectural Library 1836), pp. 463–4.
82 *Illustrations*, pp. 284–85.
83 Ibid., p. 287.
84 Oral statement by D.B. Reid on 26 July, HC 536 1842, *Report from the Select Committee on Ventilation of the New Houses of Parliament*, Q53–54.
85 Reid, *Illustrations*, p. 284–93.
86 Oral statement by D.B. Reid on 23 February, HL 719 1846, *Reports from the Select Committee of the House of Lords appointed to inquire into the progress of the building of the Houses of Parliament*, Q74–75; 1841 Ventilation Committee, Q315–16; Reid, *Illustrations*, pp. 274–6.
87 Reid, *Illustrations*, p. 284–93.
88 Reid, D.B. Letter to Phillips, 23 December 1839 (Work 11/12, nr 25).
89 Office of Woods. Letter to D.B. Reid, 8 May 1840 (Work 11/12, nr 33).
90 Oral statement on 29 April, HC 402 1852, Q3533.
91 *Illustrations*, p. 328
92 Oral statement by D.B. Reid on 26 July, SC HC 1841, Q59.
93 Oral statement by D.B. Reid on 10 March, HL SC 1843 Q203–4.
94 Oral statement by D.B. Reid on 17 June, HC 448 1844, Q317–23.
95 Ibid, Q322
96 Bruno Latour, 'On actor-network theory: A few clarifications', *Soziale Welt*, 47 (1996), pp. 369–381; 'Networks, Societies, Spheres: Reflections of an Actor-Network Theorist', *International Journal of Communication*, 5 (2011), pp. 796–810.

3 Testing the theory under real-life conditions

He did not know what expense the experiments now going on would entail, or who authorised them. Fortunately, the Gallant Member had both leisure and taste to pursue these things, if all the 658 Members took similar liberties with the House it would be converted into an arena of experiments.

Chancellor of the Exchequer[1]

The debating chamber as a field for experimentation

The temporary Houses of Parliament enabled MPs and Lords to be part of a long-term demonstration of the principles of environmental control that were to be applied in the Palace of Westminster. This demonstration began three years before Reid was formally appointed to collaborate with Barry, but continued for another 11 years whilst he developed his plans for the new Palace. During this period the debating chambers became a site for environmental field studies, in which technical, operational and social aspects of environmental control were investigated. The annual reports of HM Treasury from 1838 to 52 provide details about the operational budget – covering the cost of materials, fuel and labour – and additional funds granted for experimental investigations and technical alterations.[2]

Members of Parliament and Lords were not passive observers. As seen in the previous chapter, formal procedures were introduced to enable MPs and other users of the debating chambers to provide feedback about their experience of the environment. Members could also intervene in more substantive ways: they could scrutinise the performance of the chambers during debates; participate in the trialling of different technologies or various operational modes; or appoint an ad hoc Select Committee for more formal inquiries. In these, technologies were assessed primarily from the perspective of a Member's experience and expectations. This gave MPs a degree of agency over which neither Reid himself nor the Commissioners of Woods had much control. Discontented Members became the driving force behind the commissioning of further experiments within the chamber. Some of their scrutiny contributed towards the gradual refinement, whilst others were disruptive, preventing Reid from implementing his ideas in full.

In this chapter, it will be shown that MPs and Lords saw the lighting and ventilation of the temporary chambers as highly important as they directly affected the quality of their working environment. Their eyewitness accounts, although subject to a negative bias, provided insight into how they experienced the environment including observations on its luminous, acoustic and thermal qualities as well as air purity. Although interventions in the workings of Reid's system were largely the result of action taken by a small number of outspoken critics, this process offers critical insights into the way in which Members of Parliament exercised influence over its development.

A first public demonstration: November 1836

Reid knew that it was difficult to overcome deep-seated scepticism towards his ventilation technology. As part of an effort to counter it, he briefed MPs on 4 November 1836 about its design and capabilities, giving the first physical demonstration of the new system inside the House of Commons debating chamber. This demonstration was attended by 158 MPs, Lords and journalists, including Southwark MP Daniel Havey and two former members of the Select Committee on Ventilation, Benjamin Hawes and George Clerk.[3] To replicate the conditions of a crowded debate, Reid engaged 420 Grenadier Guards as volunteers to raise the total number of people inside the chamber to 578 people. Guards sat on the rear benches on the main floor and galleries. Reid conducted three experiments to demonstrate the power of the stack, using visual and olfactory methods. For the first experiment, smoke was injected into the stack in order to visualise the flow of air. *The Times* reported that the debating chamber was filled with smoke 'so dense that it was impossible to see five yards forwards' and that it only took 90 seconds for smoke to clear. In the second and third experiments, Reid introduced ether and orange scents to demonstrate the dissipation of smells.[4] Several newspapers reported of the experiments that the air felt fresh and that the temperature did not rise by more than 2°F. According to *The Times*, air temperatures of 60°F (~15°C) to 62°F (~16°C) were recorded, whilst the current of heated air entering through the floor was 'imperceptible'. *The Champion* wrote that the air 'remained to the last without any sensible impurity' and described the experiment as 'most decisive proof of the correctness of Reid's theory'.[5]

Although reports in the newspapers suggest that the trial was successful in convincing witnesses and the press of its technical feasibility, Reid knew that it would be more challenging to satisfy its users once the chamber itself was occupied. He produced a briefing paper *Brief Outlines Illustrative of the Alterations in the House of Commons*, which was circulated within the House in February 1837 in order to introduce MPs to the design of the new system and its capabilities. In this paper, Reid explained some its limitations in order to manage MPs' expectations.[6] The post-occupancy history of the temporary Houses of Parliament, which will be explored in the next sections, will reveal that Reid's concerns were not unjustified and that his involvement with the system did not end after he delivered his first demonstration.

First criticism inside the House

On 31 January 1837, Parliament returned to the remodelled chamber. Over the following months, MPs and staff gained first-hand experience of Reid's ventilation system under real-life conditions. Some early criticisms about the insufficient humidity were voiced by MPs in the spring of 1837, leading Reid to conduct experiments on raising the relative humidity of the atmosphere. Anxious about the impact of the largely negative feedback, Reid also collaborated with Lord Sudeley to collect signatures from the 68 MPs who considered his system to be a success.[7] In December 1837, the performance of Reid's system became the subject of a first debate in the House of Commons. This debate was initiated by Thomas Wakley, MP for Finsbury, who felt that although the system had been successful in improving the ventilation, resulting in 'more freshness in the air they respire', the atmosphere was too dry and hot. He claimed that it was 'rendered so hot and dry by the process' that some MPs 'who were subject to soreness of throat were kept in a constant state of irritation'. Wakley evidently felt very strongly about the problem, and did not trust the independence of Reid's own judgement, so he privately commissioned the physician Dr Neil Arnott for an independent evaluation of the system.

In his report, dated 22 December 1837, Arnott concluded that the problems with the atmosphere were caused by the combination of insufficient humidity and floating dust particles. The latter, he argued, was caused by the use of floor inlets, as the incoming air transformed any mud on the carpet 'into a fine and subtle dust, and is carried up to be breathed by Members'.[8] The dust, combined with the dry air, was responsible for 'dry throats, cough and feverish heat complaint of by Members'. To address these problems, Arnott advised that humidification be introduced, and that Reid's perforated floors be abandoned. He proposed a new configuration that allowed fresh air to be introduced at a higher position through openings within the walls. This would minimise rising dust and protect MPs from exposure to the rising currents.

Wakley's concerns were shared by four other MPs: Sir Edward Codrington, MP for Devonport, who said that the system was 'so imperfect that he every night suffered from it'; Andrew White, MP for Sunderland, who reported that he 'suffered from cough a very great degree'; John Hobhouse, MP for Nottingham, who thought that it was 'attended with great inconvenience and danger'; and Henry Warburton, MP for Bridport. Hobhouse and Warburton asked for improvements to be made to Reid's system. Hawes, former chairman of the Select Committee in 1835, defended Reid's scheme, and argued that it had yielded 'great improvements' and that changes should not be made without Reid's consent.

Reid was hurt by this criticism and rejected Wakley's proposal, arguing that the problems could be resolved by switching the supply from the floor to the ceiling, thereby reversing the direction of the air flow. This had been part of his original 1836 proposal (but was not authorised by the Commissions of the Woods) and required a different method of artificial illumination, gas lighting, than what was used in the House. Since February 1835, the chamber had been illuminated using wax candles that were mounted on three chandeliers.[9] As they were hanging from the ceiling, the downward supply of air was exposed to the heat and fumes produced by the candles, with the heat and exhaust fumes mostly escaping through the ceiling by the power of the ascending air current. If the ventilation were to be reversed, the current would be carried downwards into the body of the House. Reid persuaded the Commissioners of Woods to trial his original plans for an integrated system of gas lighting, similar to his model in Edinburgh.

As the criticism by MPs was reported in the national press, Reid became anxious about what would happen to his professional reputation and his system if MPs were to reject it. To counter the largely negative press, he wrote letters to the editors of several papers, enclosing copies of positive testimonials from various MPs.[10] Reid's anxiety is understandable as the criticism arrived at a time when his system was still being treated as a temporary, largely experimental, installation, and at risk of being abandoned if the MPs considered it unsuccessful. Francis Baring, Chancellor of the Exchequer, noted that the 'government acted upon the feeling that they were trying an experiment, and that, before, they ought not to incur any large expenses until he there was some fair reason to suppose that the experiment would succeed'.[11]

After the debate on 23 December 1837, Viscount Duncannon summoned Reid to London for a special meeting, attended by Arnott and also several MPs, including Baring and Hawes. During this meeting, they agreed that Reid should be given the opportunity to improve his system before any alternative arrangements be considered. Over the Christmas Parliamentary recess, Reid made a series of alterations to his system, involving the introduction of new facilities for humidification, the addition of 10,000 apertures within the floor, and the adoption of a new cleaning regime involving a weekly lifting and beating of the horse hair carpets.[12] From March 1838, an annual sum of £1,150 was provided to cover the cost for labour and materials.[13]

Reid's idea of integrated lighting system remains unrealised

Reid also collaborated with the Commissioners of Woods on a series of short demonstrations of his gas lighting, following the principle of 'exclusive lighting'.

He had previously demonstrated this on a small scale using his model in Edinburgh.[14] Reid installed a simple mock-up to demonstrate the general principle:[15] three rows of gas pipes, which had simple holes instead of gas burners, were installed above the glazed ceiling panels on both sides.[16] The light was generated by 10,000 open gas flames.[17] This simple arrangement of flames produced a bright, yet unsteady light. The first series of demonstrations were undertaken during the recess. The first of these, which took place on 30 December 1837, was attended by the three Commissioners of Woods, and several politicians including Hawes, Lord Howick, Lord Melbourne and Lord Durham.[18] During these trials, the heat of the lights was closely monitored with four additional thermometers installed above the glazed panels. The temperature above the glass reached 109°F (~43°C) but the test lights did not have a noticeable effect on the air temperature inside the debating chamber.[19] In their final report, dated 8 January 1838, the Commissioners wrote that was it more 'equally diffused perhaps than any light which could be obtained under the system at present in use' and that it was 'entirely free from shadow'.[20] However, they rejected its permanent adoption due to the risk of fire given the wide use of timber and plaster in the debating chambers.

At Wakley's request, two further trials were undertaken during actual sittings on 12 and 16 February 1838, which enabled MPs to experience the new lighting.[21] The light was reportedly very intense, compared to the more gentle light of the wax candles. The *Bristol Mercury* reported that the 'vivid glare darted down upon the body of the House' and the entire interior was 'flooded with light'.[22] Even so, MPs rejected it because it produced an unsteady light.[23] The experimental gas-lighting system was decommissioned and the House returned to using candlelight. Reid's involvement in the development of the artificial lighting used in the House had come to an end.

On 3 March 1838, Frederick Trench, MP for Scarborough, wrote to Viscount Duncannon advocating the retention of candles for the lighting, and rejecting Reid's installation, arguing that it produced 'a heavy, dull, flickering and disagreeable light'.[24] Between March and April, Trench coordinated a series of trials to improve the candlelight: the number of candles was increased from 150 to 276 to provide better illumination in the body of the House, with green lamp shades introduced in order to protect those seated in the galleries from the glare of the open flames.[25] Although Reid was no longer in charge of lighting, he continued his inquiries into alternative methods of artificial illumination, and its impact on the internal climate and air quality, for another two years.

Reid's principles challenged

Trench not only opposed the use of gas light but also became one of the most outspoken critics of Reid's system, making several pleas for substantial alterations of Reid's system and even for its outright replacement.

On 1 March 1838 he initiated another debate in the House of Commons about Reid's system.[26] In his speech Trench complained about the insufficient lighting and claimed that the recent changes to the ventilation system had not resolved problems with rising dust, and that the incoming air carried up 'great quantities of impalpable power or dust'. To eliminate this problem he recommended that large parts of the floor apertures be sealed and that the air be introduced horizontally above the head of the MPs. Similar to Arnott's scheme of 1837, he proposed that air inlets be introduced within the walls below the galleries. Trench's proposal did not receive support from the House. During the debate on 1 March, five MPs – including Reid's former critics Wakley and Warburton – opposed the idea of interfering with Reid's system, and argued that the changes made by Reid over the winter recess were successful. Sir George Strickland, MP for West Riding of Yorkshire, found that the changes were 'in a very great degree, if not wholly, successful'. Henry Goulburn, MP for the University of Cambridge, argued that it was 'well-ventilated'. Warburton believed that the House should preserve Reid's system and Wakley felt that the air was 'now almost as wholesome as anything that could be produced'. Joseph Hume, MP for Kilkenny City, opposed Trench's plans on the grounds that the new inlets would result in the backs of MPs being exposed to cold draughts, whilst Lord John Russell, Home Secretary, took the position that it was a technical issue, which should be referred to the Commissioners of Woods.

Two days later, on 3 March, Trench wrote a long letter to Viscount Duncannon describing his proposal in detail and providing more about the effect of the floating dust in the debating chamber: it 'rests upon their faces, enters their eyes, and the nostrils and mouths'.[27] Viscount Duncannon's reply to this letter makes clear that no changes would be authorised without Reid's involvement as it would be unfair if Reid, as the designer of the system, were not given the opportunity to write a response to the criticisms and to share his own view with the House.[28] In another speech on 26 March, Trench offered additional objections to Reid's system, this time opposing the very idea of a hermetically-sealed chamber on the grounds that 'natural atmosphere was preferable to any artificial means of ventilation'. Instead, he proposed and demanded a trial of a simpler system of natural ventilation through operable windows. This idea did not receive wider support in the House.

Viscount Duncannon forwarded Trench's letter and speech to Edinburgh and asked Reid for a written reply. On 28 March, Reid wrote a long and detailed response to Trench's criticisms. Clearly hurt by Trench's claims against his system, Reid became highly protective of it.[29] Reid abdicated any responsibility for the dust problem by arguing that it was the fault of the cleaning staff, whose job it was to ensure that the floor was regularly cleaned. He also stressed that the dust problem could only be resolved if, as he had proposed, the air supply were moved to the ceiling.

Reid succeeded in persuading the Commissioners of his proposal to switch the air supply, and physical alterations were made over the Easter recess of 1838.[30] The alterations involved the introduction of a second connection between the House and the chimney, thereby allowing the air extraction to be switched from the ceiling to the floor. On 5 May 1838, Reid conducted a series of demonstrations of the ceiling supply with MPs, including Trench, invited as witnesses. In the first three demonstrations, Reid traced the direction of the air currents by introducing different scents into the atmosphere. Trench reported that the scents gradually moved from the ceiling towards the floor and finally into the air chambers below, where a 'gale of wind hurried them off to feed the furnace'.[31] This olfactory method of tracing air currents was complemented by a visual demonstration: gunpowder was lit inside the air channel above the ceiling, allowing witnesses to briefly see it 'descending regularly till it was down through the carpet into the regions below'.[32] Satisfied with the outcome of these demonstrations, Trench withdrew his earlier objections. The use of the downward ceiling supply, however, was not continued as it was incompatible with the continued use of candles as a source of illumination in the chamber.[33]

Goldsworthy Gurney's intervention

The lighting issue was revisited in the autumn of 1838. Following the suggestion of Hume,[34] trials of a lighting method developed by Goldsworthy Gurney, a Cornish physician, were undertaken. As chairman of the Select Committee on State and Management of Lighthouses, Hume was familiar with the 'Bude Light', which Gurney had successfully trialled in a lighthouse operated by Trinity House, the lighthouse authority.[35] The light from the Bude Light was generated by argand oil burners, which were brighter than ordinary oil lamps due to the injection of pure oxygen into the flame. Hume secured support from the Speaker of the House and the Chancellor of the Exchequer for a trial of the Bude Light in the House of Commons. Over the Christmas Parliamentary recess of 1838 the Office of Woods commissioned Gurney to demonstrate his method of lighting in the House and also devise arrangements that would keep the atmosphere protected from any heat and exhaust fumes produced by the lighting irrespective of the direction of the air flow. The development of the new lighting system, which lasted from 1839 to 1840, incurred an additional cost of £777.[36]

A first series of trials, which were conducted inside the House between 18 January and 6 February 1839, examined various ways of integrating the Bude Light into Reid's ceiling.[37] Gurney tested various arrangements inside the House,[38] asking MPs to assess their own visual and thermal comfort in each of them.[39] In the first demonstration, the oil burners were installed above the glazed section of the ceiling (where Reid had previously tested his gas-lighting system). Although the oil burners illuminated the entire glass surface, they consumed a large quantity of fuel and the ensuing brightness also made MPs feel uncomfortable.[40] In another series of demonstrations, Gurney reduced the brightness of the light by confining the burners to seven glass panels on each side of the House. However, even this did not satisfy the MPs.

In May 1839, the House appointed an ad hoc Select Committee, chaired by Hume, to undertake a formal inquiry into whether Gurney's Bude Light technology could be made compatible with the existing ventilation system. The Committee sat between June and August 1839 and engaged several scientists – including the physicists Michael Faraday, David Brewster and Charles Wheatstone, and two physicians, Andrew Ure and Neil Arnott – to provide a range of specialist skills and knowledge. A critical review of Gurney's Bude Light and other lighting technologies was undertaken, examining their visual qualities, relative efficiency as lighting sources and the relative impact on both the atmosphere of the chamber and the health and comfort of the legislature. Experimental studies were conducted to determine the quantity of heat generated by the different lighting sources relative to their lighting intensity (luminous flux), compared to the use of wax candles with oil and gas burners.[41],[42]

At the same time Gurney also examined the effectiveness of the stack system in extracting the fumes emitted by burning wax candles. On 3 July 1839, Gurney undertook a series of smoke tests to study the behaviour of internal air currents given different ventilation rates. The ventilating fire was lit and gunpowder was ignited on the main floor of the debating chamber under each of the three chandeliers. The sliding valve at the top of the descending shaft was opened 48 in (4 ft), with the test repeated with the valve at 9 in and 3 in (the minimum ventilation rate).[43] In an oral statement made by Gurney on 4 July 1839, Gurney reports that the smoke plume ascended to the ceiling but was reflected downwards into the galleries instead of passing straight through the ceiling apertures. Additionally, he observed that hot air accumulated at gallery level. To Gurney, this was evidence that the stack did not remove enough of the candle fumes at the ceiling nor did it prevent overheating at the galleries. To overcome these issues he proposed to increase the size of the ceiling apertures, and argued that a new system, in which flames were isolated from the atmosphere of the chamber, was needed to maintain the purity of the air irrespective of its direction (i.e. ascending or descending).

In its final report of 7 August 1839, the Select Committee concluded that Gurney's experimental lighting arrangements had been successful in yielding a lighting system that was compatible with the ventilation requirements of the House: 'lamps are entirely insulated from the body of the House, so that the warming and ventilation of the House are not interfered with'.[44]

Despite this, the Committee agreed that an alternative arrangement of lights was needed in order to achieve a better distribution of light within the debating chamber and that additional trials should be conducted over the summer recess. However, these recommendations were not implemented.

For the next six months, Trench, also an outspoken critic of Gurney's system, campaigned for a return to the use of wax candles.[45] On 6 February 1849 Trench tabled a successful motion to reinstate the wax candles, arguing that it was preferable to Gurney's Bude Light. Although several MPs, including Hume, Warburton and MP for Cornwall Eastern Lord Eliot, argued that Gurney should be given the opportunity to refine his lighting system, the motion received support with a small minority of 19 votes, with 69 and 50 MPs voting, respectively, for and against it.[46] The House temporarily returned to the use of wax candles but the decision was successfully reversed on 12 March 1840 by another motion tabled by Lord Eliot. In this motion, which won a significant majority of 50 votes (136 votes for and 86 against), Lord Eliot asked the House to endorse the recommendations of the Select Committee for the continuation of Gurney's investigations.[47] He argued that further experiments were needed to realise Reid's plan for a downward supply. He said that:

> *there was a great necessity for adopting some means to improve the ventilation of the House, and that this light would afford facilities for that object, by giving opportunity for having a descending current: at all events, an improvement of the ventilation would prevent the annoyance of dust through the chinks in the floor of the House.*

Over the following months Gurney devised a different type of Bude Light, the 'Atmospheric Bude Light', which utilised coal gas and natural air instead of oil and pure oxygen.[48] Gas burners were installed above the glass ceiling to provide general background illumination, whilst a row of pendant lights were introduced below the central ceiling to illuminate the space under the galleries[49] (see Figure 2.2 in Chapter 2). The pendant lights consisted of concentric rings of gas flames and had pipes connected to a discharge shaft allowing hot fumes to be discharged.[50] The use of the downward supply was thereby enabled by isolating the flames from the atmosphere of the House.

The results of these changes were the subject of a short debate held on 8 August 1840. Notwithstanding Trench, who continued to advocate the use of wax candles, the MPs shared the view that Gurney's intervention had improved the quality of the light and the ventilation. Hume stated that Atmospheric Bude Light produced a 'rich mellow light ... in no way offensive to the sight'. Hawes also found it to be 'greatly improved' but felt that 'much credit was due to the gentleman who had undertaken the ventilation of the house, and whose plan had been so successful'. Warburton argued that the new lighting 'tended also greatly to improve the ventilation, because it facilitated the creation of a down current of air in the house, which it would be impossible to have as long as it was lighted with candles'.[51]

Charles Burrell, MP for Shoreham, thought that the 'light was now very good; and, taking the light and the improved ventilation together, he would say, that the present house was a paradise compared to the old House of Commons'.

Burrell, however, had some concerns about the use of the downwards supply. He argued that 'if a strong down current was formed, it would be necessary for him and others who, like him, had not hair to protect their heads, to wear their hats, or, like an illustrious personage in another house, wear a black cap'.

Although the changes succeeded in overcoming conflicts between the design of the ventilation and the lighting arrangements, there does not appear to be any evidence that the downward supply was subsequently used after the system was completed. Having received the approval of the majority of MPs,[52] the Atmospheric Bude lighting remained in use until the closure of the House in 1851. Gurney's system was subsequently extended to large parts of the temporary Houses of Parliament and was operated and maintained under Gurney's supervision, independently from Reid's ventilation system.

The early technical problems with Reid's ventilation system that had led to complaints from MPs in 1837 were largely resolved by Reid's initial alterations and Guerney's intervention. The investigations undertaken between 1837 and 1839 were largely based on the experience of a small number of MPs, who criticised its performance and were dissatisfied with the quality of the thermal conditions. The *Hansard* reports show that after the autumn of 1839, negative criticism of the system inside the debating chambers of the House of Commons ceased to be voiced by MPs. Indeed, MPs subsequently expressed their views on the improved system in several speeches between 1846 and 1847, whilst in September 1843 Reid had received written testimonials from Lord Campbell, MP for Edinburgh from 1834 to 1841, Lord Sudeley, Hawes and the Speaker of the House of Commons on the improved performance of the debating chamber.[53] In a debate held in the House of Lords on 25 June 1846, several former MPs reported their experience of Reid's system in the House of Commons over the past six years.[54] Lord Sudeley 'asserted that the ventilation of the House of Commons was most complete and perfect, and was the first plan of systematic ventilation ever carried out in this or any other country'.

Earl Grey, who was MP for Sunderland up until July 1845, and Lord Campbell both described Reid's system as highly successful, whilst the latter argued that Reid 'had conducted the process of ventilation more successfully than any modern philosopher' and that he 'had been for many years a Member of the other House, and could bear his testimony to the fact that Dr. Reid's success in the House of Commons had been unbounded'.

In another short debate, held at the House of Commons on 21 July 1847, three MPs had voiced their approval.[55] Redhead York, MP for York, stated that:

> *he had nowhere met with so close an approximation to a triumph as the present system of ventilation adopted in that House by Dr Reid;*

> *and hon. Members who, during the present sultry weather, occasionally went into the House of Lords must be of the same opinion as himself thereupon.*

Sir Robert Inglis, MP for Oxford University, 'begged to express the same gratitude to Dr Reid, in which he should think the whole House must concur', whilst Viscount Morpeth, MP for Yorkshire West Riding, noted that he 'for one, cordially concurred in the approval bestowed on the ventilation which they had recently experienced'.

The views expressed in these and earlier parliamentary speeches only provide insight into the views of a small minority of outspoken MPs. As such, it was not representative of a shared experience across the membership of the House nor an objective measure of building performance. Nevertheless, it suggests that after the initial problems of Reid's system had been resolved, it was widely portrayed as successful. The changes undertaken between 1837 and 1839 were instrumental in increasing MPs' confidence in the viability of Reid's ventilation system. By the end of 1840, Reid's experimental ventilation system had received the approval of MPs. Reid's original aim of creating a climate-controlled environment was still to be realised and he therefore still had much to do.

Evaluating the performance of Reid's system

Records of the original monitoring data and reports from Reid and his staff offer critical insights into the quality of the thermal environment and the practical challenges of achieving a climate-controlled environment using 19th-century technologies. The performance of the ventilation system was the subject of Select Committee reviews in each of the Houses between 1839 and 1846. Gurney also contributed to this process of evaluation, testing in 1839 how air movement affected thermal comfort, and in 1846 giving evidence on its performance to the convened Select Committee of the House of Commons.

These records show that cooling and the control of the internal air movement were the main problems. Although several methods of cooling were deployed – evaporation, exposing air to ice, or using pipes with cold water – the system relied largely on increased ventilation, which was used to lower the internal air temperature and to enhance thermal comfort by exposing MPs to air currents. Reid reported that the volume of the air passing through the house during one sitting in 1841 had to be increased to 70,000 ft^3/min, which could only be achieved by doubling the amount of coke burned inside the shaft.[56] In an oral statement given to a Select Committee in 1842, Reid testified that the stack ventilation was not always able to prevent the chamber from overheating during very hot weather. Indeed, he said, the draughts inside the chimney were observed to become weak whenever the difference between the internal and external temperature was too small. Moreover, sufficient air movement was difficult to maintain during hot weather, even when the ventilating fires were raised to artificially boost the convection currents within the shaft.[57] As a result, the thermal ventilation was most energy intensive during the summer as the air flow rate had to be boosted for cooling purposes.

Reid's objective was to keep the internal air temperature constantly below 67°F. However, historical data show that this was not always achieved. For instance, between 12 June and 11 July 1839, the temperature on the main floor was recorded as ranging from 62°F (~17°C) to 74°F (~23°C).[58] Even higher temperatures were recorded at the gallery level due to the stratification of hot air in the chamber. Complaints were made that the galleries were too hot, whilst MPs on the main floor felt it too cold.[59] Readings taken between June and July 1839 show a difference of between 1°F to 6°F at any given sitting. On 11 July, for instance, a temperature of 70°F was measured on the main floor whilst 76°F was measured in the west gallery.

Despite these problems, the recorded temperature readings suggest that the system was reasonably effective in preventing a sharp rise in indoor temperatures either during hot weather or when densely occupied. On 12 June 1839, when the outdoor temperatures was 77°F and the chamber was filled with 430 people, the indoor temperature ranged between 64.5°F (~18°C) and 70°F (~21°C). The total number of persons inside the debating chamber at any one time could fluctuate drastically even over the course of a single sitting. On 24 June 1839, the number of people in the chamber ranged from as low as 50 to as high as 750. Despite this, the indoor temperature remained relatively stable. Data for 19 June 1839 show that the number of persons increased from 100 to 700 over the first four hours of the sitting but the temperature only rose by 6°F from 65°F to 71°F. The temperature fell again by 5°F when numbers declined to 200. By the end of the sitting the number had risen to 600 people but the temperature only rose a mere 3°F.

The temperature readings themselves, however, were not sufficient to completely understand MPs' experience, as it was also strongly affected by humidity and air movement. The House of Commons debating chamber could overheat during summer, and, if not well managed, the use of air currents for cooling could also cause uncomfortable draughts.[60] During an interview on 30 September 1841, Reid testified that Members 'continually inquired into the possibility of cooling the air directly instead of producing the equivalent cooling effect by increasing the supply'.[61]

The use of ventilative cooling methods was limited by physiological factors. This issue of cooling in the chamber was investigated by a Select Committee in 1846. In May 1846, Gurney explained that the strong air currents had made people feel uncomfortably cold, despite the high indoor air temperatures. He said that air currents 'produce an increased rate of evaporation upon the skin, which produces a sensation of cold. Although the temperature of the air in motion itself is warm, say 70°F or 80°F, it will produce a temperature very considerably lower on evaporating surfaces'.[62]

In order to determine the cooling effect of currents on the human skin, Gurney recommended the use of 'evaporative thermometers' in the House instead of dry-bulb thermometers (which were only able to measure air temperature).[63] In addition to the currents around the feet and legs, MPs were exposed to horizontal cross-currents, which occurred whenever the doors of the chamber were opened. Gurney observed in July 1839 that the doors cause air to rush in 'mechanically' and that 'it takes a horizontal direction, and disturbs the whole quiescent atmosphere of the House, before it gets into the flue'.[64] These cross-currents also caused candle wax to drip into the body of the House.[65]

The House of Lords and the challenge with sustaining a social feedback loop

Reid's experimental ventilation system in the temporary House of Lords was perceived as less successful than that of the House of Commons. In several debates (discussed in Chapter 5), the Lords voiced their dissatisfaction with the working environment,[66] and its performance was also reviewed by Select Committees in 1843 and 1844. Reid did not deny that there were problems. On 2 and 10 March 1843 Reid admitted that the main issues were insufficient heating and difficulties with the diffusion of the incoming air.[67] The latter, he argued, was caused by the fact that the inlets were not evenly distributed across the floor but instead were confined to smaller areas. He noted that the air was admitted, 'principally by the sides and at those places marked under the seats'.[68] Being a very tall and narrow space the interior also had issues with internal convections currents, which, according to Reid, became most intense when Members gathered in one area leading to an 'ascent of warm air where the crowd is congregated, and a descent at the other part'.[69]

For Reid himself, however, the main issue with his ventilation system was not technological but a lack of participation by the Lords in the feedback process. In a statement given to a Select Committee in 1844 he claimed that his technology had the capacity to provide full control over the temperature or velocity, but believed that it would perform better if the Lords would report their experiences regularly and on time. Reid had, at the beginning of the project in 1837, briefed both MPs and Lords about the importance of active participation, but their participation had declined over time. This problem was particularly pronounced in the House of Lords.[70] Feedback was scarce, even at times when Lords felt uncomfortable, and problems were often reported retrospectively. On 10 March 1843 Reid reported that 'our great regret always is, want of complaints; those complaints never being made to us'.[71] It might appear paradoxical that he was inviting Lords to be more critical, but Reid clearly thought it essential to enable the system to become responsive to their experiences. At times Reid himself had to be proactive in approaching individual Lords and MPs in order to gain a better understanding of their experiences.[72] It is evident from these accounts how much the user-responsive system depended on the quantity of personal feedback that could be physically gathered. Environmental control was a sociotechnical process that relied heavily on active participation. Despite these difficulties, Reid continued to pursue inquiries into climate control strategies that enabled accommodating individual differences in the two debating chambers of the Palace of Westminster (see Chapters 5 and 7).

Conclusion: an emerging model of evidence-based practice

The post-occupancy history of the temporary Houses of Parliament offers an intimate insight into the processes by which Reid's system was tested and developed over the course of several years after initially implementing the foundational infrastructure for the ventilation and climate control. At the beginning of the process in 1836, neither the technologies nor the operational procedures were fully developed (whilst parts for the House of Lords were also compromised by time constraints and a small budget). The final system in the House of Commons, more sophisticated than the basic system installed in 1836, was not completed until 1840 and was developed through both trial and error and experiments (both under Reid's own directions and others led by external advisors under the supervision of the Office of Woods or a Select Committee). Gurney's substantive involvement resulted in the resolution of the problems caused by integrating the ventilation system with the type of artificial lighting used in the chamber.

This refining of the system was a highly collaborative endeavour, which involved multiple stakeholders within Parliament. In addition to the participation of MPs as its primary users, it involved working with the Office of Woods and Department of the Sergeant-at-Arms, which were responsible for the maintenance and operation of the temporary Houses. The ventilation system's performance was also dependent on the continual stewardship of Reid, who was employed to ensure it was operating as intended and its delivery depended on wider institutional participation. If viewed through the lens of modern organisational learning theory, this level of participation facilitated a process of organisational learning, which is the 'product of organisational members' involvement in the interaction and sharing of experience and knowledge'.[73] This learning formed an important part of Reid's transition from conducting experiments in a laboratory to conducting them in the field under real-life conditions; it also illuminated the complexity of managing buildings in use and enabled staff and MPs as the main users to participate in the process of evaluating and refining the technical arrangements.

As seen in Chapter 2, formal procedures were developed to facilitate user involvement in the day-to-day operation. However, as seen in this chapter, the involvement of MPs was not confined to these planned programmes of engagement. MPs participated in various technical trials and also acted as agents, who drove the commissioning of further investigations. The technical experiments and the changes made to improve Reid's system were both undertaken in response to pressure from MPs who exercised their influence by voicing their views during debates, tabling motions that demanded improvements, or appointing an ad hoc Select Committee. The Select Committees were charged with coordinating formal inquiries in which MPs collaborated with external scientific advisors. Several additional grants were also awarded by HM Treasury to cover the cost of consultants, physical improvements and technical trials. The investment into the development of Reid's system, however, was not solely a means to resolving the immediate problems within the temporary chambers. It was also considered a way to test and refine technical solutions before they were applied to the new Palace of Westminster. As such the temporary Houses also fulfilled the function of a building research laboratory.

Through their interventions MPs gained direct influence over the development of technologies. Their direct participation was also significant as it resulted in technologies being assessed primarily on the basis of MPs' own personal experience of the environment with respect to the thermal, luminous and auditory qualities of the debating chamber. Whilst this level of participation enabled user experience to inform design, it also gave MPs a degree of agency over which neither Reid nor the Commissioners of Woods had much control. This shows that MPs were unusual occupants in that they had the agency and financial resources to shape and reshape their working environment according to their own expectations. MPs attempted to take ownership over their environment.

The disruption caused by MPs' interventions became of increasing concern to the Office of Woods and the government of the time. It must be remembered that these interventions were not the initiative of the whole House, but of a small minority of MPs. Although criticism was voiced by only a few outspoken individuals, they had a strong influence on debates about the performance of Reid's system. Despite the small number of dissenters, the House of Commons and the Office of Woods nevertheless agreed that the issues with lighting and ventilation were important enough to justify formal inquiries and the expense of further experiments. The Office of Woods devoted some time to addressing criticisms and commissioned several physical alterations and trials of technical solutions in an effort to improve the quality of the environment for its users.

In retrospect, these investigations can be seen as part of an extended design process in which empirical evidence was used to evaluate and refine technical solutions based on their actual performance. Empirical evidence, which was in the form of measured data, direct observations and self-reported experience, revealed a gap between Reid's original design intentions and its actual performance in use, leading to refinements of the system. If viewed through the lens of design methodology research this could be viewed as an example of evidence-based practice in early 19th-century

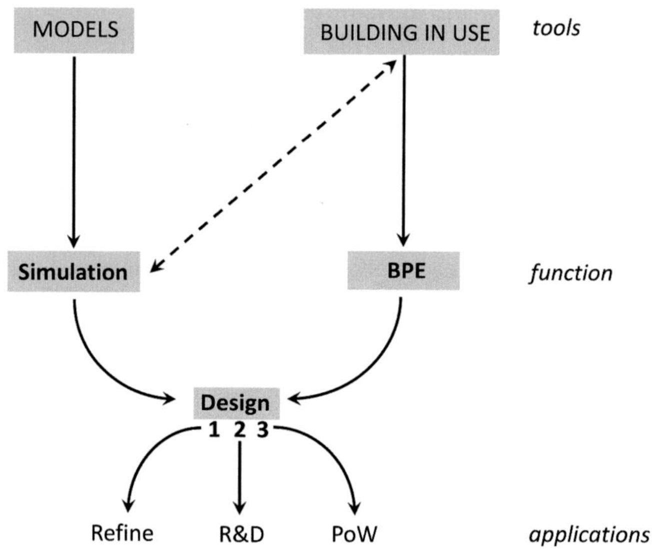

Figure 3.1 A model of evidence-based practice (author's own drawing). Key: BPE: Building performance evaluation. Three main applications of new knowledge gained through BPE and simulations: 1. Refinement of Reid's systems inside the temporary Houses of Parliament, 2. Development of new technologies, 3. Generating new knowledge to underpin the design of the new Palace of Westminster.

architecture. Similar to Reid's earlier mock-up, the temporary Houses were used as a design tool that facilitated a more evidence-based approach to design. Reid's practice involved processes of simulation, undertaken within physical mock-ups with the aim of testing refining solutions before they were applied in buildings, as well as evaluations and improvements of their performance through field testing within buildings in active use. The latter could be interpreted as an early example of building performance evaluation through 'post-occupancy evaluation' (POE).[74] POE is a modern methodology used to assess the gap between the design intention and measured performance of a building in use.[75] Similar to modern POEs, Reid combined physical measurements, such as climate monitoring and air flow tests, with the review of operational procedures and occupant satisfaction.

From an epistemological perspective the early mock-up of the debating chamber and temporary Houses fulfilled similar functions to the computational simulations tools of current architectural practice, such as computational fluid dynamics (CFD). Without access to such tools Reid and his contemporaries relied largely on physical trials to verify and optimise design solutions. Although they were clearly buildings in use, the temporary Houses of Parliament retained some of the qualities of Reid's earlier physical mock-ups. The technology and fabric of the Houses underwent various and substantial physical changes throughout their operational life. Some were more long-term alterations that were undertaken to improve the ventilation, while others were transient changes made to enable brief trials of different technologies. The temporary Houses could therefore be interpreted as a third typology – a form of occupied prototype – situated somewhere between model and actual building, and in terms of their function, as a design tool, somewhere between field test (POE) and simulation. This is important to understand the changes in design methodology that Reid would later encounter in his transition from the temporary Houses towards the design of the new Palace of Westminster.

Notes

1 There is no *Hansard* entry, but the debate was reported in 'Imperial Parliament', *The Morning Post*, 23 May 1838, p. 2.
2 HC 277 1837-1838 'Estimates', 23 March 1837, p. 10; HC 310 1839 'Estimates', 11 April 1838, p. 10; HC 14-1 1841 'Estimates', 8 September 1841, pp. 11-12; HC 211–1 1851 'Estimates', 14 April 1851, p. 12.
3 'House of Commons', *Sheffield Independent*, 12 November 1836, p. 1.
4 'Alterations in the House of Commons', *The Times*, 5 November 1836, p. 3.
5 'Alterations in the House of Commons', *The Champion*, 13 November 1836, p. 66.
6 David Boswell Reid, *Brief outlines illustrative of the alterations in the House of Commons* (Edinburgh: Neil & Company, 1837), pp.9–28.
7 All three letters were reproduced in *The Morning Post*, 29 December 1837, p. 1: Tracy, H. Letter to D.B. Reid, 10 May 1837; Chancellor of the Exchequer, Letter to D.B. Reid, 6 May 1837; Viscount Duncannon, Letter to D.B. Reid, 17 June 1837.
8 Arnott, N. Letter to T. Wakley, 22 December 1837, reported in House of Commons, *Hansard's Parliamentary Debates: The Official Report* (23 December 1837, vol. 39, cols. 1517-20) (London: Hansard).
9 HC 251 1842, *Report from the Select Committee on Lighting the House*.
10 'Ventilation of the House of Commons', *The Morning Post*, 29 December 1837, p. 1.
11 Ventilation of the House, House of Commons, *Hansard's Parliamentary Debates: The Official Report* (23 December 1837, vol. 39. cols. 1517–20) (London: Hansard).
12 Viscount Duncannon. Letter to F. Trench, 6 March 1838, in HC 204 1838, p. 4.
13 HC 277 1838 'Estimates', 31 March 1838, p. 10.
14 'Imperial Parliament', *Freeman's Journal*, 15 February 1838, p. 3; 'Imperial Parliament', *The Morning Post*, 2 March 1938, p. 2.
15 HC 725 1837-1838, *Return of the detailed expenses incurred in experiments for improving the Ventilation of the House of Commons, in the experiment of lighting with gas, also in the lighting with candles, ending with the present lustres and shades*.
16 David Boswell Reid, *Illustrations of the Theory and Practice of Ventilation* (London: Longman, 1844), pp. 301–2.
17 Viscount Duncannon, Stephenson, B. and Milne, A. Letter to H.M. Treasury, 8 January 1838 (National Archives: Work 11/26/1, no. 5)
18 'Ventilation of the House of Commons', *The Morning Chronicle*, 1 January 1838, p. 4.

19 Statement of the thermometer in the space between the present and old ceiling and in the roof of the House of Commons during the lighting of the House with gas on the 5th January 1838, 8 January 1838 (National Archives: Work 11/26/1 no. 1–4).
20 Report by Duncannon, Stephenson and Milne to the Lords of H.M. Treasury, 8 January 1838 (National Archives: Work 11/26/1, no. 1-4).
21 'Parliamentary Intelligent', *Bristol Mercury*, 17 February 1838, p. 4.
22 'Lighting the House of Commons with Gas', *Bristol Mercury*, 17 February 1838, p. 4.
23 Oral statement by D.B. Reid on 25 March, HC 402 1852, *Second report from Select Committee on ventilation and lighting of the House*, Q224.
24 Trench, F. Letter to Viscount Duncannon, 3 March 1838, in HC 204 1838, p. 3.
25 'House Of Commons, Wednesday, April 11', *The Times*, 12 April 1838, p. 3; 'Miscellaneous', *Derby Mercury*, 5 September 1838, p. 1;
Lighting the House, House of Commons, *Hansard's Parliamentary Debates: The Official Report (06 February 1840, vol. 51 cols. 1343–6)* (London: Hansard).
26 House of Commons, *Hansard's Parliamentary Debates: The Official Report (01 March 1838, vol. 41, cols. 329–32)* (London: Hansard).
27 Trench, F. Letter to Viscount Duncannon, 3 March 1838, in HC 204 1838, p. 3.
28 Viscount Duncannon, Letter to F. Trench, 6 March 1838, in HC 204 1838, p.4.
29 Reid, D.B. Letter to Viscount Duncannon, 28 March 1838, in HC 277 1838, pp. 1–8.
30 Reid, *Illustrations*, pp. 284–90.
31 Trench, F. Letter to Viscount Duncannon, 6 May 1838, in HC 358 1837–38, p.5.
32 Ibid.
33 House of Commons, *Hansard's Parliamentary Debates: The Official Report (08 August 1840, vol. 55, cols. 1396–8)* (London: Hansard).
34 HC 251 1842, Select Committee to inquiry into renumeration to Mr Gurney for Services in lighting House of Commons, p. 3
35 HC 590 1834, Select Committee to inquire into State and Management of Lighthouses. Report, Minutes of Evidence.
36 HC 251 1839–1840, Bude Light (Lighting the House): Return of the expenses connected with the experiments of the Bude light, from the commencement to present day.
37 Ibid.
38 HC 251 1842, Select Committee, p. 9.
39 Trench, F. 'Letter to John Bull', *John Bull*, 8 September 1839, p. 424; Ventilation and lighting of the House, House of Commons, *Hansard's Parliamentary Debates: The Official Report (01 March 1838, vol. 41, cols. 329–32)* (London: Hansard); Lighting the House, House of Commons, *Hansard's Parliamentary Debates: The Official Report (06 February 1840, vol. 51, cols. 1343–6)* (London: Hansard).
40 Oral statement by Gurney, in HC 501 1839, Q53–56.
41 HC 251 1842, Select Committee, p. 33; Oral statement by Charles Wheatstone, 4 June 1839, pp. 20–23. In modern building science this relationship between heat and light is referred to as 'luminous efficacy'. Wheatstone was engaged to measure the light intensity using a photometer:
42 Oral statement by G. Gurney, 4 July 1839, in HC 501 1839, *Select Committee*, Q912–14. For more on the experiments with lighting, see Edward Gillin, *The Victorian Palace of Science – Scientific knowledge and the building of the Houses of Parliament* (Cambridge: Cambridge University Press, 2017), pp. 184–213.
43 Oral statement by G. Gurney, in HC 501 1839, *Select Committee*, Q868–71.

44 HC 501 1839, *Select Committee,* Report, p. iii.
45 The Bude Light, House of Commons, *Hansard's Parliamentary Debates: The Official Report (19 August 1839, vol. 50, col. 420)* (London: Hansard); Lighting the House, House of Commons, *Hansard's Parliamentary Debates: The Official Report (13 August 1839, vol. 50, cols. 2601–1)* (London: Hansard).
46 Lighting the House. House of Commons, *Hansard's Parliamentary Debates: The Official Report (06 February 1840, vol. 51, cols. 1343–6)* (London: Hansard).
47 Lighting the House. House of Commons, *Hansard's Parliamentary Debates: The Official Report (12 March 1840, vol. 52, cols. 1158—61)* (London: Hansard).
48 John Bethell, 'On the Atmospheric Bude light', *Transactions of the Society for the Encouragement of Arts,* 54 (1843–1844), pp. 198–200.
49 Oral statement by G. Gurney on 26 April 1852, HC 402 1852, *Second report from Select Committee on ventilation and lighting of the House,* Q2966–76.
50 Bethell, 'On the Atmospheric Bude light'.
51 Lighting the House. House of Commons, *Hansard's Parliamentary Debates: The Official Report (08 August 1840, vol. 55, cols. 1396–8)* (London: Hansard).
52 HC 251 1842, *Select Committee.*
53 These letters, dated 7 to 11 September 1843, were reproduced in the document 'Extracts from Official Documents, Reports, and Papers, Referring to the, pp. 7–9.
54 The new Houses of Parliament. House of Lords, *Hansard's Parliamentary Debates: The Official Report (26 June 1846, vol. 87, cols. 1033–5)* (London: Hansard).
55 Dr Reid's ventilation. House of Commons, *Hansard's Parliamentary Debates: The Official Report (21 July 1847, vol. 94, cols. 663–4)* (London: Hansard).
56 Oral statement by D.B Reid, 26 July 1842, in HC 536 1842 Committee on Ventilation, Q55.
57 Ibid, Q49–55.
58 B. Riches, 'State of the thermometer in June and July', in HC 501 1839, *Select Committee,* p.76.
59 Oral statement by W. Gossett, 24 June 1839, in HC 501 1839, *Select Committee,* Q 744–7.
60 Oral statement by D.B Reid, 30 September 1841, in HC 51 1841, *Report from the Select Committee on Ventilation of the New Houses of Parliament; with the minutes of evidence, and appendix,* Q161.
61 Ibid, Q 164.
62 Oral statement by G. Gurney, 11 May 1846, in HL 719 1846, *Reports from the Select Committee of the House of Lords appointed to inquire into the progress of the building of the Houses of Parliament,* Q15.
63 Ibid, Q16–19.
64 Oral statements by G. Gurney, 4 July 1839, in HC 501 1839, *Select Committee,* Q891.
65 Ibid, Q892–94.
66 Lord Campbell. Letter to Reid, 11 September 1843, in 'Extracts from official documents, reports' (Hume Tracts, UCL Library); House of Lords, *Hansard's Parliamentary Debates: The Official Report (21 February 1843, vol. 66, cols. 1033–6)* (London: Hansard); House of Lords, *Hansard (05 June 1845, vol. 81, cols. 120–2)* (London: Hansard); House of Lords, *Hansard (24 April 1846, vol. 85, cols. 970–6)* (London: Hansard); House of Lords, *Hansard (26 June 1846, vol. 87, cols. 1033–5)* (London: Hansard).
67 Oral statements by D.B Reid on 2 March, in HL SC 1843 Q203, Q107–8.
68 Oral statement by D.B. Reid on 10 March, in HL SC 1843, Q200.
69 Ibid., Q185.

70 Oral statement by D.B. Reid, 1844 Committee on Houses, Q317–28, 387; Reid, *Illustrations* pp. 292–3.
71 Oral statement on 10 March, in HL SC 1843 Q201, 203.
72 Oral statement by D.B. Reid on 17 June, HC SC 1844, Q386.
73 Carla Curato, 'Organisational Learning theory and organisational design', *The Learning Organisation*, 13 (2006), pp. 25–48.
74 Tim Sharpe, 'Building Performance Evaluation', in *Innovations for Sustainable Building Design and Refurbishment in Scotland, Green Energy and Technology*, ed. by Branka Dimitrijevic (Cham, Switzerland: Springer, 2013), pp. 127–146.
75 Wolfgang Preiser & Andrea Hardy, 'Historic Review of Building Performance Evaluation', in *Architecture Beyond Criticism: Expert Judgment and Performance Evaluation*, ed. by Wolfgang F.E. Preiser and others (London: Routledge, 2015), pp. 146–59.

PART II

Attempted application: 1840–46

I will reserve speculation but tell your Lordships that the Parliamentary Buildings (Restoration and Renewal) Act 2019 established a sponsor body and a delivery authority to ensure independent oversight and management of delivery, accountability and costs. That was very important, because we can think back to some projects—I have the construction costs of the Scottish Parliament in mind. In 1997, the White Paper estimated these at between £10 million and £40 million. In 1998, when Holyrood was named as a site, the cost was £50 million. In 1999, the Scottish Parliament voted to continue the project at £109 million, and in 2004, when it was completed, the report of the Holyrood inquiry, led by the late Lord Fraser of Carmyllie, estimated the final cost at £414.4 million. That is why we have established the sponsor body and the delivery authority and built in independent oversight.

Lord McFall of Alcluith, Senior Deputy Speaker, March 2020[1]

Note

1 Palace of Westminster: Restoration and Renewal, House of Lords, *Hansard's Parliamentary Debates: The Official Report* (12 March 2020. vol 802, Col. 1094).

4 A tentative master plan for ventilating the Palace of Westminster

The right adjustments of the relative claims of the fine and useful arts, where they meet or interfere, forms the very root and basis of all that is most important, more especially in public buildings. Nothing can attain the highest order of beauty that attempts to do so at the expense of an admitted want.[1]

David Boswell Reid, 'The revision of architecture in connection with the useful arts' 1855

The paths convergence

The extension of David Boswell Reid's experimental ventilation system to the temporary House of Lords occurred four years after he had participated in the first inquiry into the ventilation of the House in 1836. His experiments within the temporary Houses of Commons were instrumental in enabling MPs to gain experience of a hermetically- sealed and climate-controlled environment and succeeded in overcoming some of the scepticism expressed in 1835 when it was still only theoretical. The prospect of his ideas being applied to the design of the new Palace of Westminster, however, remained uncertain in 1839. Although the various Select Committees that had investigated the ventilation and lighting of the Houses had alerted the Commissioners of the Woods to the importance of integrating these two strands from the outset, they remained separate until April 1840 when the Commissioners finally agreed to engage Reid in the design of the new Palace. This marked the beginning of a new phase in which Reid had to work with Charles Barry's project team.

Barry's architectural design for the new Palace had to be substantially altered in light of the late adoption of Reid's system. Although Barry's own scheme was not finalised, this late adoption limited the degree to which the architecture itself could be adapted to accommodate the functional requirements of Reid's system. Whilst some aspects of the architectural plans were changed to accommodate the requirements of the system, in other cases Reid's system had to be adapted to fit Barry's plans. This involved a complex, and at times difficult, process of amalgamating two different schemes. Indeed, as John Darwin – a former resident engineer at the Palace of Westminster – has written, it was the 'first time in history that the architectural style of the building was fundamentally altered on account of the needs of the mechanical system'.[2]

However, it is not strictly true that it changed the new Palace's architectural style, which was required to be Gothic from the outset, as it proved to be sufficiently adaptable to accommodate the physical changes required to implement the system without undermining the integrity of Barry's existing design.

This chapter focuses on the early phase of Reid's work with Barry's project team from 1840 to 1842. During this period, inquiries into the application of Reid's principles to the new Palace were limited to the development and approval of the general design concept – a 'masterplan' – which would guide the development of Reid's scheme until 1846. These early investigations also provide us with some insight into the difficulties of establishing a successful working relationship between a scientist and a professional architect. These difficulties are critical to understanding the impact of organisational factors, which intensified in the later stages of the project, on the development of the new Palace's design.

The appointment of the 'ventilator'

The first conversations about applying Reid's system to the new Palace of Westminster were in the autumn of 1839. On 3 October, Charles Barry wrote a letter to Viscount Duncannon, First Commissioner of the Woods, informing him that it was time to begin designing the ventilation and central heating systems and that an engineer should be employed to provide assistance to him in addressing technical questions.[3] Barry also voiced some reservations about involving Reid as the 'ventilator', a term that Barry had used to describe the role of the building services engineer. Although he acknowledged that Reid's system in the temporary House of Commons was successful and that Reid himself was a person 'who in all respects are eminently qualified to be employed upon this business',[4] Barry argued that 'he does not profess to be thoroughly acquainted with the practical details of building and machinery'.[5] Moreover, he claimed that Reid would be unable to take on this new role due to Reid's own professional commitments in Edinburgh. Instead, Barry recommended that Charles Manby, a civil engineer, be appointed and that an alternative ventilation and central heating system be used, one that he and Henry Cruger Price, his business partner, had successfully applied in other public buildings. Barry specifically cited his and Price's work on the British Museum (1823–47),[6] and the fact that they had recently begun a collaboration with Manby designing the ventilation and central heating of one of Barry's other architectural projects – the Reform Club, situated in Pall Mall (1837–41).[7] As such, Manby was familiar with their alternative system and the potential design implications of adopting it. It is not clear from Barry's letter, nor indeed subsequent correspondence, if Barry himself objected to the involvement of Reid in any way. However, it is clear that he believed that the main role should be held by someone from an engineering background and who had practical experience in construction, and that Reid did not meet either of these criteria.

Reid was quick to challenge Barry's assumptions about his availability. On 5 October, he wrote to Viscount Duncannon expressing an interest in taking on the role of 'ventilator',[8] and subsequently expressed his willingness to end his teaching career in Edinburgh to move with his wife and children to London, if given adequate remuneration.[9] Although Viscount Duncannon was not prepared to make any firm commitments, he agreed to engage Reid temporarily for a preliminary investigation.[10] The aim of this investigation was to determine the general implications of adopting Reid's ventilation system for the design and construction of the new Palace of Westminster. Reid had several meetings with Barry in October 1839, undertook site visits and examined the drawings held by the architect's office in Westminster. In a joint statement, dated 22 October 1839, Reid and Barry concluded that the changes required for accommodating Reid's ventilation system would not affect the first phase of the Palace's construction.[11] However, in another letter to Viscount Duncannon, Reid warned that the next phase of the construction should be postponed until the scheme for the ventilation system was completed because it would have a significant impact on the Palace's architectural fabric. Indeed, it would 'necessarily modify architectural arrangements'.[12] Reid also highlighted the fact that it would take time to complete the additional experiments required to demonstrate the feasibility of applying stack ventilation at a much larger scale than that with which he had experience of, which was limited to relatively small and simple structures. With the Palace of Westminster, Reid proposed to apply stack ventilation at the scale of an urban block: since the Palace was 300 m (984ft) long, and between 120m(393ft) and 70m (229ft) wide, the ventilation shaft would have to draw air over long distances (over 200 m(656ft) in some areas).

The uncertainty about Reid's involvement in the design of the Palace ended in January 1840. Despite Barry's earlier advice, the Commissioners of the Woods decided to adopt Reid's system and employ him as a specialist consultant responsible for overseeing its implementation.[13] Although Barry had originally asked for a technical assistant, Reid did not work under his direct supervision and was neither his employee, nor embedded within Barry's team. Instead, he was employed by the Office of the Woods and was given his own office in Westminster. A wage of £500 per annum, and the costs for laboratory facilities and new staff employed to assist Reid in his work, were covered through additional grants from HM Treasury.[14] The Commissioners – Viscount Duncannon, Sir Benjamin Stephenson and Alexander Milne – also set out the terms of Reid and Barry's collaboration and their respective roles and responsibilities. Milne, Third Commissioner of the Woods, made clear that Reid was in charge of the ventilation and heating system and that Barry was expected to assist in him in accomplishing his plans. Barry was responsible for supplying copies of drawings and also should 'offer him every required facility and guidance' needed to realise his plans.[15] In return, Reid was asked to acquiesce to Barry's views 'upon all point affecting either the solidity or the architectural character of his building'.[16] Milne did not provide an explanation as to why such an arrangement was advocated, but the result was that Reid was given a significant level of authority and independence. This ensured that the requirements of the ventilation system were given a prominent role in the subsequent development of Barry's scheme. The appointment of Reid had the (perhaps unwitting) consequence that Barry, the architect, was the person assigned responsibility for the integration of building services, something that the 1836 Select Committee had recommended in their final report. Today, this has been described by the contemporary historian Moritz Gleich as a 'historically unique constellation' in which the architect was 'placed alongside an independent service architect on the construction site'.[17]

Transition: embedding the cultural practices of science

The appointment of Reid as 'ventilator' marked the end of the phase of isolated experimentation and the beginning of a new phase in which the findings of his earlier experiments could be directly applied to the design of the new Palace of Westminster. In retrospect, the development of the temporary Houses of Parliament, which was the focus of Chapters 2 and 3, can be understood as part of a slow transition from entirely speculative design studies using specially-constructed rooms towards the design of system for actual buildings. Reid retained his role as superintendent of the ventilation system inside the temporary Houses, which enabled him to gather empirical data on its performance and involve its operational staff in the study of atmospheric pollution. This experience directly informed the design of the more sophisticated arrangements for the two permanent chambers in the new Palace of Westminster.

However, this was not the end of the use of scientific experimentation to empirically assess design proposals. Reid set up his own office at the Palace of Westminster, initially within one of the committee rooms, but in Spring 1841 he moved into a permanent office replete with an 'experimental laboratory'. This laboratory, erected with funding from the Office of the Woods, provided him with the facilities, special equipment and instruments he needed to undertake the technical experiments required to underpin the development of the new Palace's design.[18] In a letter to Milne, Reid noted that parts of the design 'can be determined solely by actual experiments'.[19]

Reid also used the experimental laboratory for educational purposes, demonstrating new technologies and general principles using scale models and life-size mock-ups to interested parties including, in 1841 and 1842, MPs and Lords from the various pertinent Select Committees. For example,

Reid used a simplified glass model of the Central Tower system, which with the aid of the smoke of a lit spirit lamp enabled the air flow within the building to be seen.[20] Reid reported that he had used this model for 'experiments on the subject' but also to undertake live demonstrations of his principles. Reid reported that on the occasion of Queen Victoria's and Prince Albert's visit, he and Barry 'worked that model, and moved the valves'.[21]

Between 1841 and 1842, Reid undertook various experiments and also asked his staff to monitor the outdoor air quality within the near vicinity of the site of the new Palace. On 26 July 1841 he reported that his trials focused on the development of a mechanical system for the pressurisation of interior spaces, ventilation flues and cleaner fuels that would eliminate the need for the sweeping of chimney flues.[22] The latter was a response to recent changes in the legislation of child labour, which prevented the use of chimneys requiring young boys for their maintenance.[23] Reid conducted further experiments after 1842 including testing new methods of thermal insulation for smoke and air channels, which would prevent the thermal convection from being compromised by heat loss.[24] He also planned to conduct more large-scale experiments (for example, by constructing a full-scale mock-up of a sample Committee room). However, due to the cost, these plans were never implemented.

Formulating a general design concept: 1840–41

Reid begun to develop an outline for his proposed ventilation system in the new Palace in April 1840, using copies of Barry's original architectural drawings. He also frequently shared and discussed initial ideas with the Commissioners, Barry himself and Barry's staff within the architect's office. Some of these discussions were conducted through letters, which offer detailed insight into the early phase of Reid's scheme. This phase, which lasted from Spring 1840 to Autumn 1842, focused on the development of the general design concept, but also involved lengthy reviews of the costs involved and its practical implications.

Reid had received a first set of architectural drawings from Barry in January 1840, which Barry believed contained sufficient information for Reid himself to explain his proposed ventilation system. According to Reid, this set of drawings comprised floor plans and cross-sections that provided a general overview and also had 'red marks indicating certain spaces which Mr Barry considered might be at liberty for ventilation'.[25] Reid received a set of more detailed drawings of the River Front,[26] which was one of the first sections of the Palace of Westminster to be built once the foundations and river wall were completed, in June 1840. After their initial correspondence in January 1840, Barry had asked Reid to produce a set of drawings and to explain his proposal 'in the clearest and most definite manner'.[27] Barry expected Reid, the 'ventilator', to be able to work independently and supply the architectural team with drawings for a complete scheme in advance, and argued that these drawings should be sufficient to enable his own team to implement the system.[28]

On 7 July 1840, three months after his appointment as 'ventilator', Reid provided a first written account of his ideas for a central air supply. He proposed to introduce all of the fresh air for the new Palace through tall shafts and to distribute it across the several hundred room sites (including the two debating chambers) through a network of basement passages.[29] Speaking about his own observations of the atmospheric pollution in the surrounding neighbourhood, Reid claimed that the purity of the air was greater at a height of several hundred feet than at one near the ground. His aim was to protect the interior from 'offensive odours from the river' as well as from various types of atmospheric pollution such as the smoke from surrounding chimneys and the emissions from factories. Reid did not explain in the letter how exactly the shafts would be accommodated, but Barry suggested that one of the architectural towers in his plans could be adapted for Reid's purposes. On 14 August 1840, Barry proposed to transform three of the four corner turrets of the 'Victoria Tower' – a large square tower that was to be built at the south end of the Palace to house the Parliamentary Archives – into ventilation shafts, and claimed that it would allow air to be sourced from a height of 240 ft (~73 m).[30] This adaptation required a significant physical intervention as the original architectural plans had not included a basement level.[31] Barry said that this was an 'entirely new creation'.[32] The lowest level in Barry's original scheme was the ground floor, which acted as functional level, containing spaces for kitchen, workshops, storages and also some limited space for building services.

On 17 July, ten days after his first letter, Reid introduced Barry to the idea of adding a massive central discharge shaft, which would act as a central outlet for smoke and vitiated air. Unlike the fresh air shafts, which were accommodated by adapting the turrets of the Victoria Tower already included in Barry's plan, this discharge shaft would have to be accommodated within a large new tower above the Central Lobby. This third tower, which became known as the 'Central Tower', would eventually be added to the plans for the new Palace specifically for the ventilation system.

Reid set about producing a set of schematic drawings after receiving Barry's request of 14 August for further details. This took Reid eight months and throughout this period he continually engaged Barry and his staff in extensive conversations about the functional requirements of the system and their architectural implications. Reid undertook calculations to determine the requirements for the ventilation system and supplied Barry with schematic sketches and written specifications. When face to face, Reid and Barry would discuss and use sketches to explore how the requirements of the ventilation system could be integrated into the existing architectural plans. Amongst the issues explored in the early meetings between Reid and Barry were the dimensions and architectural form of the Central Tower.[33] Although Reid supplied Barry with diagrams and written specifications for its required height and width, they also considered the possibility of using alternatives tower forms. This gave Barry, whose role was to translate Reid's ideas into an architectural feature, the flexibility to develop a form that was acceptable from an architectural perspective. On 14 August 1846, Reid noted that they 'often talked of having it at various altitudes, and he asked me what modifications could be made. I said "if you increase the altitude, you may diminish the diameter"'.[34]

In a letter, dated 24 August 1840, Reid discusses the specification for an enormous shaft, which would rise 150 ft (~45 m) above the ceiling of the Central Lobby.[35] Several alternative designs are also explored, including a proposal by Barry to create an even taller shaft of more slender proportions with a height of 194 ft (~59 m). Figure 4.1 is an example of the early explorative sketches produced in 1840: it shows a diagrammatic cross-section drawn by Reid, and Barry's outline for an architectural tower, which he had traced over the top of this during one of their meetings.[36] Barry transformed this new functional feature into a highly-ornate Gothic spire, which harmonised with the exterior architecture but which concealed its functional nature. This shows that the discussion about the form and scale of the ventilation shaft was shaped as much by formal architectural, as by functional, considerations.[37]

On 2 April 1841 Reid presented his first set of drawings to Barry, comprising six sheets of paper with schematic floor plans and sections. However, these only provided an outline of his general design concept. Unfortunately, these original drawings could not be located in the relevant archives, but according to Reid they comprised diagrammatic floor plans showing the location of the 'leading' horizontal channels within the roof space and basement, as well as a schematic cross-section illustrating the general principle by which these horizontal channels would be connected to each room. The first set of drawings also contained specifications for the size of main channels and tower.[38]

In another letter, dated 28 April, Reid provided a written statement about his design objectives, focusing primarily on his strategies for responding to the problem of severe air pollution in Westminster and surrounding environs.[39] He proposed various measures to reduce and displace emissions from the Palace itself and also to protect the interior atmosphere from external sources of pollution. To prevent smoke from the several hundred fireplaces in the Palace from polluting the courtyards and adjacent streets, Reid recommended the use of cleaner fuels and that Barry's plan for traditional chimneys be abandoned. In their place, he proposed a central extraction system that would convey smoke into one central shaft, to be located within a proposed new tower above the Central Lobby, and eject it at a high level. To protect the interior atmosphere from external sources of pollution, both of which were affected by changes in the direction of the wind, he proposed several high-level intakes inside the proposed 'Victoria Tower' and 'Elizabeth Tower' (also known as the 'Clock Tower', containing the hour

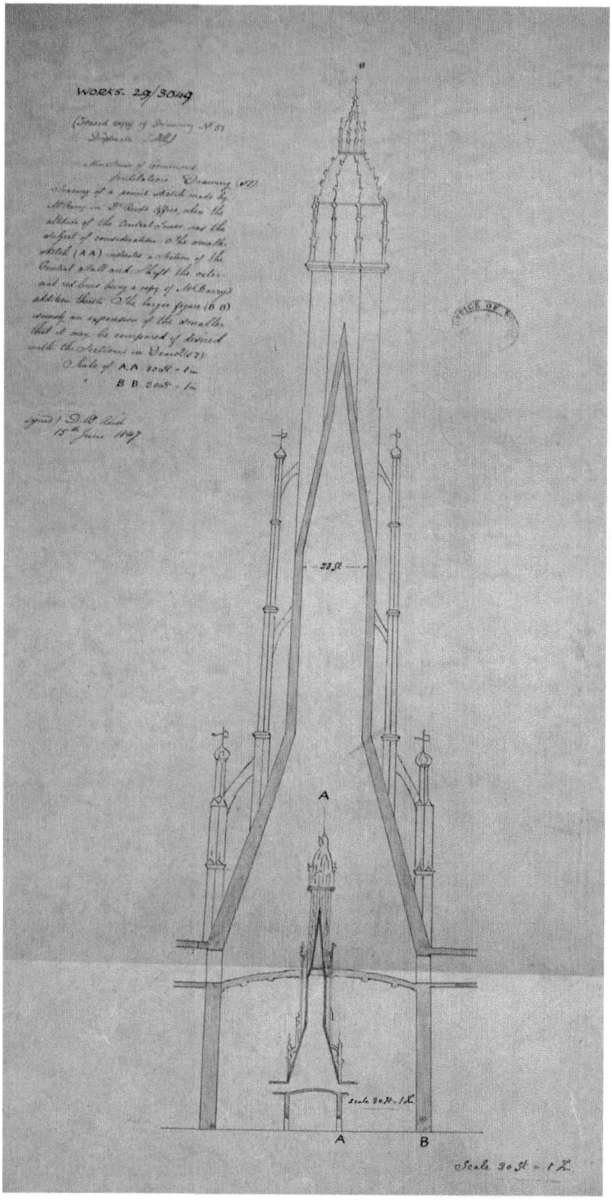

Figure 4.1 Sketch produced during early conversations between Barry and Reid about the architectural form of the Central Tower. Tracing of a pencil sketch made by Mr Barry in Dr Reid's office, 15 June 1847 (National Archive: Work 29/3049).

bell 'Big Ben'). Reid was aware that even with the provision of a central tower, the inlets inside these two towers could be exposed to smoke emitted from the Palace. Reid thus proposed to switch the fresh air supply between the two towers as needed.

Aware of its significant architectural impact, the Commissioners of the Woods consulted Barry about Reid's proposal. On 27 April 1841 Barry confirmed that he did not have 'practical or other objections' to the adoption of Reid's scheme.[40] This approval was important as Reid's proposal entailed substantial architectural intervention. The removal of the large number of individual chimneys (which if retained would have resulted in a silhouette characteristic of many early 19th-century buildings) and the introduction of a third central tower, marked the beginning of an architectural transformation driven by environmental considerations.

Early drawings, produced in Barry's office before the adoption of Reid's scheme, provide an indication of the architectural qualities of his original plans replete with individual chimneys.[41] Despite this, Barry became an outspoken advocate of the third tower on architectural, rather than functional, grounds. In a letter to Viscount Duncannon, the First Commissioner of the Woods, he wrote that 'I have not hesitation in expressing my opinion that the adoption of such feature ... would considerably improve the general effect and importance of the intended building'.[42]

On 28 May 1841, Barry and Reid provided a first estimate of £65,000 for the cost of the proposed ventilation system and associated fireproofing measures, which were not covered by the original budget agreed in 1837. This equated to an increase of almost 10% of the original budget.[43] The estimate included a budget of £20,000 for the construction of the Central Tower and £12,300 for physical alterations needed to create an internal network of 'air and chimney flues in the walls under all the floors and roofs; vaulting of the basement story, and various alterations of the building' for

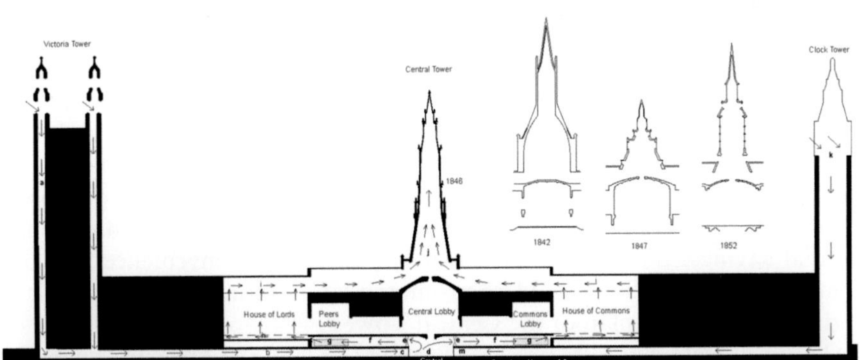

Figure 4.2 Diagram of Reid's concept for a centralised ventilation system (author's own drawing).

the handling of smoke and air. A sum of £20,650 was also included for additional fire barriers, which were necessary to protect the building against the increased fire risks associated with the use of the central smoke extraction system. The estimates show that over half of the budget covered the cost of changing the architectural fabric for the ventilation system, whilst the mechanical plant (which assisted the system), estimated at £12,000, was a comparatively small part of the scheme. The low cost of the latter was due to the fact that most of the ventilation infrastructure was constructed as an integral part of the structural masonry: a concealed network of voids embedded within the depth of the structural walls, buttresses and floors, combined with large architectural features, such as towers and corner turrets. By contrast, the network of steel ducts used in modern mechanical ventilation and air-conditioning systems are typically independent from the structural fabric of the building.

The new estimate of £65,000 was submitted to the Lords of the Treasury, but was not approved by them as the cost and scale of the proposed changes to Barry's original plans required formal approval from Parliament.[44] This delayed the new proposal by over a year. Parliament was also unable to review the matter before September 1841[45] and parts of the construction relating to the foundations were also intermitted for two months.[46] On 2 July 1841, and again on 15 July, Reid urged the Commissioners to intermit the parts of the ongoing construction that would likely be affected by the changes in the original design.[47] The Commissioners of the Woods were also anxious that the proposed changes might be rejected on the grounds of costs and so asked both Barry and Reid to explore ways in which costs could be reduced.[48] Two alternative proposals were investigated: integrate the central extraction shaft into the Victoria Tower, thereby eliminating the need for a third tower, or reduce the size and simplify the design of the Central Tower. The first proposal was rejected due to the close proximity of the central shaft to the inlet shafts,[49] which would eject pollution near the inlet shaft; the second proposal was rejected by Barry on architectural grounds.[50] Writing of this second proposal, he states that 'it will be necessarily become an important feature of the building, and should therefore in my opinion accord in architectural character with the exterior generally'.[51] It took Barry and Reid three weeks to complete their investigation and produce a joint statement in which they made the case for retaining a central tower for its architectural and functional merits, and that the proposed alternative use of the Victoria Tower as an outlet would not achieve substantial savings. Indeed, the 'architectural effect, the convenience and the comfort of the building would be promoted by incurring the larger outlay, and resorting to the plan of the Central Tower'.[52]

The joint statement was presented to the Lords of the Treasury, together with a letter of support from Viscount Duncannon.[53] On 20 September 1841, Lord Wellington, Leader of the House of Lords, was successful in tabling a motion in the House of Lords for the appointment of a Joint

Select Committee to be in charge of reviewing Reid's proposed ventilation scheme.[54] Viscount Duncannon was seriously concerned about the risk of further delays and that the scheme could be rejected. He wrote to the Chancellor of the Exchequer, who was chair of the Joint Committee, arguing that the high cost could not be avoided if the Palace was to be equipped with modern building services. He noted that it was necessary 'if the houses are to be warmed and ventilation on the system adopted by Reid in the temporary House of Commons and which appear to have given general satisfaction'.[55]

This statement is important as it illustrates that important members of the Select Committee had become persuaded that modern building services of a type and scale proposed by Reid and Barry were needed.

Barry's original design assumed that the new Palace, with the exception of the two debating chambers, would be heated with conventional fireplaces.[56] Reid's new scheme, in contrast, serviced the entire building using an integrated system of ventilation and warm air central heating. Whilst the general principles of his system were well established, his proposal was up until this time unprecedented in terms of the scale of its application. It foreshadowed a type of fully-serviced public building that would only became common in Europe from the 1860s onwards.[57] The Commissioners of the Woods were aware of this, and in a note to HM Treasury they admitted that 'such scientific and comprehensive system of warming and ventilating has not henceforth been adopted in any building of the same magnitude'.[58]

The Joint Select Committee interviewed Barry and Reid, but did not conclude its inquiry within the allotted time. In its final 1841 report, dated 1 October, the Joint Select Committee recommended that the inquiry be resumed in the next Parliamentary session. To avoid further delays in the construction of the new Palace's foundations, it agreed to sanction the disbursal of funds required to strengthen the foundations below the Central Lobby. Consequently, the development of Reid's scheme was paused for another year, leading him to endure another period of uncertainty about its future. This uncertainty lasted until August 1842, when the Joint Select Committee, reappointed in July, concluded their inquiry.[59] This review was brief: it only held one sitting in which they interviewed Reid and Barry again, and the final results were presented two weeks later.

Defending the idea of a central tower: 1841–42

The 1842 Joint Select Committee comprised 19 MPs and Lords, half of which were familiar with the subject due to their involvement in the previous 1835 and 1839 Select Committees that had looked at, respectively, ventilation and artificial lighting. This Committee focused solely on questioning Barry and Reid, and did not call on external witnesses to provide independent technical advice. The MPs scrutinised the proposed scheme and explored ways in which costs could be reduced, focusing on alternatives to the proposed Central Tower. In addition to revisiting the earlier proposals for the Victorian Tower, the MPs discussed with Reid and Barry four other possible scenarios.

The first scenario was to reduce the size of the proposed Central Tower and limit its use as an outlet for the two debating chambers and adjacent lobbies only. The second scenario was to substitute the ornamental tower in the centre of the building for a more utilitarian chimney to be located outside the site boundary. The third scenario was to abandon the large central extraction shaft and replace it with an array of smaller shafts.[60] In this case, the committee rooms and the two debating chambers were to have their own separate shafts.[61] The fourth scenario was to use an entirely mechanically-driven system.

None of the four alternative proposals were successful as either they were considered impractical or they did not yield sufficient savings to justify a departure from Reid's original scheme. In 1841, and again in 1842, Reid defended the need for a central tower on functional and economic grounds. He argued that the use of thermal convection and natural wind pressure as the primary motive powers of his ventilation system provided it with the capacity to both reduce air pollution as well as operational costs: 'the natural impulse of the wind, aided by the action of the shaft, will generally provide sufficient power on all ordinary occasions'.[62] In his statements Reid voiced strong objections against the use of mechanically-driven ventilation, arguing that they had higher operational costs and were less reliable than his system due to the risk of mechanical failure.

Although Reid and Barry were unable to quantify the potential savings in operational costs, both believed that the additional capital cost for erecting the proposed Central Tower would be recovered in the long term through the savings in operational costs.[63] Similarly, Reid hoped to use the natural wind pressure to reduce the use of steam-powered fans for fresh air supply. He believed that air could be introduced naturally if the intakes, equipped with rotating cowls, were turned in the direction of the oncoming wind.[64] This illustrates that his scheme relied on the use of architectural features rather than mechanical technology, and his objective was to devise physical arrangements that would enable natural air flow. Reid noted the need for mechanical assistance could be minimised if the 'structure be so arranged that there is no impediment introduced to the process of natural ventilation, it is so certain to be insured by means connected with the laws that regulate the movement of air'.[65]

This represented a clear a departure from the design of Reid's earlier stack system inside the temporary Houses of Parliament, which had relied heavily on the use of coke furnaces due to the location of the shaft on the ground next to the House of Commons. In that system, the vitiated air, which was collected above the ceiling, was forced down through a vertical shaft and into a subterranean tunnel before reaching the base of the ventilation shaft. By contrast, in his scheme for the new Palace of Westminster, his objective was to devise a way for the air to rise naturally from the two debating chambers into the discharge shaft.[66] The tower was thus moved to roof level, which allowed the bottom of the shaft to connect directly with the roof space above the two Houses.[67]

Although Reid's statements offer insights into his aspirations to avoid mechanically-driven ventilation, he was aware of the limitation of natural ventilation and that it would still require some mechanical assistance. His system was a hybrid system, similar to many modern examples of naturally ventilated public buildings that combined natural and mechanical methods. Reid's objective was not to exclude mechanical technologies or ventilating fires from the system altogether but to limit their use to periods in which natural processes on their own were unable to meet demand. He claimed that 'mechanical power or heat' was required during periods when rooms were crowded and during hot weather.[68] Reid claimed that the Palace's users, its several hundred fireplaces, Atmospheric Bude gas lighting and the central heating itself could provide most of the heat needed to create the stack effect.[69] He wanted to enhance the natural stack effect within the tower by harvesting the waste heat of the smoke fumes emitted by boiler furnaces and kitchen stoves and several hundred fireplaces. These fumes could be conveyed through a large stack in the middle of the central ventilation shaft, exchanging its heat with the surrounding air. In the summer, however, when the central heating and fireplaces were not in use, the convection had to be sustained artificially with the aid of a coke furnace at the base of the shaft. Most of the mechanical assistance was mostly required for the fresh air supply, which was a separate system from the stack system used for the extraction of smoke and foul air. Reid's proposal was to deploy steam powered fans to draw fresh air from the top of the two towers into the basement, move large volumes of air through the underground passages and several hundred vertical flues and to keep the hermetically-sealed interior of Palace under a uniform atmospheric pressure. The latter, Reid argued, was necessary to prevent draughts at the doors in-between the individual rooms.[70]

Monumental versus technical functions

During this period, Barry focused primarily on the design of Reid's ventilation system from an architectural perspective, as well as that of cost and fire safety, leaving it largely to Reid to defend his proposal from a technical and operational perspective. Barry was reluctant to engage with technical questions in detail. In his responses to the Joint Select Committee Barry noted that he was neither aware of any precedent to Reid's system nor did he feel qualified to evaluate its technical viability, which he considered to be Reid's domain.[71] Indeed, the subject has 'been altogether referred to Dr Reid, I have not considered it necessary or within my province, to give serious attention to it'.[72] Barry continued to be an advocate of the Central

Tower, which he perceived as a functional necessity that also offered new architectural opportunities.[73]

The proposed form for the Central Tower was not entirely the outcome of functional considerations: it also reflected Barry's architectural ambitions. As mentioned above, the alternative designs explored between 1840 and 1841 included Barry's proposal to increase the height of the Central Tower to 194 ft (~59 m).[74] This was one-third higher than Reid, who was aware of the cost implications, had originally specified. Reid also distanced himself from the architect's proposal to further increase the height of the Central Tower for architectural effect. In September 1841, when the Joint Committee queried the proposal of a taller tower, Barry claimed it was essential to enhance the stack ventilation, but also admitted that his 'object in putting it in that form was to make the Central Tower differ as much as possible in outline from the two other towers, by which a more picturesque effect will be produced'.[75]

In another oral statement he noted that 'the appearance of the tower, not only from the River, but from a number of other points, will in my opinion improve very materially the general effect of the building'.[76]

Barry's aspiration was to transform the Central Tower into a monumental Gothic spire. This scheme was scrutinised by the Committee, who revisited the possibility of eliminating the additional cost of such architectural elaborations by adopting a more economic and functional tower design. However, this design was acceptable to Barry only if it were moved to a more remote location outside the site boundary as the large monumental feature would be incompatible with the refined gothic architecture required of the new Palace. Moving the tower off-site would require linking it with the new Palace, which Barry suggested could be achieved by the use of a functional shaft that linked them through a subterranean tunnel.[77] Although this idea was never fully realised, a dramatic depiction of what the tower might have looked like can be seen in an 1844 oil painting.[78]

A final decision and the acknowledgement of risks

In 1842 the Joint Select Committee also challenged Reid's design from a technical perspective by objecting to the use of a central smoke shaft. Concerned about the lack of evidence for its technical feasibility, and to avoid the risk of failure, MPs discussed with Reid the possibility of separating the smoke extract from the ventilation and returning to a more conventional arrangement of local chimneys. They also objected to a central smoke shaft for aesthetic reasons as they feared that it would emit a large plume of visible smoke.[79] For example, Sir Robert Inglis, MP for Oxford University, asked Reid if the stack could be 'disfiguring the effect of the whole [building] by the column of smoke'.[80] Although Reid made the case that visible smoke could be prevented through the use of cleaner varieties of coke, this did not persuade his critics. In its final report, published on 5 August 1842,

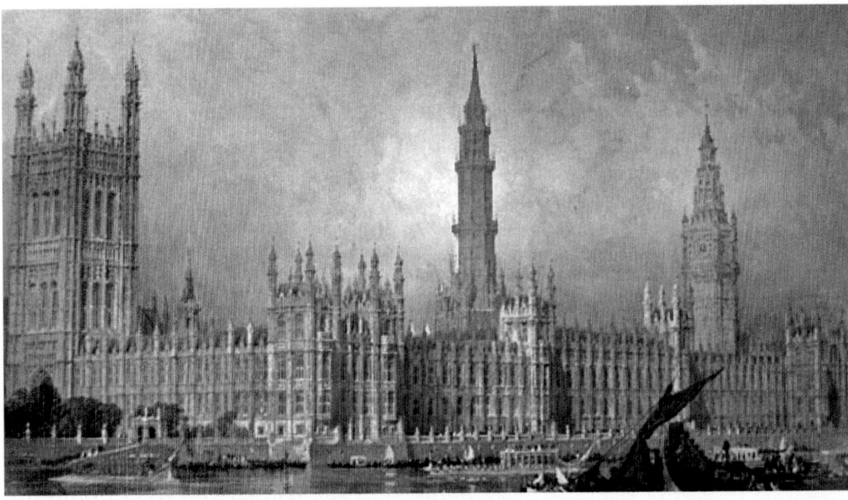

Figure 4.3 Perspective of the Palace of Westminster showing Barry's vision for transforming the central tower into an architectural feature. (View of the Houses of Parliament from south-east. A watercolour by Thomas Allom, 1844, Parliamentary Archives: Arc/var/279.)

the Committee approved the Central Tower as an important functional feature, but reiterated its doubts about the central smoke extract. It described its adoption as a seriously risky experiment and argued that there has 'not yet been any adequate trial in any building of corresponding magnitude'.[81] Although it did not reject the scheme, the Committee insisted on installing a parallel infrastructure, composed of conventional chimneys, as a back-up system in case Reid's system failed. Barry's original drawings, produced between 1843 and 1846, show that he included a set of flues and conventional chimneys alongside the flues for the central extract.[82]

Recent radar impulse scans of the walls have also shown that these back-up flues were built, but never saw active use. These features are physical evidence of the lasting influence that the involvement of MPs and Lords had on the technical design of the Palace. As will be seen in Chapter 5, this influence would further grow in the following four years, and ultimately culminate in the decision to partially abandon Reid's original scheme.

Establishing collaborative working practices: 1841–42

The publication of the Joint Select Committee's final report represented the end of a lengthy inquiry into Reid's scheme, focusing on its architectural, functional and economic aspects. The decision to adopt his system, however, also necessitated operational changes within Barry's office. Barry realised during the early phase that a new framework for collaboration would have to be adopted in order to effectively integrate Reid, and his specialist skills and knowledge, into the wider project team.

Although the Office of Woods expected Barry to offer Reid guidance and assistance, it was more extensive than they had anticipated. In a letter addressed to Barry, dated 16 January 1840, the Commissioners of the Woods wrote 'that Dr Reid may receive from you, from time to time, as the architect of the building, such assistance as he may require' and that Barry himself 'will, in all matters of detail connected with the structure and arrangement of the new buildings, afford him every requisite facility and assistance in carrying it into effect'.[83]

In another letter sent to Reid on the same day, the Commissioners wrote that they:

> *impress upon you the expediency, from time to time, of making such arrangements with Mr Barry, on points involving any interference with the structure of the building, as shall at the same time secure the best means of giving effect to your plans, and render any recourse to alterations and extra works unecessary.*[84]

The use of the words 'from time to time' implies that both parties were expected to work independently. The extent of support necessary to enable Reid to develop his ventilation system also went well beyond what Barry had originally considered to be part of his responsibility as the architect. Since Reid was not able to offer the skills that Barry himself expected from the 'ventilator', the architect had to adjust his expectations and accept a more collaborative design approach in which Reid's skills were complemented by those offered by the architects, engineers and draftsmen of Barry's office. Barry had initially intended to follow a different model of collaboration, one in which Reid was treated as an external consultant and was expected to have the competence to produce the required drawings for the ventilation scheme independently and with minimal assistance from the architect's staff. In this, Barry's expectations were not dissimilar to the working relationship that currently exists between architects and external consultants in many modern architectural practices.

However, it became apparent within the first year (i.e. 1840) that Reid was unable to satisfy Barry's expectations without support from more experienced staff within Barry's own office. As Reid had no formal training in technical drawing and did not employ his own draftsmen, he was unable to produce drawings to a satisfactory standard. The first set of drawings that Reid provided Barry in April 1841 were not produced with professional help. As such, Barry was unhappy with the quality of the drawings as they only provided a schematic outline of the general design concept and did not go into sufficient detail about its application within each individual room of the new Palace. According to Milne, one of the Commissioners of the Woods, Barry considered them imperfect and unintelligible.[85] Indeed,

Joseph Gwilt, the architect, examined Reid's drawings in 1845 and reported them to be of the 'rudest description' and 'very vague'.[86]

Since Reid was not able to meet Barry's original expectations, Barry had to accept a more collaborative approach. This represented a departure from the clear division between the roles of the architect and the 'service engineer'. Even in the development of the early design concept, Reid had relied heavily on practical support from Barry and his chief clerk. Barry was dissatisfied with this arrangement as it took up a considerable amount of time. In May 1841 he informed the Commissioners of the Woods that it was causing serious disruption to his practice's work and demanded additional resources to support Reid's work.[87]

Reid did not deny this dependence on Barry and his staff. In March 1841 he had written to the Commissioners complaining that that the level of support he was receiving was insufficient, and requesting that two hours of additional help per day be provided.[88] Subsequent discussions between the Office of the Woods and Barry resulted in Barry advocating a more formal arrangement. He proposed to appoint a full-time clerk to act as Reid's personal assistant and liaison with Barry's office. The responsibilities of this clerk would be to manage the communication with Reid and to assist him with any technical drawings, whilst also limiting his interaction with other members of staff. Barry wrote that this new arrangement was to furnish Reid with the 'preparation of all the drawings and instructions that will be necessary'.

To cover the additional workload, Barry estimated that he would need to employ another clerk at a wage of between £180 and £200 per annum; he also expected to receive a payment of between £250 and £300 to cover the additional time he and his chief clerk of works would need to invest in assisting Reid with the scheme.[89] In June 1841 the Office of the Woods agreed that Barry's staff would create the drawings for Reid and that the costs for another full-time post in Barry's office would be covered. The Office of the Woods noted that the purpose of the role was to assist 'Reid in the drawings and communicated to him all the information he required' from Barry's office.[90] As already mentioned, their collaboration did not begin in earnest for another year due to the delays caused by the Parliamentary inquiry. The earliest set of ventilation drawings – referred to as 'ventilation sets' – that was produced following this new model of collaboration was completed between April and July 1842.[91] From 1842 until 1846 nearly all of Reid's ventilation drawings were produced by staff in Barry's office. The funds approved by the Office of the Woods initially only covered a period of two years, but Barry secured an extension of this agreement for another two years in 1844 when the collaboration became more intense in the detailed design and construction of the Palace.[92]

On 3 January 1844 Barry wrote to the Commissioners of Woods that the design of the ventilation was 'very complex and perplexing' and also that it will 'continue to occupy much of his own attention, and will require the whole of the time of one clerk at the least as they have hitherto done, up to the completion of the work'.

His staff were still expected to execute drawings for Reid that exhibited the 'arrangements for warming and ventilating in all its details'.[93]

Reid himself did not see a problem with his lack of drawings skills and considered it reasonable that Barry's staff produced drawings on his behalf. On 14 August 1846, he noted that as he did not 'profess drawing, Mr Barry was to exhibit in all their details, my arrangement' and that 'I give the data, as to all Architects, and he prepares the drawings, and received payment for that preparation'.[94]

These changes represented the first attempt to formally organise the collaboration between Reid and Barry, and this marked a departure from the latter's preferred model of divided responsibility to one of increasing interdependence. The design of the ventilation system became a shared responsibility – an increasingly collaborative endeavour – which Barry and his staff had a central role in facilitating. The function of collaboration was threefold. First, to resolve complex design problems. Second, to facilitate an effective exchange of design information. The third function of the collaboration was necessitated by Reid's lack of experience working on architectural projects and was to support Reid in fulfilling his role as the 'ventilator'. The process of designing the system, let alone integrating it into the architectural fabric of the new Palace, would have constituted a significant challenge in its own right. However, Barry was faced with the additional pressure of establishing and sustaining an effective collaboration with a scientist, who, whilst offering important knowledge and skills, did not have sufficient experience of working on architectural projects.

For Reid, his involvement in the new Palace of Westminster represented a significant cultural shift. Hitherto his work had focused exclusively on the design of experimental structures, and he followed a largely functionalist approach to design, one that was primarily concerned with the satisfaction of environmental requirements. The outcomes of his work were highly utilitarian designs that were developed without consideration of architectural style, form or other, potential conflicting, functional requirements. In the new Palace, these aspects of environmental design could no longer be treated in isolation or even given priority over all other aspects. Whilst the temporary Houses of Parliament represented an important first step towards enabling a transfer of scientific working practices into the field of architecture, their application within the Palace required Reid to adapt to a new working practice, one characterised by a focus on reconciliation and collaboration. Indeed, Reid had to be in constant dialogue with Barry in order to arrive at a mutually agreeable solution for the ventilation arrangements themselves.[95] The influence of this collaboration on Reid can be seen when he notes that he can no longer treat a physical space as a 'plain room that you may cut and carve upon as you like'.[96]

The organisational architecture of a collaborative design process

The archival records allow us to gain an insight into the organisational architecture underlying the collaborative design process. At the beginning of the early phase, Reid's principal clerk in Barry's office was a 'Mr Fletcher', who was succeeded by a 'Mr Grove' (first names are not known); from 1844 to 1846 the role was held by a 'Mr McDougall'. Reid received additional assistance from several other clerks in Barry's team, including Robert Banks and a 'Mr Paulby'. These clerks were based within Barry's office and remained under the architect's supervision. He also discussed his proposals with Barry's chief engineer, Alfred Meeson.[97] Very few drawings were produced by Reid's own office, and those that were produced were primarily diagrams and sketches used in discussions about complex design problems. Most of the drawings were produced by clerks in Barry's office following Reid's instructions, which were communicated verbally and rarely through the medium of letters or sketches. Barry reported that the clerks produced ventilation drawings under his 'constant superintendence' and following his verbal instruction.[98] Reid visited Barry's office to supervise the production of new drawings and review them after their completion. Barry's staff members were also sometimes sent to Reid's office to ask for needed information.[99] The ventilation sets produced under Reid's supervision were only advisory drawings produced for the purpose of communicating Reid's ideas to Barry and his wider team. They comprised schematic floor plans and cross-sections showing the arrangements of vertical shafts within the walls and buttresses and the horizontal air chambers within the floors; they were also annotated and colour coded, explaining the function of the different elements, as well as arrows indicating the direction of air flow.

A first set of ventilation drawing were produced for the design of the River Front between 1842 and 1844.[100] Barry's team, who were responsible for producing the final working drawings and tender documents, used these drawings for guidance, and if necessary, made adaptations to Reid's proposals to ensure their compatibility with other design requirements (such as fire safety, structural integrity or architectural design). The arrangements in the final working drawings and the actual built infrastructure itself often differ significantly from those shown in the ventilation sets produced under Reid's direction.

In addition to producing the final working drawings, Barry's staff developed the technical details of Reid's system and integrated them into the architectural fabric of the new Palace. The working drawings showed, amongst others, how air flues were to be constructed and embedded within the masonry walls and floors, and how these could be connected to inlets within the architectural interiors. Staff also discussed with Reid the selection of suitable materials and the construction of various ventilation features. As such, the contribution of Barry's office went well beyond the task of supporting him in the execution of drawings. It actively participated in developing Reid's ideas on a technical level. By illustrating how design information about the ventilation was produced, shared and finally integrated into working drawings, these procedures also show that the collaboration was characterised by a clear division of roles.

From the autumn of 1842, Reid was provided with an assistant inside his own office. His name was John Imray and he worked in Reid's office from 1842 to 1850.[101] Reid had written to Milne in September 1842 to ask for a technical draughtsman who could assist him with the production of technical drawings in his own office without having to rely on support from Barry's office. He requested a person who was 'acquainted more particularly with machinery' and that he should 'be placed exclusively under my direction'.[102] Imray was paid by the Office of Woods but was not permitted by Barry to work directly with his own architectural staff.[103] His involvement between 1842 and 1846 was limited to the development of some of the mechanical engineering details.

Conclusion: a transition towards synthesis

The approval of the Central Tower in 1842 marked the end of the conceptual phase in the development of Reid's scheme. This process lasted almost two years as the design and budget underwent lengthy Parliamentary reviews before it was formally approved. Over this period, however, Barry and Reid were only able to finalise the key design features. The development of the more detailed plans and their integration in Barry's architectural drawings for the Palace of Westminster was delayed due to the uncertainty caused by the involvement of Parliament in the approval of the ventilation system.

The first year of developing the general design concept of the ventilation system illustrated the practical reality of engaging a scientist in its design. The impact of this collaboration, however, was not solely physical: it also affected the organisation of the wider design team and its working practices. These were important steps towards the establishment of a more collaborative model of architectural practice, in which the skills of the scientist were complemented by those of architects, draftsmen and engineers in Barry's office.

According to Gleich, Reid was given the same amount of influence over the design as Barry despite being 'completely foreign to the architectural profession'.[104] Furthermore, Reid was 'empowered to interfere on an equal basis in the floor plan and construction of the building designed' by the architect. Gleich bases this interpretation mainly on a study of the original letters of January 1840, which describe how the Commissioners of the Woods envisaged the working arrangements to be.

However, as demonstrated in this chapter, the design process was much more collaborative than the intended original working agreement. Reid and Barry's working relationship evolved significantly over the first two years, leading to a different model of collaboration.

The full scale of the impact of this new model of collaboration, however, would only become fully apparent after 1842. This model became an increasingly important factor in the next phase of the project where the focus was on developing a detailed design and construction of the new Palace. From 1842 to 1846 Barry's practice collaborated with Reid to develop and integrate his scheme into the architectural fabric. However, as will be shown in Chapter 5, efforts to establish a framework for their collaboration were unsuccessful, resulting in major disruptions. The project became increasingly dependent on collaboration in order to resolve complex design problems. These issues are not dissimilar to those associated with establishing interdisciplinary design practices. In its modern form, organisations such as Arup, RWDI or BDP, and people from multiple disciplines, including those working in very specialist fields, are required to work within large project teams. Such models of practices were unheard of in the early 19th century, as were many of the specialist disciplines now involved in such collaborations. In the context of the Palace of Westminster, however, Barry and Reid were confronted with the task of defining their own roles and how they would interact with each other.

Notes

1 David Boswell Reid, 'The revision of architecture in connection with the useful arts', *Builder*, 5 (May, 1855), pp. 208f.
2 John Darwin, 'Yesterday in Parliament',' *Building Services*, March (1980), p. 21.
3 Letter from C. Barry to Lord Duncannon, 3 October 1839, in HC 574 1846, *Third report from the Select Committee on Westminster Bridge and new palace; together with the minutes of evidence*, p. 22
4 Ibid.
5 Ibid.
6 Marjorie Caygill & Christopher Date, Building the British Museum (London: British Museum Press, 1999), p. 72; Howard Colvin, History of King's Work: 1782-1851 (London: HMSO, 1963), p. 413.
7 Dan Cruikshank, Timeless architecture (London: Architectural Press, 1985), pp. 37-38; John Olley, 'The Reform Club', *Architects Journal*, 27 February 1985, pp. 34–60.
8 Letter from D.B. Reid to Viscount Duncannon, 5 October 1839 (National Archives: Work 11/12, no. 12).
9 Letter from D.B. Reid to Viscount Duncannon, undated. (National Archives: Work 11/12, no. 17).
10 Gore, C. Letter to D.B. Reid, 19 October 1839 (National Archive: Work 11/12, no. 13).
11 Letter from C. Barry and D.B. Reid to C. Gore, 22 October 1839 (National Archives: Work 11/12, no. 15).
12 Letter from D.B. Reid to Viscount Duncannon, undated. (National Archives: Work 11/12, no. 17).
13 Letter from Office of Woods to C. Barry, 16 January 1840 (National Archives: Work 11/12, no. 29).
14 Letter from A. Milne to D.B. Reid, 24 January 1840 (National Archives: Work 11/12, no. 30-32).

15 Letter from A. Milne to C. Barry, 16 January 1840 (National Archives: Work 11/12, no. 29).
16 Letter from A. Milne to D.B. Reid, 24 January 1840 (National Archives: Work 11/12, no. 30-32).
17 Moritz Gleich, 'Architect and Service Architect: The Quarrel between Charles Barry and David Boswell Reid', *Interdisciplinary Science Reviews*, 37:4 (2012), 332-344 (p. 340).
18 Letter from the D.B. Reid to Office Woods, 28 April 1841 (National Archives: Work 11/12, no. 45).
19 Letter from D.B. Reid to A. Milne, 16 May 1841 (National Archives: Work 11/12, no. 49).
20 David Boswell Reid, 'The Study of Ventilation, to the editor of The Lancet', *The Lancet*, 2 May 1903, 1264–65.
21 Oral statement by D.B Reid, 14 August 1846 in HL 719 1846, *Reports from the Select Committee of the House of Lords appointed to inquire into the progress of the building of the Houses of Parliament, and to report thereon to the house: Fourth Report*, Q26.
22 Oral statement by D.B Reid, 26 July 1842, in HC 536 1842, *Committee on Ventilation*, Q1-2. Note: The Chimney Sweeper Act of 1834 prohibited the employment of boys below the age of 14, in 1840 the 'Chimney Sweeper and Chimneys regulation Act' raised the minimum age to 16. See Kathleen H. Strange, *Climbing Boys: A Study of Sweeps' Apprentices 1772-1875* (London/ Busby: Allison & Busby, 1982).
23 The claim that the flues of the Palace of Westminster might contain the corpses of Victorian chimney boys is not correct. Reid, aware of the changes in legislature, adopted an arrangement of smoke flues that dispensed with the need to enter vertical smoke flues for sweeping. The depth of flues varied between 20 and 30 cm.
24 Oral statement by D.B. Reid, 17 March 1846, in HC 1846 (574) *Third report from the Select Committee on Westminster Bridge and new palace; together with the minutes of evidence*, Q730–31.
25 Oral Statement by D.B. Reid, 10 March 1846, in HC 574 1846, *Third report*, Q318.
26 Oral Statement by C. Barry, 23 February 1846, in HL 719 1846, *Select Committee on progress: First Report* Q17–20.
27 Letter from C. Barry to A. Milne , 14 August 1840 (National Archives: Work 11/12, no. 37)
28 Oral Statement by C. Barry, 23 February 1846, in HL 719 1846, *Select Committee on progress: First Report*, Q27.
29 Letter from D.B. Reid to Viscount Duncannon, 7 July 1840 (National Archives: Work 11/12, no. 34)
30 Letter from C. Barry to A. Milne , 14 August 1840 (National Archives: Work 11/12, no. 37)
31 Oral Statement by D.B. Reid, 1 October 1841, in HC 51 1841, *Report from the Select Committee on Ventilation of the New Houses of Parliament; with the minutes of evidence, and appendix*, Q324.
32 Oral Statement by C. Barry, 30 September 1841, in HC 51 1841, *Select Committee on ventilation*, Q129.
33 Letter from Reid to Office of Woods, 15 June 1847, in Copies of correspondence relative to the new Houses of Parliament, (HC 1847–48 (205)) pp. 2f; 'Section of Central Tower traced from a drawing furnished to Dr Reid in 1840', 13 June 1847 (National Archives Work 29/3052).
34 Oral Statement by D.B. Reid, 14 August 1846, in HL 719 1846, *Select Committee on progress: Fourth Report*, Q26.

35 Letter from D.B. Reid to C. Barry, 24 August 1840, in HL 719 1846, *Select Committee on progress, Fourth Report,* p. 43.
36 'Tracing of a pencil sketch made by Mr Barry in Dr Reid's office', 15 June 1847 (National Archives: Work 29/3049).
37 This approach was not unique to Westminster. It exemplified an approach to reconciling form and function that was compatible with an era of architectural historicism where historical forms (e.g. Italian campanile, Gothic church spires) were re-appropriated as a means of accommodating new technical requirements. This approach became characteristic of many public buildings in the latter part of the 19th century. Prominent examples are the University of Glasgow in Scotland, and the Natural History Museum and the Royal Courts of Justice, both in London.
38 Oral statement by D.B. Reid, 17 March 1846, in HC 1846 (574) *Third report,* Q808–10.
39 Letter from D.B. Reid to Viscount Duncannon, 7 July 1840 (National Archives: Work 11/12, no. 34).
40 Letter from C. Barry to D.B. Reid, 27 April 1841 in 'Dr Reid's reply to the report of the referees appointed to consider the warming and ventilating arrangements for the new Houses of Parliament, and Dr Reid's accompanying letter to the Lord Viscount Morpeth. (HC 555 1846), p. 10.
41 Examples of original drawings showing arrangement of individual chimneys are: 'North Wing, Sections shewing flues, 5 April 1841 (Work 29/2012); South Wing, 'Sections shewing flues of fires places', 1841. (National Archives: Work 29/2052).
42 Letter from C. Barry to Viscount Duncannon, 21 August 1841 (National Archives: Work 111/12, no. 79–80).
43 C. Barry, 'Abstract of Estimate for warming and ventilating', 26 May 1841 (National Archives: Work 11/12 no. 54–55).
44 Letter from Office of Woods to the Lords of H. M. Treasury, 7 June 1841 (National Archives: Work 11/12, no. 57).
45 Letter from H. H. Treasury to the Commissioners of Woods, 24 June 1841 (National Archives: Work 11/12, no. 59).
46 Letter from Alexander Milne to Charles Barry, 16 July 1841 (National Archives: Work 11/12, no. 66).
47 Letter from D.B. Reid to A. Milne, 15 July 1841 (National Archives: Work 11/12, no. 65).
48 Letter from Office of Woods to D.B. Reid, 8 July 1841 (National Archives: Work 11/12, no. 64).
49 Letter from C. Gore to C. Barry, 17 July 1841 (National Archives: Work 11/12, no number); Letter from A. Milne to D.B. Reid, 17 July 1841 (National Archives: Work 11/12, no. 67).
50 Letter from C. Barry to C. Gore, 31 July 1841 (National Archives: Work 11/12, no number).
51 Letter from Charles Barry to Charles Gore, 27 July 1841 (National Archives: Work 11/12, no. 68).
52 Ibid.
53 Letter from Viscount Duncannon (Woods) to the Treasury, 10 August 1841 (Work 11/12, no. 76); New Houses of Parliament. Copy of letter from the First Commissioner of Woods to the Chancellor of the Exchequer on the subject of warming, ventilating and securing from fire the new Houses of Parliament (HC 23 1841–2).
54 New Houses of Parliament – Ventilation, House of Lords, *Hansard's Parliamentary Debates: The Official Report* (20 September 1841. vol 59. cols. 611).

55 Letter from Viscount Duncannon to Chancellor of the Exchequer, 27 August 1841 (National Archives: Work 11/12 no. 82).
56 Note: Central heating was widely used in many notable public buildings from the early 19th century. The architect John Soane adopted central heating in his scheme for the Westminster Law Courts (1827) and Dulwich Picture Gallery (1817) (Dean Hawkes, *Architecture and Climate* (Abington: Routledge, 2010).
57 Notable examples of public building with sophisticated building services are the British Library Reading Room (1857), Royal Courts of Justice (1882) Natural History Museum (1873) and St Thomas Hospital, Westminster (1871).
58 Letter from the Office of Woods to H. M. Treasury, 7 June 1841 (National Archives: Work 11/12, no. 57).
59 HC 536 1842, *Committee on ventilation*, Report, pp. 3f.
60 Oral Statement by D.B. Reid, 24 September, in HC 51 1841, *Select Committee on ventilation*, Q10.
61 Oral statement by D.B. Reid, SC 1841, Q56-58, Q191.
62 Letter from D.B. Reid to Chancellor of the Exchequer, 30 September 1841, in HC 51 1841, *Select Committee on ventilation*, pp. 36–37.
63 Oral Statement by C. Barry, 24 September 1841, in HC 51 1841, *Select Committee on ventilation*, Q95; Oral Statement by D.B. Reid, 10 October 1841, in HC 51 1841, *Select Committee on ventilation*, Q289.
64 Oral Statement by D.B. Reid, 26 July 1842, HC 536 1842, *Select Committee on ventilation*, Q8.
65 Ibid, Q4.
66 Oral Statement by D.B. Reid, 24 September 1841, in HC 51 1841, *Select Committee on ventilation*, Q2–7.
67 Oral Statement by D.B. Reid, 26 July 1842, in HC 536 1842, *Select Committee on ventilation*, Q56.
68 Oral Statement by D.B. Reid, 26 July 1842, HC 536 1842, *Select Committee on ventilation*, Q4.
69 Despite Reid's argument about energy efficiency, the inherent problem with his system was the fact that it encouraged the loss of heating energy. The stack effect relied extensively on the warm air produced by the heating systems. As such his approach stands in stark contrast to modern approaches of energy conservation, which utilises air tightness and heat recovery as a means to minimise the loss of heat to the external atmosphere.
70 Oral Statement by D.B. Reid, 24 September 1841, in HC 51 1841, *Select Committee on ventilation*, Q67.
71 Oral Statement by C. Barry, 26 July 1842, HC 536 1842, *Select Committee on ventilation*, Q88–89, Q99.
72 Ibid, Q104.
73 Oral Statement by C. Barry, 1 October 1841, in HC 51 1841, *Select Committee on ventilation*, Q263–87.
74 The question if a tower of this height was required for functional reason was the subject of a dispute between Barry in June 1847. Letter from Reid to the Commissioners of Woods, 15 June 1847, in New Houses of Parliament. Copies of correspondence relative to the new Houses of Parliament (1847–48 (205)) pp. 2f.
75 Oral Statement by C. Barry, 1 October 1841, in HC 51 1841, *Select Committee on ventilation*, Q88.
76 Ibid, Q97.
77 A similar idea had been proposed by Reid for the ventilation of St George's Hall in Liverpool, but this was considered impractical for the new Palace of Westminster.

78 'View of the Houses of Parliament from South-East'. A watercolour by Thomas Allom, 1844 (Parliamentary Archives: Arc/var/279).
79 Oral Statement by C. Barry, 26 July 1842, in HC 536 1842, *Select Committee on ventilation*, Q93; Oral statement by C. Barry on 26 July, in HC 536 1842, *Select Committee on ventilation*, Q93.
80 Oral Statement by D.B. Reid, 26 July 1842, in HC 536 1842, *Select Committee on ventilation*, Q72.
81 HC 536 1842, *Select Committee on ventilation*, Report, pp. 3–4.
82 'South Curtain. Elevation and section shewing the flues', 22 November 1844 (National Archives: Work 29/2048).
83 Letter from A. Milne to C. Barry, 16 January 1840, in HC 574 1846, *Third report*, pp. 182.
84 Ibid.
85 Oral Statement by A. Milne, 17 March 1846, in HC 574 1846, *Third report*, Q713.
86 Oral Statement by J. Gwilt, 13 March 1846, in HL 719 1846, *Select Committee on progress: First Report*, Q121–25.
87 Letter from D.B. Reid to A. Milne, 16 May 1841 (National Archives: Work 11/12, no. 49).
88 Letter from the Office of Woods to D. B. Reid, 3 May 1841 (National Archives: Work 11/12, no. 47).
89 Letter from C. Barry to D.B. Reid, 12 May 1841, in HC 574 1846, *Third Report*, p. 71.
90 Letter from the Office of Woods to Charles Barry, 1 June 1841 (National Archives: Work 11/12, no. 56).
91 E.g. 'Plan of north front, first floor from north east pavilion to centre', 22 April 1842 (National Archives: Work 29/2016).
92 Letter from A. Milne to C. Barry, 9 January 1844 (National Archives: Work 11/12, no. 96).
93 Letter from C. Barry to A. Milne, 3 January 1844, in HC 574 1846, *Third Report*, pp. 180.
94 Oral Statement by DB. Reid, 14 August 1846, in HL 719 1846, *Select Committee on progress: First Report*, Q26.
95 Oral statement by 10 March, HC SC 1846, Q331, Q338, Q362.
96 Oral Statement by D.B. Reid, 10 March, in HC 574 1846, *Third Report*, Q344.
97 Oral Statement by D.B. Reid, 17 March 1846, in HC 574 1846, *Third Report*, Q825–29; Oral Statement by D.B. Reid, 24 March 1846, in HC 574 1846, *Third Report*, Q1109–1113; Oral Statement by D. B. Reid, 19 March 1846, in HC 574 1846, *Third Report*, Q900–5.
98 Oral Statement by C. Barry, 23 February 1846, in HL 719 1846, *Select Committee on progress: First Report* Q17–20.
99 Oral Statement by D.B. Reid, 10 March 1846, in HC 574 1846, *Third Report*, Q381–90, 331.
100 E.g. Plan of the first floor, centre portion, showing the chambers and flues through which the vitiated air passed into the ceiling', 1843 (National Archives: Work 29/2846); 'South front: sections through southern section showing flues marked 'unfinished' 1843 (National Archives: Work 29/2859), River front, south section: plan and section of ground floor showing the position of the air chambers, 18 March 1844 (Work 29/2875); No 36 House of Commons: plan of the libraries showing the supply and discharge of air, 20 July 1843 (Work 29/2857).
101 Obituries: John Imray, Proceedings of the Institute of Mechanical Engineers (London: 1902), pp.1026–28.

102 Letter from D.B. Reid to A. Milne, 3 September 1842 (National Archives: Work 11/12, no. 92).
103 Obituries: John Imray, Proceedings of the Institute of Mechanical Engineers (London: Institute of Mechanical Engineers, 1902), pp. 1026–28.
104 Moritz Gleich, 'Architect and Service Architect: The Quarrel between Charles Barry and David Boswell Reid', *Interdisciplinary Science Reviews*, 37:4 (2012), pp. 332–344.

5 Reid's original proposals for the two debating chambers: 1842–46

From concept to implementation

The approval of the Central Tower in August 1842 marked the end of the conceptual phase in the development of the ventilation system. Reid's involvement in the early phase of development (1840–42) was limited to design investigations that focused exclusively on formulating the general design concept and reviewing its cost implications. Hitherto Reid's scheme only existed in outline as a set of schematic drawings, which were separate from Barry's original architectural plans. As seen in the previous chapter, the process of developing more detailed plans and accommodating the ventilation system within the existing architectural design was delayed due to the political uncertainty caused by the Parliamentary inquiries, led by Select Committees in 1841 and 1842 and it did not begin in earnest until the scheme and associated budget was approved.

In this chapter, the development of Reid's original scheme is retraced, focusing on the design of the House of Lords, which was the first chamber to be designed and completed. Between 1840 and 1846, when Reid was in charge of the ventilation of the entire Palace, the focus was on designing and constructing the basement, the River Front and the House of Lords. These were the first parts of the Palace to be completed and occupied whilst other parts were not completed for several more years. For example, the design of the ventilation system for the House of Commons, which is discussed in Chapter 7, did not advance beyond the arrangement of the horizontal air chambers, the vertical shafts and the wall apertures that formed part of the structural carcass until after the autumn of 1846. As the design for the House of Lords was almost complete before Reid's departure in 1846, it illustrates most clearly his original vision for the ventilation of the debating chambers. The proposed system was more complex and sophisticated than the experimental systems tested inside the temporary Houses of Parliament, and was also different from the final arrangements completed under Barry's direction in 1847.

The shift towards detailed design and integration posed new challenges. Reid's and Barry's earlier difficulties establishing a successful working relationship resurfaced and intensified, ultimately culminating in its breakdown. Although the difficulties between Barry and Reid have been mentioned in previous studies on the history of the Palace of Westminster, the causes of these interpersonal tensions and their impact on the design of the final ventilation system have not been examined in any great detail (see Introduction). The original project correspondence, drawings, sketches and Parliamentary records, the detailed study of which is documented in this book, offers critical insights into the nature of Reid's collaboration with Barry himself and Barry's own staff.

As shown in the last chapter, the design of the ventilation system was the outcome of a highly collaborative process and that Barry and Reid held an equal share of responsibility for its successful delivery. From the outset of the ventilation project, however, it was difficult to draw clear boundaries between their roles. As will be seen in this chapter, the breakdown in their relationship in 1846 was the result of a failure both to fully understand and accept the limitations of Reid's scientific background and to devise a mutually-agreeable arrangement that used the skills and knowledge of the wider project team to compensate for these. The working arrangements between Reid and Barry specified by the Office of the Woods ultimately caused serious disruption, and became the subject of multiple Parliamentary inquiries between 1845 and 1846, but these were unsuccessful in resolving them.

A problem of strategic planning

The fact that the requirements of Reid's ventilation system were not considered an integral part of the architecture of the new Palace was problematic from the very beginning. As has been seen in the previous chapters, the late introduction of the ventilation system represented a major disruption to the development of Barry's architectural plans at a time when parts of it and the design brief were still not finalised. The architectural plans for the new Palace evolved into a much more complex scheme than was originally conceived, not only technologically but also in terms of its spatial and functional organisation. For example, plans for important areas, such as the ceremonial approach from the Victoria Tower to the House of Lords, were not finalised until 1845. The scope of the original design brief, which had only set out the basic space requirements, was also expanded. Barry reported in 1844 that it was necessary to refine the brief due to the 'absence of correct information in the original instructions or sufficiently detailed information'.[1] In order to gain a deeper understanding of the functional requirements of Parliament, Barry consulted various key stakeholders in the House of Commons, such as the Clerk, Librarians, the Speaker and the Sergeant-at-Arms. Then, as now, it was not unusual for project briefs to continue to be developed in the early stages of an architectural project, but in the case of the new Palace of Westminster interference by the Office, and the Commissioners, of the Woods gave rise to further complications.

The adoption of Reid's ventilation system for the new Palace was only one of several alterations that the Office of the Woods had imposed on Barry. Indeed, between 1836 and 1844 the various amendments that they were instructed to implement led to the cost of the project rising by 50% and by the end of this period it had reached over £1,000,000.[2] Although these costs also covered changes that Barry himself had initiated – including alterations to the foundations, river wall and fire proofing – by far the largest increase was due to the changes instructed by the Commissioners. These interventions were disruptive as the Office of the Woods was following a largely reactive approach. Over the first eight years, the Office had given Barry various ad hoc instructions, something over which neither the architect himself nor Reid had any control over. The late adoption of Reid's system should therefore be understood as the result of the failure to maintain sufficient strategic oversight in the management of the project. New requirements, added to the brief by the Office of the Woods after 1842, adversely affected the implementation of the ventilation systems as they occupied space needed for the ventilation. Between 1845 and 1846 these changes contributed to increasing the tension between Reid and Barry. Some spaces that had been initially adapted for the ventilation system were reclaimed by Barry to accommodate the new functions required by the Office of the Woods, sometimes even without consulting Reid about the possible implications this might have for the ventilation system. For example, in 1844 the Office asked Barry to amend the design of the roof to accommodate a new repository for the Public Records Office, which was to be relocated from its old premises at Carlton Ride to the new Palace of Westminster.[3] Although these plans would be abandoned in 1845, they led to further delays in the development of Reid's scheme.[4]

A mandate for physical change

The scale of the proposed physical changes to the original 1836 design of the Palace was significant and in 1844 the governance of the project became the subject of two separate inquiries by Select Committees from both the House of Lords and the House of Commons.[5] These Committees considered some of the proposed alterations to the original design to be unacceptable. Although the inquiries were primarily concerned with the proposed architectural alterations, such as changes to the design of the Sovereign's entrance at the base of the Victoria Tower, the Committees also questioned Barry about the reasons for the scale of modifications he had made to accommodate Reid's ventilation system. Barry successfully defended his actions in these inquiries by arguing that he was acting on instructions he had received from the Commissioners of the Wood and that the formal approval of Reid's scheme in 1842 had included Barry's mandate to make any needed changes to the architectural design.

In a written statement presented to the House of Commons Select Committee on 29 April 1844, Barry writes that the system alone has 'caused many serious and important changes'.[6] A few days later, on 2 May 1844, Barry provided them with a list of the numerous changes made to the original design. Amongst these, Barry lists three directly associated with the ventilation system: first, alterations to the 'Victoria' and 'Elizabeth' Towers; second, the addition of a basement and a central tower; and third, 'numerous alterations in the sections of the building for the warming and ventilating arrangements'.[7] This third change was significant as it explicitly recognises that other physical modifications were necessary to accommodate the implementation of Reid's system. Whilst major changes to the original design, such as the addition of a central tower and basement, had been reviewed by Parliament as part of the approval process between 1841 and 1842, many other, more minor, alterations were made after 1842 without the explicit approval of the Office of the Woods. Barry argued that the mandate for these minor changes was implied rather than direct and that they were necessary to accommodate Reid's system. This was necessary as the plans approved in 1842 only outlined the general design concept of the new Palace and the full scale of the impact of the ventilation system on the architectural form had not been fully considered.

The stack ventilation system as a driver of wider changes

Amongst the proposed changes to the original design of the new Palace of Westminster introduced between 1843 and 1844 were alterations to the roof and elevations of both Houses. The roofs above the debating chambers, Royal Gallery and the Prince's Chamber were to be raised to the same level as the floor of the Central Tower to provide space for the horizontal channels that were to convey smoke and vitiated air to the central extraction shaft. In case of the Royal Gallery, the ceiling was to be raised by 30 ft (~9 m).[8] These changes transformed the roof space into one continuous 394 ft (~120 m) long space, extending the whole length from the Central Tower to the south end of the Palace.[9] Reid confirmed to the Select Committees that these drastic changes were necessary to implement the stack system. In an oral statement on 29 April 1844 he argued that the level roof would make the stack ventilation more efficient, as it minimised the level obstruction to the natural flow of air and smoke towards to the Central Tower.[10] The 'vitiated atmosphere', he explained, would drain 'naturally upwards, as the water would flow downwards'. He argued that if these geometric requirements could not be met, the ventilation system would become more dependent on the use of coke fires and steam-driven fans to maintain the air flow artificially. He noted that 'a special power, or an increase of power in the shaft, would be required to counteract loss incurred by local descent'. In Barry's original drawings these spaces were covered by separate roofs at different heights, which was to create the illusion that Parliament was an organically-evolved group of buildings reminiscent of Medieval architecture.[11]

These interventions illustrate the degree to which Barry had changed the external form of the building to accommodate Reid's system. The changes affected the roof of the entire Palace of Westminster as the Central Tower was connected to a much larger network of four main smoke and air channels. In addition to a main channel that lead south from the Central Tower towards the House of Lords, there were main channels linking it east to the River Front, north to the House of Commons, and west to St Stephen's area (adjacent to Westminster Hall). These channels terminated in four pointed arches at the base of the Central Tower and allowed it to be connected to several hundred rooms across the large site.[12]

Reid knew that the central stack system could not service all the rooms simultaneously and that its effectiveness would be dependent on its intelligent management:[13] the ventilation rate in each space was to be regulated with the aid of manually-operated valves, and adjustments were to be made to respond to changes in occupancy and weather conditions and to carefully manage the allocation of the central shaft to the different spaces. Reid noted that the 'particular power which, under ordinary circumstances, was operating upon one room, should be diverted and turned on another'.[14] Thus, the ventilation for specific spaces, such as the debating chambers of the House of Lords and House of Commons, could be controlled by connecting or disconnecting them to the central tower as required depending on whether they were occupied or not.[15] Reid estimated that between 12 and 16 attendants would be required to constantly operate the ventilation system and heating, and claimed that it would be comparable to the amount of staff required in a conventional building for the opening and closing of windows or maintenance of fireplaces.[16] These staff would be responsible for the 'purpose of watching and taking care of those valves were always in order'.[17]

The intelligent management required to operate the ventilation system of the (according to Reid) 767 individual rooms in the new Palace, with its manual controls for each individual room equipped with a set of supply and extract flues, would have been complex.[18] This design allowed each vertical flue to be connected to or disconnected from the horizontal channels in the roof of the central tower network. New surveys, undertaken by the author over the past four years, have shown that this network, including the system of flues and valves, had by 1846 (when the idea of using a central tower as a main outlet was abandoned) been completed according to Reid's original plans and that large parts of it have survived to this day. This original network, large parts of which have survived to this day, was ultimately retained and subsequently connected to an array of multiple ventilation shafts (see Chapters 6 and 7).

Increasing political pressure

Reid's proposed ventilation scheme followed a centralised approach to stack ventilation that was to serve the entire Palace of Westminster. However, this approach required that the Central Tower be constructed before other parts of the system could become operational. The potential implications of this level of dependence were first examined by the House of Lords in Spring 1843. Indeed, the construction of the Central Tower was only completed in 1855, three and eight years after the Houses of Commons and the Lords, respectively, were completed and occupied.

The Lords were highly dissatisfied with their temporary facilities and exerted pressure on Barry to complete their debating chamber.[19] Barry had in 1836 originally estimated that the Lords chamber would be completed within six to eight years alongside the House of Commons. However, by 1843 it was evident that its construction was already several years behind schedule. Indeed, Reid and Barry had only just began producing their first set of ventilation drawings for the House of Lords. During a debate on 21 February 1843, Ulick de Burgh, 1st Marquess of Clanricarde, successfully tabled a motion for the appointment of an ad hoc Select Committee in the House of Lords to lead a formal inquiry into the delay.[20] This Committee reviewed two options: first, to improve the ventilation within the temporary debating chamber; and second, to outfit the shell of the new House of Lords with a temporary interior. On 2 and 10 March 1843 it discussed with Barry and Reid the possibility of installing a temporary interior with mock furniture and introducing a small temporary ventilation shaft, which would enable the Lords to vacate their temporary House by Spring 1844.[21] Barry was not supportive of the proposal and warned that he could not guarantee that the chamber would be 'fit for occupation' even with a temporary interior or that it would result in additional cost. A temporary ventilation tower, he argued, could only be realised at 'considerable expense' and also had to be demolished once the permanent Central Tower was completed.[22]

Despite this, the ad hoc Committee recommended in its final report, published on 6 May 1843, that the temporary interior for the House of Lords be installed. This recommendation, however, was not implemented because the Office of the Woods was never formally instructed to realise them, and in March 1844 the Select Committee was reappointed by the House of Lords to revisit matters.[23] In their second report, published in August 1844, the ad hoc Committee agreed to abandon the plans for a temporary interior for their debating chamber. Barry revised his original estimate of when the new House of Lords debating chamber would be completed: February 1846. Since neither the Central Tower nor the Elizabeth or Victoria Towers would be completed at this time, the temporary ventilation shaft and inlets would still be required.[24] On 6 August 1844 Reid noted that a temporary boiler would have to be fitted and arrangements made for the 'discharge as well as the ingress of air'.[25] Barry estimated that the Central Tower could be completed by February 1847 at the earliest, but emphasised that it would only be achievable 'with great exertion' and additional funding to cover the cost of a larger on-site workforce.[26] The House of Lords approved the recommendations of this reconvened ad hoc Select Committee on 9 August 1844.[27]

The earliest drawings outlining the temporary extraction system were produced in 1845.[28] This temporary system was composed of a short stone tower located on the roof of the Peers' House Lobby and connected to the vitiated air chamber above the ceiling of the House of Lords through large passages.[29] As with his original estimates, Barry's predictions regarding the completion date of the House of Lords debating chamber and the Central Tower itself were overoptimistic and it would take another three years before the Lords would be able to formally move into their new chamber.

These developments, which were the result of political pressure from the Lords, are important as the temporary shaft, known as 'the Peers' Shaft', became a permanent feature of the House of Lords ventilation system (see Chapter 6 for further details). In contrast to the planned use of the Central Tower as the main ventilation shaft for the House of Lords (which was never realised), the Peers' Shaft remained in active use for over 70 years (see Chapter 6).

The ventilation system as a driver of structural intricacy

As seen in the previous section, the scale of changes that Barry made to the exterior architecture of the new Palace in order to accommodate the central ventilation system was not insubstantial. Although the addition of a central tower and remodelling of the roof were the most obvious alterations, the adoption of Reid's system had an even greater impact internally as it motivated changes in the planned construction, spatial layout and interior architecture. When it was completed, the ventilation infrastructure occupied 25–30% of the entire volume inside the Palace. The full extent of changes on Barry's plans, however, is not easily discernible as most of these spaces are well concealed within the depths of the structure. The entire fabric of the building was permeated by a network of horizontal and vertical voids for the ventilation and warm air central heating, as well as by flues for the extraction of smoke and gas fumes. The vertical shafts were embedded within the depth of walls, columns and external buttresses, whilst horizontal voids were embedded within the structure of floors and ceilings.

Reid's system also motivated the adoption of more sophisticated methods of masonry construction and the extensive use of structural iron, including the adoption of iron frames and perforated iron floors, both of which were used for the admission of fresh air into the two debating chambers and adjacent (east and west) division lobbies. Reid and Barry introduced hollow masonry floors to accommodate horizontal voids, examples of which can be found in the Central Lobby, the Royal Gallery, St Stephen's Hall and on the River Front (see Figure 5.1). Large parts of the vaulted ceilings in the basement were also not solid masonry, but hollow shells containing a network of horizontal air channels. Andrew Saint, the architectural historian, has noted that this made the new Palace of Westminster the 'most intricate building yet assembled'.[30]

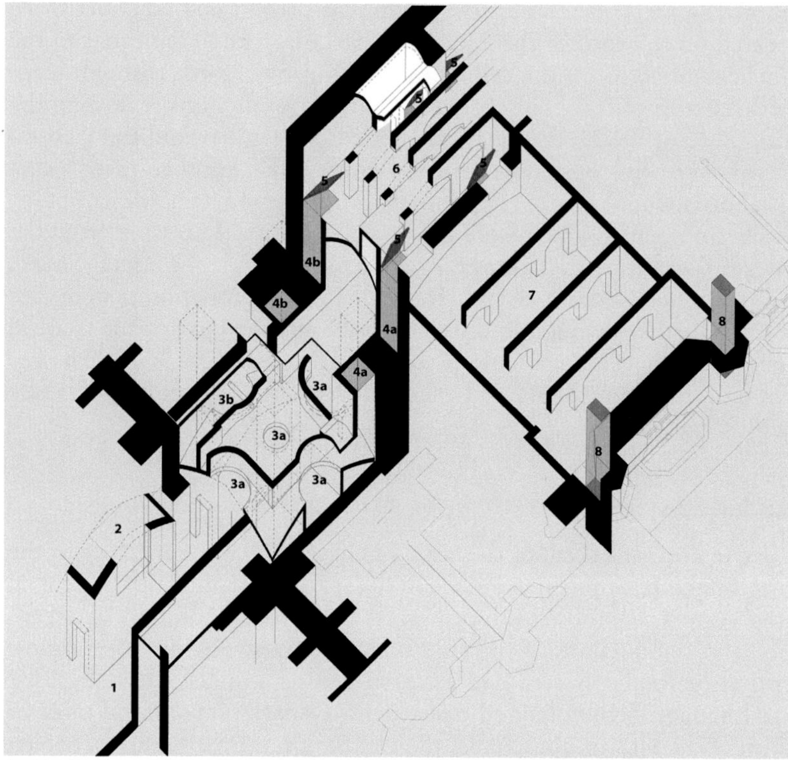

Figure 5.1 Axonometric projection showing network of flues in bay of River Front (author's own drawing). Key: 1. Main fresh air channel for River Front (basement), 2. Heating chamber inside fresh air channel, 3. Air valves in basement vaults: A. for cool air, B. for heated air, 4. Vertical flues carrying cool (a) and warm air (b) from basement to mixing chamber, 5. Square timber valve for controlling air supply to mixing chamber, 6. Mixing chamber above ceiling of ground floor corridor, 7. Fresh air plenum linking mixing chamber to vertical flues inside buttresses (8).

As a result of these substantive structural changes, more time was required for the design and construction of the new Palace, which contributed to the delays in the delivery of the project. Indeed in August 1843, only one year after the formal approval of the plans, Barry informed the Office of the Woods that there would be serious delays in the construction of the River Front: in addition to strikes by the stonemasons, these were caused by difficulties with resolving the intricate technical details of the ventilation and fireproofing. He wrote that 'adapting it to the mode of warming and ventilation prescribed by Dr Reid, which in order to avoid interference with the sightliness, accommodation and convenience of the building, is continual sources of difficulty'.[31]

The House of Lords and Reid's original plan for a central air supply

The ventilation network inside the House of Lords was developed over a period of three years. It began with the design of the primary infrastructure for the central air supply, which was intimately intertwined with the construction of the masonry structure on the ground floor and basement level. The first set of specialist ventilation drawings for the central air supply was completed between March 1843 and December 1844 and shows that the debating chamber was linked to the central air chamber of the main fresh air network.[32] This 'Central Chamber' was an octagonal space below the Central Lobby and connected to the Elizabeth and Victoria Towers through large air passages in the basement. It extended over the basement and ground floor levels and had two functions: first, it enabled the air supply to be switched from one tower to the other to minimise contamination by air pollution, which was highly susceptible to changes in the wind conditions; second, it provided a space for tempering the air (for example, in the winter the entire body of fresh air for the Palace was to be pre-warmed to a mild temperature). Reid noted that the air was 'to partake of the character of a mild climate before it is introduced, instead of being thrown in cool currents'.[33] In summer the basement was to be used to naturally cool the incoming air, utilising the large thermal mass of its masonry floors and vaults.[34]

The Central Chamber was also connected to a series of air mains, which could distribute the tempered air across the entire basement. The air main for the House of Lords was located on the south side of the Central Chamber, whilst for the House of Commons it was located on the north side. The quantity of air admitted into each of these mains could be regulated with the aid of large valves. Although the basement allowed air to be distributed horizontally across the large site without interfering with the layout of the principal floor, large spaces on the ground floor were still required to connect the two Houses to the basement network.[35]

The continuing development by Reid of the proposed ventilation system necessitated modifications to Barry's architectural plans with the former negotiating with the latter over securing access to spaces that had already been reserved for other uses. Two floor plans from 1843 show that Reid had introduced a network of passages and chambers on the ground floor to convey air horizontally to the House of Lords.[36] The Central Chamber was connected to a heating chamber below the Peers' Lobby through a vaulted passage (which in the drawings is referred to as a 'distribution flue'). This latter chamber would enable the temperature of the pre-warmed air to be raised to over 70°F (~21°C) if needed,[37] before distribution to four separate plenum air chambers, which were situated below the House, the Prince's Chamber and each Division Lobby. Another drawing, dated 28 April 1843, shows that Reid introduced large circular apertures inside the load-bearing walls to connect these plenum chambers to the heating chamber.[38] The air was to enter the interior of the debating chamber through an array of different inlets at various heights. Vertical shafts were also introduced within the masonry walls to convey fresh air to two sets of inlets, one below the gallery and another inside the ceiling.[39]

The final drawings for the ceiling arrangements were finalised in July 1846, shortly before Reid's involvement in the design of the House of Lords was stopped. These show that the roof space above the ceiling was divided into three longitudinal compartments. Two of these were to be used to supply fresh air downwards into the galleries on the east and west sides of the debating chamber and were connected to the central air supply by six vertical shafts at the south end of the Prince's Chamber.[40] The third compartment, which was located above the central row of ceiling panels, was for the extraction of vitiated air. Whilst the earlier plans for the House of Lords showed that Reid had originally intended to link the main smoke and air channels of the chamber and Royal Gallery to the Central Tower itself, the final drawings show an arrangement with a temporary ventilation shaft. Another array of inlets, which integrated into the furniture and floor of the chamber, allowed air to be introduced from below. These openings formed part of a complex system for the control of the ventilation and internal climate.

The proposed system for the House of Lords, which was to be mirrored in the House of Commons, was considerably more complex than the experimental systems deployed inside the temporary Houses of Parliament. Instead of replicating the same principles within the new Palace of Westminster, Reid (like any good scientist) saw it as an opportunity to use the knowledge that he had gained from his original experimental systems to develop a more sophisticated arrangement. Speaking of his ambitions, on 19 March 1846 Reid said that his objective was to 'make as great an advance on the ventilation of the new houses beyond that of the temporary houses as the temporary houses were before anything that been done previously'.[41]

Reid's plans for this complex system were never fully completed and were ultimately abandoned in 1846, but the archival material allows his original vision to be reconstructed. Letters, reports and the transcripts of oral statements that Reid provided to the various Select Committees between 1843 and 1846 offer insights into the underlying concepts and objectives underlying his vision, whilst the original drawings and sketches illustrate how he intended to realise them.

Personal and collectivist approaches to thermal comfort

As in his earlier inquiries, Reid's new ventilation scheme reflected a strong interest in the human aspects of environmental control such as thermal comfort and user satisfaction. The design of this system was informed by his earlier observations of the thermal environment and how it was affecting MPs and Lords, both individually and collectively. In his oral statements, Reid refers to reviews of user feedback collected inside the temporary chambers from Lords and MPs. Although Reid knew that it was impossible to produce a thermal environment that met the expectations of every individual user, he believed that differences in user experience could be accommodated by introducing some level of local control. His main objective was thus to realise a system that combined a personal and collectivist approach to climate control. He proposed two different strategies to do so.

The first strategy was to divide the debating chamber into multiple zones so that the condition of the atmosphere could be adjusted in each zone separately, thereby accounting for the often uneven distribution of Lords within the physical space of the House. Reid claimed that the indoor temperature would vary locally by several degrees if the air were to be introduced at a uniform temperature throughout the House.[42] To solve this problem, Reid proposed that the temperature of the supply air was lowered in the more crowded areas, which had a greater tendency to overheat, and raised in the less crowded areas, which had the opposite tendency to cool. Reid proposed to introduce five zones, located around (1) the throne, (2) the bar, the blocks of (3) ministerial and (4) opposition benches, and in (5) the central floor area.

The second strategy was to provide facilities for personal control. Reid argued that thermal comfort could not be treated as a uniform standard that applied equally to all Members. Referring to its application in the House of Commons, he identified two groups of users, which, due to the nature of their roles, had different needs. The first group, which accounted for the majority of users, comprised MPs who attended sittings for relatively brief periods. The second group – including the Speaker of the House, the Sergeant-at-Arms and Cabinet Ministers – constituted a small number of senior MPs and officers that had to regularly attend sittings in the debating chamber for long periods. Due to this, this group was more likely to be adversely affected by the interior climate than the first group.[43] In the temporary House of Commons, the Speaker and the Sergeant-at-Arms had complained that they felt uncomfortable even when the temperature satisfied the majority of MPs. Reid had thus undertaken trials in which he fit their chairs with individual controls. However, the customised chairs left the Speaker and Sergeant 'subject to an amount of annoyance of which they bitterly complained, from the state of the atmosphere being more congenial to those around them than to themselves'.[44]

Reid's plan for the 'permanent' Houses of Lords and Commons was to overcome the potential differences in users' thermal comfort by equipping each individual bench with their own separate controls and air supplies controlled by attendents below the debating chamber. Each bench was intended to accommodate the needs of 'members who may be forced by public business to occupy the same seat for numerous successive hours, at almost every day, for many consecutive months'.[45]

Equipped with valves, the air supplies were to be manually regulated by attendants stationed inside the air chamber below the House. Using instruments such as thermometers and hygrometers, the attendants were also expected to continuously monitor the 'precise constitution of the air proceeding to the individual benches'.[46]

Integration of the ventilation system into the interior architecture: 1844–46

Between 1844 and 1845 Reid himself and Barry and his staff explored in detail how this new concept could be implemented inside the House of Lords without disfiguring the Gothic interior. The ornamental details of Pugin's interior had to be adapted to accommodate the apertures needed for the ventilation system. In Reid's new scheme the air was no longer to be introduced in vertical currents through openings that covered the entire floor. To protect peers from rising dust particles and exposure to rising currents, Reid proposed to move the main apertures to locations furthest away from the benches. This required Reid and Barry to carefully plan the positioning of the apertures, utilising any available surfaces, such as skirting boards, wall panels, covings or tables. These apertures also had to be large enough to enable the air to enter the chamber without causing strong currents. The air, he argued, was to 'flow in and rise as gently as the movement of the air on a mild summer's evening'.[47] A second set of outlets was located below both the benches and the risers within the gangways.[48] As the air entered the apertures through the back of the benches near the Lords' feet, the use of these inlets was to be limited to periods when the main apertures were not sufficient to provide the required ventilation or if was it requested specifically by the Lords themselves. On 6 August 1844 and again in 14 August 1846 Reid testified that these inlets were only intended to be activated during hot and sultry weather or when the chamber was exceptionally crowded.[49] He said:

> *On extreme nights when Members of the House of Peers might complain of the state of the atmosphere, and wish a still greater portion than could be supplied by those means, then I would propose to bring in the back of the seats and using the steps as surfaces of diffuse ventilation under such control as the Peers themselves might desire.*[50]

The plans for this complex arrangement of apertures were never fully finalised as Reid and Barry could not agree on the final arrangement. A first set of working drawings was completed between November and December 1844.[51] These were produced by Barry's clerks following Reid's verbal instructions and included details for perforated cast iron floor plates and special iron plates with rectangular apertures for the supply to the benches and steps inside the gangways.[52] A floor plan, dated 11 December 1844, also shows the layout of the proposed configuration of inlets: the floor inlets were to be confined to a small area with apertures located within the floor of the gangways and dais, single rows of apertures were to be located along the walls, and the remaining inlets were to be integrated into the wooden screen of the bar and behind the throne as well as within the furniture. In this latter case, included the individual inlets were located underneath the benches, table, the Woolsack and even the Judges' Woolsack.[53]

Reconciling tensions between Reid and Barry

In addition to the immediate challenge of resolving complex design problems, the development of the ventilation system was seriously disrupted by the complexity of the working relationship between Reid and Barry's project team. Both of these increased as the project entered the detailed design phase. Reid and Barry had to cooperate with one another in order to reconcile the requirements of the ventilation with the architecture, fire safety or construction. The tensions between the two caused serious delays to the delivery of the House of Lords and became the subject of separate formal reviews by the Office of the Woods and several Select Committees, between 1844 and 1846. Although letters and reports from the time largely describe the relevant parties' personal experiences, they offer detailed insights into Barry and Reid's perspectives on this problem, as well as that of the three Commissioners of the Woods. Problems with Reid's and Barry's collaboration were apparent from the outset, but in the first three years they were not seen as sufficiently serious to justify a formal complaint. Over this period Reid's view was that:

> *so long as I considered that Mr Barry gave what was desirable for ventilation, or that the matter was so clear that there could be no doubt about it, and I apprehend no difficulty, I considered that would have been uncalled for on my part to have made any communication.*[54]

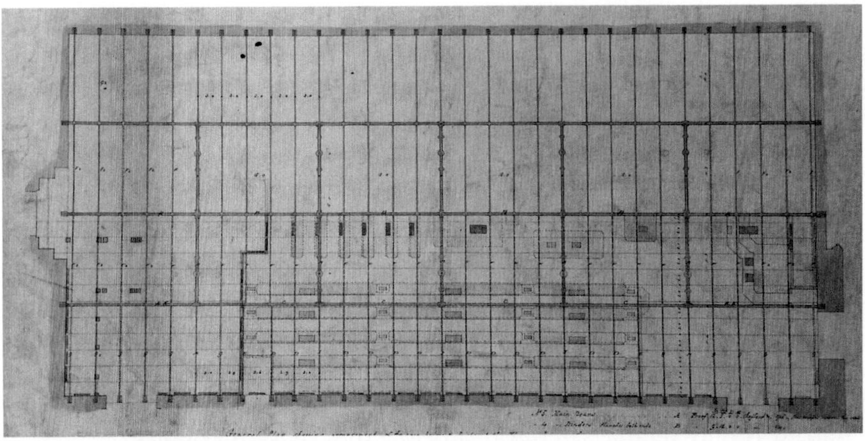

Figure 5.2 Floor plan showing the distribution of inlets incorporated within the furniture, skirting boards and the bar, dated 11 December 1844 (National Archives: Work 29/192).

The first signs that the tensions between Reid and Barry were more serious and were causing disruption to the delivery of the House of Lords began to surface in 1844. On 17 June, Reid voiced his frustration to a House of Commons Select Committee about the impact that various changes Barry had made to the architectural plans without his consent. He also complained about the pressure to repeatedly alter his own drawings – sometimes four or five times – in response to changes that Barry was making to his own plans. Moreover, some shafts had been 'rendered useless' as a result of changes that Barry had made to the architectural plans.[55] Barry, he alleged, had retrospectively reduced the size of available spaces for ventilation below and above the debating chambers in order to accommodate other functions within the envelope of the building. Reid had written a formal complaint to the Commissioners of Woods six days before his statement to the Select Committee on 17 June, notifying them that the air chamber underneath the House of Lords had been compromised by Barry's new alternative plan to introduce a sub-hall for horses and carriages in its stead below the two Houses on the ground floor.[56] This hall, which was covered with masonry groin vaults resting on columns,[57] occupied a space that was reserved for the ventilation. Barry's changes had reduced the height of the air chamber to 2.6 ft (~0.80 m). Reid demanded that this space be increased to at least 7 ft (~2 m), which was needed for the attendants to access the ventilation controls, arguing that the need for a sub-hall was 'secondary with the desirableness of promoting the personal comfort'[58] of the Palace's users.

On 6 August 1844, he testified to the Lords that attendants should be allowed 'to go freely and apply any instrument and all times to know the conditions of the air upon each individual bench'.[59] In another oral statement given to a House of Commons Select Committee in June 1844 he explained his objective to give 'any attendant the power of knowing personally what is the nature of the air that supplies every particular bench, without that, if called on to alter the character of atmosphere supplied to it, he cannot do it precisely'.[60]

Following a review of this issue during the sitting of the House of Commons Select Committee on 17 June, Barry agreed to abandon the proposed sub-hall even though construction on it had already started.[61] Barry subsequently increased the height of the air chamber to 6 ft (~1.8 m)[62] by lowering the ceiling and replacing the ornamental vaulting with a more shallow and utilitarian ceiling composed of jack arches between iron beams.[63] An original cross-section, dated 28 April 1843, shows the original groin vaults as well as pencil lines added later to show the position of the lowered ceiling.[64] The original groin vaulting, which is clearly visible from the Peers' Court, was only retained in the area below the Division Lobbies.

Although the conflict over the height of the air chamber beneath the House of Lords debating chamber was resolved after the meeting of the Select Committee in June 1844, tensions between Reid and Barry resurfaced and intensified from February 1845. In a period of just three months they quarrelled about multiple design issues, which resulted in the work on the House of Lords being intermitted by more than a year. These delays were a threat to Barry's own professional reputation and forced the Office of the Woods to both act as moderators in the disputes between the two and revisit the existing working arrangements.

The cast iron roof of the House of Lords was completed in January 1845.[65] After this, Reid began working on the details for the air and smoke channel that were intended to link the Royal Gallery and debating chamber to the Central Tower. The clerks working under Reid's supervision produced a series of schematic cross sections and floor plans that outlined several alternative proposals for the arrangements of air chambers, gas lighting and smoke channels above the ceiling. On 10 February 1845, however, Reid wrote to the Commissioners of the Woods complaining that Barry had claimed space again without being consulted, having decided to transform parts of the roof space above the House of Lords into an archive for the Public Records Office.[66] According to sketches from 27 January 1845 the proposed archive would have occupied the upper level of the roof space and significantly reduced the area available for ventilation. The upper level was to be separated from the lower level, which contained a chamber for the vitiated air, by brick vaults that would have acted as a fire barrier.[67] These changes made by Barry would have forced Reid to move the main smoke channel to the lower level, thus limiting the space available for the air chambers. In May 1845, three months after Reid raised the issue, Barry agreed to remove the archive spaces.[68]

Questions of architectural integration

Although a set of working drawings had been completed in 1844, Reid was not satisfied. In March 1845 he insisted that construction be paused to allow the earlier plans to be revisited and amended. Over the following three months Reid and Barry discussed several alternative schemes through the medium of letters and drawings. Reid reports that, over this period, drawings were passed:

> *backwards and forwards between Mr. Barry and myself, defining every point on the floor where air was to enter; what was to be the amount of entrance at the throne, the air at the bar, the air at the floor, the air at each individual bench; and in what manner the ventilation was to be worked at the prorogation, at an ordinary attendance, at a crowded attendance.*[69]

According to Barry, no fewer than eight different proposals were considered without settling on a final scheme.[70] In one of the proposals, he said, the air was to be introduced though openings within the skirting boards and wall

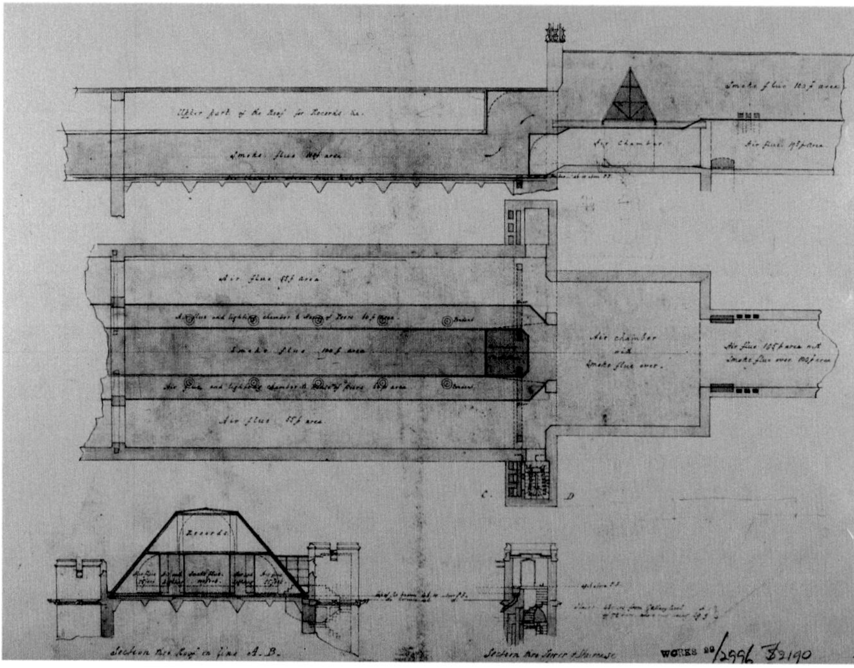

Figure 5.3 Cross section showing archive space encroaching on Reid's space, 27 January 1845, National Archives (Work 29/2996).

panels at opposite ends of the House. In another, the air was to be injected horizontally through perforated panels, which Barry described as 'breathing walls' and which were to line the walls to a height of 13 ft (~4 m).[71] In the last of the alternative proposals, air was to be admitted through the backs of the raised seats, rows of perforated floor plates along the wall, and through a series of 'air fountains in the form tables'. Despite the search for an alternative scheme, a final decision was not made, and the construction of the floor was delayed for more than a year.

Barry and Reid struggled to agree on a final plan because they failed to fully collaborate to resolve the closely-intertwined and complex architectural design and technical problems. This failure led to increasing tension between them. Barry accused Reid of constantly changing his plans and failing to agree on a final plan; Reid claimed that Barry's team had not provided him with enough of the architectural details to successfully incorporate the inlets into the interior woodwork, replete with its intricate Gothic details, and was therefore unable to complete his system as the inlets 'might come in a shape and in a manner with an appearance that he might not approve of'.[72] In addition, the flues were 'utterly useless, unless they work in according with the furnishings of the ceiling, of the sides, off all the details of furniture, may be considered, as it were, the terminations of the ventilation'.[73]

Reid wrote that the impasse between Barry and himself could only be resolved through a 'compromise', one that that reconciled technical and architectural requirements.[74] These issues were never fully resolved, and Reid's plan was abandoned after 1846 (see Chapter 6).

Contested responsibilities

The tensions between Reid and Barry illustrate that both parties felt that they did not adequately communicate their plans to one another and failed to come to a shared understanding of the boundaries between their respective roles and responsibilities in the design of the ventilation system. However, the records suggest that the perception of Reid as indecisive by Barry was, at least in part, due to the different approach he took to arriving at integrated design solutions. Reid defended his approach arguing that he presented Barry with the choice of multiple alternative solutions rather than just one single proposal. He believed that this would give the architect the flexibility to choose the most acceptable to him. As Reid struggled to understand which arrangements Barry was prepared to accept, due to the latter's lack of both feedback and design information, he saw the presentation of multiple options as a way of resolving potential conflicts.

Reid had taken a similar approach when planning the network of vertical air flues. In this case Reid only outlined his requirements, specifying the size or direction of flues, instead of presenting fully developed and detailed plans. It was Reid's understanding that he was only responsible for specifying the direction and the 'area or content of the flues' and that Barry was responsible for determining how these requirements could be met.[75] He left it to Barry to decide whether flues were to be embedded within the walls or buttresses, or whether the required quantities were supplied through single or across multiple flues. Reid saw these arrangements as equivalent and argued that 'it was matter of perfect indifference to the ventilation in what shape it came, provided it would equally cleaned, equally managed and controlled, and provided with valves'.[76]

Barry himself was not content with this approach. He expected Reid to be directly involved in the practical implementation of his scheme and asked him to be more precise and specific in his instructions.

The lack of shared understanding between Reid and Barry regarding their working relationship manifested itself in another major quarrel about fire safety, which began in April 1845.[77] This was triggered by the realisation that Reid's ventilation system was not compatible with Barry's fire safety strategy, which followed well-established principles of physical compartmentation that were originally developed to protect cotton mills in the later 18th century.[78] In the Palace of Westminster, brick arches were introduced above every ceiling to prevent the vertical spread of fire[79] and continuous party walls with cast iron doors were inserted to contain fire horizontally.[80] The use of timber was confined to internal finishes, such as the wall and ceiling panels.[81] A large number of the original 'fireproof' doors have survived and examples can still be found in roof spaces above the House of Lords, the Prince's Chamber and various Committee Rooms.

Barry alerted Reid that the vertical risers on the River Front penetrated the horizontal brick floors, and thus that there was a risk that fire could bypass the barriers. A heated epistolic disagreement between Reid and Barry, played out in a series of letters, ensued about who held the responsibility for resolving the conflicts between the requirements of ventilation and fire protection. In his letters Reid makes clear that he considered himself only partially responsible for their successful integration as it was, in his view, an integral part of the structural design, which was Barry's responsibility. In a letter to Barry, dated 23 April 1845, he writes that 'isolation of all apartments between the great channels of fresh and vitiated air is a matter of structure alone'.[82] Moreover, Reid claims that the ventilation system could be easily adapted to comply with Barry's principles of fire compartmentation by using fire resistant materials for the construction of the flues, equipping the shafts with iron dampers, and constructing the ornate panelled ceilings using iron instead of timber.

In another letter, dated 26 April 1845, Reid reiterated his belief that his role was to provide the specifications for the quantities of space required for the supply and extraction of air, and that Barry's team was responsible for both the construction of the physical channels and ensuring these did not compromise the architecture, fire safety or structural integrity of the building itself.[83]

Barry refused to accept Reid's characterisation of his responsibilities and in a letter, dated 24 April 1845, asserts both that Reid refuses to take responsibility for his own work and that he himself cannot take responsibility for 'the principle and application of the warming and ventilating arrangements ... which you [Reid] alone have devised and directed'.[84]

These fire safety issues were serious concerns, and as they were never fully resolved, the ventilation network has remained a fire risk to this day.[85] Indeed, the risk posed by historic but still extant ventilation shafts has recently been illustrated by the devastating fire in 2014 in another important public building: the Glasgow School of Art.[86]

Revisiting the mechanisms of collaborations: 1845–46

The conflict over who was responsible for reconciling the requirements of the ventilation system and fire protection was never fully resolved.[87] By May 1845, their working relationship had collapsed and the development of the ventilation system for the House of Lords was intermitted for an entire year.[88] Barry had lost his patience with Reid and had imposed new restrictions (which Reid considered unacceptable) on the way that they were to communicate: Reid was not to speak to him in person and all communications were to be conducted directly through the medium of drawings and letters or indirectly through Barry's clerks, who acted as liaisons between the two.[89] Barry wanted to keep a better record of their communications and to force Reid to communicate his own proposals more clearly through the medium of drawings. These restrictions heralded a significant, though not necessarily positive, change in their working relationship. In the previous five years, Reid and Barry had primarily talked in person, and also used simple sketches and written specifications to communicate their ideas.[90] However, Barry had become increasingly frustrated with Reid, claiming that he sometimes interpreted conversations incorrectly and that he made false accusations.

In addition to the aforementioned restrictions on the way that Reid could communicate with Barry, Reid was prohibited from borrowing drawings from Barry's architectural office and was only permitted to view and produce copies of them.[91] Although these ventilation drawings were produced under Reid's guidance, Barry viewed them as the property of his architectural practice. Some of the original drawings were kept in Barry's project office at the Palace of Westminster and some in his main office (that assisted the project team) in George Street.[92] Barry believed that this was necessary to prevent the loss of the original ventilation drawings produced by Barry's own staff.[93] Although Reid was still permitted to work with Barry's clerks and view drawings in his office, the process of resolving design problems became more time consuming and bureaucratic in nature. The clerks continued to produce drawings for Reid but they now had to liaise with Barry himself on Reid's behalf to gain his approval of Reid's proposals.

Barry also approached the Office of the Woods to request an independent review of Reid's working practices. The Commissioners appointed the architect Joseph Gwilt to undertake this review, who between June and September 1845 undertook a thorough examination of them, examining the original drawings, sketches and written correspondence, and conducting interviews with Reid, Barry and several of the Barry's clerks.[94]

In his final report, submitted to the Office of Woods on 29 September 1845, Gwilt concluded that the risk of further delays could not be eliminated without limiting Reid's involvement in some way. He argued that Reid did not have the knowledge and skills to fulfill his role as ventilator. He was unable to produce detailed technical drawings that are 'usually furnished by ventilators',[95] but even with the assistance of Barry's office clerks, Reid could not fulfil his role adequately. Since these deficiencies were sufficient to cause serious disruptions, Gwilt recommended that responsibility for the ventilation system be completely transferred to Barry's team, and that Reid be retained only as a consultant.

Over the next four months, the Commissioners of the Woods negotiated with Barry and Reid over the possible changes that could be made to the organisation of the design team, exploring how far previous issues could be addressed by redefining Reid's roles and responsibilities within the project. In a letter to the Commissioners, dated 19 November, Barry voiced his support for the recommendations in Gwilt's final report and demanded that Reid no longer be permitted to interfere directly in the work of his clerks nor have access the construction site.[96] Gwilt's recommendations represented a radical departure from the more collaborative design approach that had been followed in the preceding five years. However, the Commissioners of the Woods were reluctant to terminate Reid's employment and, on 2 December 1845, therefore proposed an alternative structure to Reid and Barry's working relationship. The Commissioners believed that the existing difficulties could be resolved through making changes in workflow management and by redefining Reid's and Barry's roles and responsibilities. The Commissioners suggested moving Reid into a separate drawing studio with his own staff and drawing library; Reid and his staff could independently produce a separate set of ventilation drawings, which could then be submitted to a clerk in Barry's office, who could then take on responsibility for coordinating the production of full working drawings that incorporated Reid's ideas. In this way, Reid could be retained as the lead designer for the ventilation system and his capacity to interfere with the work of Barry's team would be minimised.

In his reply, dated 4 December, Barry rejected the Commissioners' proposal, claiming that it would not address the most fundamental problem, namely Reid's lack of both the technical skills and knowledge required to coordinate the production of technical details.[97] As neither Barry nor Reid could agree on an approach, the Office of the Woods re-engaged Gwilt to undertake an independent review of the plans and the responses received from Barry and Reid.[98] In his report, which was returned on 11 December, Gwilt recommended that Barry's proposal be adopted. Although he did not go as far as proposing to remove Reid entirely, at least in public, he agreed that his position should become less 'prominent'.[99] However, according to Alexander Milne, Second Commissioner of the Woods, Barry had privately lost patience with Reid and would ideally have had him removed from the project altogether.[100] Gwilt's recommendations were not formally adopted but were revisited six months later due to growing political pressure from the Lords.

The final inquiry: 1846

In the summer of 1845 the Lords continued their criticism of Barry, lamenting the delays in construction. Several speeches at the time voiced the Lords' discontent with the internal climate of the temporary debating chamber, with Lord Campbell and Lord Brougham being particular outspoken about this issue. The problem with the internal climate was raised, amongst others, after a series of long debates on the Maynooth College Bill held at the beginning of June 1845, which took place in the temporary chamber over three days each, lasting over 12 hours and involving between 200 and 400 Lords at any one time.[101] Lord Brougham reported that during these debates, the interior had been uncomfortable due to the 'imperfect ventilation'. According to Constantine Henry Phipps, the Marquess of Normanby, the ventilation was in a 'wretched state'; whilst Lord Campbell noted the he suffered severely from the 'sudden draughts of hot and cold air'.[102] During another debate on 24 July 1845, Lord Brougham claimed that 'nothing could exceed the suffering they endured in the morning sittings, sometimes from the heat, and at other times from cold', whilst Lord Campbell noted that the Lords themselves suffered from an 'unwholesome' atmosphere.[103]

In February 1846, the Lords appointed yet another ad hoc Select Committee to undertake a formal investigation into the impact of Reid's involvement in the delays with building the House of Lords debating chamber.[104] Further debates about the delays were also held between April and June 1846.[105] The Select Committee's inquiry lasted four months and produced four separate reports. A parallel inquiry, which lasted six months and yielded three reports, was also undertaken by an ad hoc House of Commons Select Committee. Reid, Barry and the Commissioners of the Woods were questioned by both of these Committees about the impact of Reid's involvement on the delays, as was Gwilt about the findings of his earlier inquiry.

In their first report, published in March 1846, the Lords asked the Commissioners of the Woods to take a more proactive role in managing and monitoring the collaboration between Reid and Barry.[106] The Commissioners subsequently instructed Barry to finish the drawings required to complete the House of Lords and supply Reid with the information he needed to finalise the arrangements for the ventilation system (including the details for the floor-level inlets). In return Reid was expected to produce final plans for the ventilation system, which would then enable Barry to incorporate it into his final working drawings. Reid received 81 additional drawings from Barry's office.[107] The renewed pressure from the Lords resulted in significant progress being made on the drawings in the ensuing three months: drawings of the ventilation arrangements above the ceiling were largely completed, including floor plans showing the layout of vitiated and fresh air compartments, and construction details of key features, such as the ceiling panels, internal partitions, air valves and access walkways for the operational staff.[108]

Technical feasibility of system is challenged

Whilst the plans for the House of Lords were being developed, the Select Committee continued its inquiry into Reid's impact on the project. Additional pressure on the project was brought to bear by Goldsworthy Gurney, who publicly questioned the technical viability of Reid's system. Gurney lobbied Members of both Houses to consider the adoption of a different system. On 21 April 1846 Dr John Bowring, MP for Bolton, presented a petition by Gurney to the House of Commons, asking that his ideas for an alternative ventilation system be considered and tested.[109] On 28 April 1846, the Earl of St Germans presented the same petition to the House of Lords.[110] Gurney's proposal was subsequently referred to the House of Commons Select Committee, who interviewed him on 26 May 1846. In this interview, Gurney questioned the technical feasibility of Reid's scheme, arguing that the building was too large to be ventilated through a single chimney.[111] Gurney proposed to divide the ventilation system into a series of smaller systems, each with their own separate shafts. However, the Committee did not feel confident that this approach was viable due to a lack of evidence for its efficacy in other buildings. As the choice between rival ventilation systems was a technical question, the proposal was subsequently referred to an Expert Advisory Panel appointed by the Commissioners of the Woods to assist them with reviewing the problems delivering Reid's scheme.[112] The Advisory Panel was composed of the engineer George Stephenson, the architect Philip Hardwick and the chemist Professor Thomas Graham.[113]

In its final report, dated 25 June 1846, the Expert Advisory Panel agreed with the tenor of Gurney's general verdict and concluded that a centralised stack system, if applied to the entire building, would be far too complex. They made three recommendations: first, that the system be simplified by confining its use to only a small part of the Palace and that it should only be connected to the two debating chambers, the public corridors, the libraries, the kitchens and the dining rooms; second, that the two chambers should be given their own separate ventilation shafts that could be operated independently from each other; and third, that the other areas of the new Palace, including the Committee rooms, be heated and ventilated through more conventional methods using fire places and operable windows.[114]

Although the endorsement of Gurney's criticisms by the Expert Advisory Panel did not lead to the adoption of his proposed alternative scheme, Gurney's ideas significantly influenced the implementation of subsequent changes to Reid's original plans after 1846. His proposal also provided the blueprint for his own later plans for remodelling the ventilation systems inside the two chambers, which were implemented in the mid-1850s.

The Lords' final verdict

Reid, who feared for his reputation, was not successful in convincing the Lords that the ventilation system could be completed under his direction without the risk of further delays. The various reviews of the progress of the new Palace reveal that the witnesses shared the view that Reid's involvement was a major cause of disruption. Alexander Milne, who had been a Commissioner of the Woods since 1835, arrived at the same verdict. Whilst acknowledging that Barry also shared some responsibility for the delays, he opined that the main problem was the disruption caused by Reid's insufficient 'practical knowledge of construction and general acquaintance with building'.[115] The Earl of Lincoln, Henry Pelham-Clinton, who had served as First Commissioner of the Woods under Robert Peel's Conservative government, noted that Reid could not adequately fulfil his role due to his 'ignorance of the rules of construction and everything appertaining to building'. Reid, he asserted was 'unable, from working drawings, sufficiently to understand the mode of carrying out his plans, and off course unable to explain'.[116] However, he was reticent to endorse Reid's removal without the approval of Parliament. As large parts of the ventilation system infrastructure had already been completed, he argued that Reid should continue to be involved in the completion of the system, and that the substantial amount of public funds invested in the project should not be wasted. Indeed, 'all subterranean flues are made; in fact they are made throughout the buildings, and the connecting flues throughout the whole of the River Front'.[117]

The final decision about Reid's future involvement thus became a political issue.

Barry persuaded the House of Lords ad hoc Select Committee to remove Reid from the project believing that the problems that had beset their attempted collaborations were insoluble. Reid had had, after all, 'every facility which any competent and reasonable man could have required to enable him to make out a complete' scheme.[118] Barry offered to take over the responsibility for ventilating the House of Lords and guaranteed that it would be completed in 1847 if Reid was not involved. He had worked closely with Reid for over six years and felt confident that he and his team had acquired sufficient knowledge of the ventilation system to complete it on their own.

The House of Lords ad hoc Select Committee accepted Barry's offer to take over responsibility for implementing of the Lords' ventilation system. In its final report, published on 23 June 1846, it stated that the risk of further delays due to Reid's involvement could only be avoided if responsibility for completing the ventilation in the House of Lords was transferred to the architect himself. On 26 June 1846 the House of Lords approved the recommendations by vote.[119] Bruceman and Prowler have interpreted this as a 'momentous' *step* 'in the architect's bid to regain full design control'.[120] It also marked the beginning of the end of what Moritz Gleich has today described as the "service architect's claim for dominance over the architect".[121]

Bicameral split – divisions between the Houses

Unlike their counterpart, the House of Commons ad hoc Select Committee was more optimistic that the problems between Reid and Barry could be resolved, concluding that Reid should be retained and efforts made to resolve the tensions between them. In its third and final report, dated 5 August 1846, it wrote that it 'anxiously desired to see the re-establishment of a cheerful and ready cooperation between the parties'.[122] To enable this it recommended establishing a new body: a third party that would perform the role of a mediator between Barry and Reid. This body was to be composed of two people, one to be selected by Barry and the other by Reid. If they were unable to resolve disagreements between the two, a third person was to be appointed by the Office of the Woods to take on the role of adjudicator. The Select Committee also expressed their support for Reid's system, arguing that the system in the temporary House of Commons had successfully demonstrated its potential and that the improvements that it had 'affected in the atmosphere ... can be appreciated by every Member'.

As a direct result of the differing recommendations of the two Select Committees, Reid's services were retained so that he could complete the ventilation system in the House of Commons debating chamber with all other parts of the Palace, including the House of Lords, coming under Barry's control. Reflecting the split between the Houses, a physical boundary was drawn between the territories controlled by Reid and Barry (see Chapters 6 and 7). The ventilation network of the House of Commons became isolated from that of the House of Lords. The original plan for using a central tower as a shared extraction shaft was also abandoned and superseded by a decentralised approach to stack ventilation involving an array of local ventilation shafts.

Conclusion: revisiting Reid's role

The tensions between Reid and Barry illustrate the serious difficulties that best the development of the plans for the ventilation system and its integration into the architectural fabric of the new Palace. Whilst the former can be clearly attributed to Reid, the latter was clearly both men's responsibility. As has been shown, the final design of the ventilation was the outcome of a close, although difficult, collaborative enterprise in which both the architect and ventilator bore a substantive share of responsibility for the realisation of the ventilation system. This was partly because its design could only be resolved through working together and partly because Reid was unable to fulfil the role of ventilator without receiving direct support from Barry's staff. Barry was forced to take on the role of a steward responsible for orchestrating a process of cross-disciplinary collaboration. However his commitment to sustain the collaboration wavered and had ended by the spring of 1845. The fundamental problem was that Barry failed to fully accept the impact of the limitations of Reid's specialist role as a building scientist (for example that Reid himself needed a higher level of support to successfully fulfil his role).

The delivery of the project was also compromised by serious problems in the management of the workflow and the exchange of information. These issues were of particular significance for a project of such an enormous scale and complexity. Barry had began introducing formal procedures to centrally manage the exchange of drawings with Reid as early as 1841, which can be understood as an early example of modern building information management. Similar to the role that Building Information Modelling (BIM) plays in contemporary construction projects, these procedures were designed to enable, monitor and manage the flow of large quantities of design information between stakeholders. However, the formal procedures put in place by Barry were not up to the task of managing the large quantity of information that had to be exchanged between Reid and Barry's team for the collaboration to be effective. A major reason, which should not be underestimated, for the failure to collaborate effectively and the consequent delays to the project, was that they became increasingly restrictive, in particular after May 1845, when Barry would not even talk to Reid in person. As will be seen in in Chapter 7, this persisted after 1846 when Reid's involvement was confined to the House of Commons. Despite the reduction in scale, Reid's and Barry's failure to collaborate effectively continued to disrupt the delivery of his system for the House of Commons

It is evident that many of the difficulties with the design and construction of the new Palace were due to the fact that the design of the ventilation scheme was not completed before the start of construction on the Palace itself. A comprehensive set of drawings that showed the final system in its entirety, was never produced. Between 1842 and 1846 large parts of its design were developed following a reactive rather than thought-out approach and without a complete set of drawings. A substantial part of the network was constructed without the aid of finalised drawings. The information about the system was also fragmented. The archives contain a large number of drawings that show only small sections of the system, whilst only a few provide an overview of the wider system to which these sections belonged. As a result, large parts of the final ventilation arrangements were not documented. The fact that parts of the infrastructure were built without any plans had, and continues to have, serious long-term implications in terms of establishing and retaining an institutional understanding of the system over several generations, which the staff of the Office of Works (later known as the Ministry of Works) would have to rely on to manage its day-to-day operation. This problem was recognised in the 1860s when John Percy, Superintending Engineer, commissioned a survey in an (ultimately unsuccessful) attempt to map the network as built. Efforts made since the late 20th century to mitigate the potential fire risk of the historic infrastructure have also been compromised by an insufficient understanding of its design.

Although the challenges detailed in this chapter caused serious delays to the construction of the House of Lords, the surveys shows that this did not prevent Barry and Reid from completing a large part of the physical infrastructure required for the latter's original ventilation scheme. This is significant as the completed infrastructure provided the physical foundation on which the final plans for the House of Lords, completed under Barry's direction after 1846, could be built. These final arrangements will be investigated in the next three chapters.

Notes

1 Oral Statement by Barry, 3 June 1844, in HC 448 1844, Report from the Select Committee on Houses of Parliament, Q80.
2 Ibid, Q31. The original cost was £707,104 and in June 1844, the cost had had risen to £1,016,924.
3 Oral Statement by Earl of Lincoln, 17 June 1844, in HC 448 1844, *Select Committee on Houses of Parliament*, Q408; Oral Statement by C. Barry, 10 June 1844, in HC 448 1844, *Select Committee on Houses of Parliament*, Q259–61.
4 To create this repository Barry reclaimed parts of the roof spaces reserved for ventilation. To mitigate the additional fire risk posed by the housing of these documents, he also introduced a system of cast iron roofs, which replaced the original plans for more traditional roofs, which was to be constructed in timber and covered with lead and slate.
5 HC 448 1844, *Select Committee on Houses of Parliament*; HL 381 1844-43, *Report of the Select Committee appointed to inquiry into the progress of the building*; HL 269 1844, *First report of the Select Committee appointed to inquiry into the progress of the building*; HL 629 1844, *Second report of the Select Committee appointed to inquiry into the progress of the building* .
6 Letter from C. Barry to Select Committee, 29 April 1844, in HL 269 1844, *Select Committee on progress, First report*, pp. 20–21.
7 Paper read by C. Barry, 2 May 1844, in HL 269 1844, *Select Committee on progress, First report*, p. 31.
8 Oral Statement by Earl of Lincoln, 17 June 1844, in HC 448 1844, *Select Committee on Houses of Parliament*, Q419.
9 Oral Statement by C. Barry, 29 April 1844, in HL 269 1844, *Select Committee on progress, First Report*, Q182; Oral Statement by C. Barry, 6 May 1844, in HL 629 1844, *Select Committee on progress, Second Report*, Q255–66.
10 Oral Statement by 29 April 1844, in HL 269 1844, *Select Committee on progress, First Report*, Q189–205.
11 Section through Peers' Court showing east elevation of House of Peers also transverse section through House of Peers', undated (National Archives: Work 29/152); Section through Peers' Court looking west showing east elevation of House and transverse section looking north , 7 February 1844 (National Archives: Work 29/165).
12 The arches inside the Central Tower are fully preserved and are still connected to the various spaces of the Palace, many of which have been adapted to house modern plant rooms.
13 David Boswell Reid, *Illustrations of the theory and practice of ventilation*, (London: Longman, Brown, Green & Longmans, 1844), 93–96.
14 Oral Statement by D.B. Reid, 17 March 1846, in HC 574 1846, *Third report from the Select Committee on Westminster Bridge and new palace; together with the minutes of evidence*, Q779–80.
15 Ibid; Reid, *Illustrations*, 287.
16 Oral Statement by D.B. Reid, 17 March 1846, in HC 574 1846, *Third Report*, Q783–4.
17 Oral Statement by D.B. Reid, 19 March 1846, in HC 574 1846, *Third Report*, Q919.
18 Ibid, Q919–20.
19 HL 381 1844–43, *Select Committee on progress, report*, p. 3.
20 The new Houses of Parliament. House of Lords, *Hansard's Parliamentary Debates: The Official Report (*21 February 1843 vol. 66 cols. 1033–6.

21 Oral Statement by C. Barry, 2 March 1843, in HL 381 1844-43, *Select Committee on progress*, Q11–96.
22 Oral Statement by C. Barry, 10 March 1843, in HL 381 1844-43, *Select Committee on progress*, Q155, 160.
23 The New Houses of Parliament. House of Lords, *Hansard's Parliamentary Debates: The Official Report (*11 March 1844 vol 73 cols. 797–8).
24 Oral Statement by C. Barry, 10 June 1844, in HC 448, 1844, *Select Committee on Houses of Parliament*, Q256–7.
25 Oral Statement by D.B. Reid, 6 August 1844, HL 629 1844, Select Committee on progress, second report, Q535–36.
26 Oral Statement by C. Barry, 27 June 1844, in HC 448, 1844, *Select Committee on Houses of Parliament*, Q637–40)
27 New Houses of Parliament. House of Lords, *Hansard's Parliamentary Debates: The Official Report (*09 August 1844 vol 76 cols. 1953–5).
28 'Plans and sections of boiler chimney and ventilating shafts', 1845 (National Archives: Work 29 678).
29 Oral Statement by D.B. Reid, 23 February 1846, in HL 1846 719, *Select Committee on progress*, Q55–58.
30 Andrew Saint, *Architect and engineer: a study in sibling rivalry* (New Haven: Yale University Press, 2007), pp. 95–99).
31 Letter from Barry to A. Milne, 15 August 1843 (National Archives: Work 11/6/1.
32 'House of Peers. Plan of ground floor', 17 March 1843 (National Archives: Work 29/133); 'Plan of central hall including Peers' Inner Court and Commons' Inner Court, ground floor', 27 March 1843 (National Archives: Work 29/101); Sections through sub-vaults of Royal Gallery showing groins and vaults, also provision for ventilation, 15 July 1843, (National Archives: Work 29/476); 'Plan of House lobby and parts adjoining', 31 July 1843 (National Archives: Work 29/188); 'Plan and section of floor of the House showing piers to carry floor and openings for ventilation', 8 June 1844 (National Archives: Work 29/2878).
33 Oral statement by D.B. Reid, 24 September 1841, in HC 51 1841, *Report from the Select Committee on Ventilation of the New Houses of Parliament; with the minutes of evidence, and appendix*, Q2–4.
34 Interview with Reid, 30 September 1841, in HC 51 1841, *Select Committee on ventilation*, Q157–67.
35 These were very large spaces and today many of these spaces have been converted into space for offices, storage and modern plant rooms.
36 'Plan of central hall including Peers' Inner Court and Commons' Inner Court, ground floor', 27 March 1843 (National Archives: Work 29/101).
37 Oral statement by D.B. Reid, 4 August 1844, in HL 629 1844, *Select committee on progress, second report*, Q557.
38 Sections showing provisions for ventilation (National Archives: Work 29/2834; Ground floor– showing provision for ventilation, Work 29/476.
39 House of Lords: ground floor plan and sections showing provision for ventilation, 23 April 1843, (National Archives: Work 29/2834); Masonry details of Peers' Lobby, east, west and north sides, 31 July 1843, (National Archives: Work 29/185).
40 'Section through Royal Gallery looking south showing ventilation flues', 1845 (National Archives: Work 29/2883); 'House of Lords: plan of principal floor showing provision for ventilation', 1844 (National Archives: Work 29/2894); 'Peers Lobby throne end of House', 22 July 1844, (National Archives: Work 29/171).

41 Oral Statement by D.B. Reid, 19 March 1846, in HC 574 1846, *Third report*, Q990.
42 Oral statement by D.B. Reid, 4 August 1844, in HL 629 1844, *Select Committee on progress, second report*, Q561, 558.
43 Note: Similar observations have also been made inside the current chamber of the House of Commons from the 1950 to the present day. The experience of MPs in the modern chamber has been the subject of another study by the author: Henrik Schoenefeldt, 'Delivering occupant satisfaction in the House of Commons, 1950–2019', *Building & Cities* (1 – 1) 141–163.
44 Oral statement by Reid, 17 June 1844, in HC 448 1844, *Select Committee on Houses of Parliament*, Q320.
45 Oral statement by D.B. Reid, 6 August 1844, in HL 629 1844, *Select Committee on progress, second report*, Q554.
46 Ibid, Q561.
47 Oral Statement by D.B. Reid, 19 March 1846, in HC 574 1846, *Third report*, Q991.
48 Oral Statement by D.B. Reid, 6 August 1844, in HL 629 1844, Select Committee on progress, second report, Q552, 554.
49 Ibid, Q556; Oral Statement by D.B. Reid, 14 August 1846, in HL 719 1846, *Select Committee on progress*, Q35.
50 Oral Statement by D.B. Reid, 6 August 1844, in HL 629 1844, *Select Committee on progress, second report*, Q552.
51 'Plan and section through floor showing cast iron columns, binders, joists and rakers to seating', 10 December 1844, (National Archives: Work 29/191; 'Detail of cast iron floor plates', 11 December 1844, (National Archives: Work 29/189).
52 Oral statement by C. Barry 23 February 1846, in HL 719 1846, *Select Committee on progress*, Q13–14.
53 Oral Statement by C. Barry, 23 February 1846, HL 719 1846, *Select Committee on progress*, Q12–14' General plan shewing arrangement of the iron beams and joists of floor', 11 December 1844 (National Archives: Work 29/192)
54 Oral statement by Reid, 17 June 1844, 17 June 1844, in HC 448 1844, *Select Committee on Houses of Parliament*, Q372.
55 Ibid, Q367. Note that surveys have revealed that the fabric contains flues, which had been constructed in accordance with an earlier plan but became redundant as a result of later changes in Barry's plans. Examples of such redundant shafts can be found on the south side of the Prince's Chamber.
56 Oral statement by D.B. Reid, 17 June 1844, in HC 448 1844, *Select Committee on Houses of Parliament*, Q372.
57 'Ground floor of sub-hall', 1843 (National Archives: Work 29/136); 'Elevation and cross-section', undated. (National Archives: Work 29/152); 'Plan and section of floor of House showing piers to carry floor and openings for ventilation', 8 June 1844 (National Archives: Work 29/2878).
58 Oral statement by D.B. Reid, 17 June 1844, in HC 448 1844, *Select Committee on Houses of Parliament*, Q317.
59 Oral statement by D.B. Reid, 6 August 1844, in HL 629 1844, *Select Committee on progress, second report*, Q562.
60 Oral statement by Reid, 17 June 1844, in HC 448 1844, *Select Committee on Houses of Parliament*, Q337.
61 Surveys undertaken by the author have revealed that part of the sub-hall was already under construction at the time these changes were undertaken, and traces of it are still clearly visible today. The ornamental stone columns of the proposed sub-hall were retained and the height of external arches linking the

cloister to the two courtyards was lowered. These features as well as the original jack arches are fully preserved.

62 Oral statement by D.B. Reid, 17 June 1844, in HC 448 1844, *Select Committee on Houses of Parliament*, Q326.
63 'Cross-sections of vaults under House', 28 April 1843 (National Archives: Work 29/2834).
64 'Sections showing provisions for ventilation' 28 April 1843 (National Archives, Work 29/2834).
65 'New Palace of Westminster – queries and answers', in HC 574 1846, *Third report*, pp. 161f.
66 Letter from D.B. Reid to Office of Woods, 10 January 1845 (National Archives: Work 11/12, no. 147).
67 'Plan and sections showing ventilation through roof and staircase', 27 January 1845 (National Archives: Work 29/2996).
68 Letter from Charles Barry to Reid, 10 May 1845 (National Archives: Work 11/12, no. 140–41).
69 Oral Statement by D.B. Reid, 19 March 1846, HC 574 1846, *Third report*, Q984.
70 Oral Statement by C. Barry, 23 February 1846, HL 719 1846, *Select Committee on progress*, Q12–16
71 Ibid, Q15.
72 Oral Statement by D.B. Reid, 19 March 1846, in HC 574 1846, *Third report*, HC 574 1846, Q1001–5, Q989).
73 Ibid, Q984.
74 Oral Statement by C. Barry, 23 February 1846, in HL 719 1846, *Select Committee on progress*, Q40–44.
75 Oral Statement by D.B. Reid, 17 March 1846, in HL 719 1846, *Select Committee on progress*, Q834–35.
76 Ibid, Q820.
77 Letter from Charles Barry to Reid, 15 April 1845 (National Archives: Work 11/12, no. 124); Letter from Reid to Charles Barry, 17 April 1845 (National Archives: Work 11/12, no. 124–5).
78 Sarah Werweil, *The development of fireproof construction in Great Britain and the United States in the nineteenth century, in Robert Thorne (ed.) Structural Iron and Steel, 1850-1900* (Aldershot: Ashgate, 2000), pp. 69-9; A.W. Skempton & H.R. Johnson, The First iron frames, in Structural Iron, R.J.M. Sutherland, Structural Iron, 1750–1850, p. 19–44; *Denis Smith, 'Techniques of the Building', in* H.M. Port, The Houses of Parliament (New Haven: YUP & Paul Mellon Centre for Studies in British Art, 1976), pp. *195–17;* John Olly, The Reform Club, *Architects Journal*, 27 February 1985, pp. 34–60.
79 The jack-arched floor inserted above the panelled ceilings are shown in the original drawings, e.g. 'River front, centre sections: transverse section through Committee Rooms and corridor looking south', 1841 (National Archives: Work 29/2833).
80 'Plan, elevation and details of fire-proof doors for first room next Peers' entrance to terrace, south section', 23 October 1853, (National Archives: Work 29/2767).
81 The application of the principle of compartmentation is most clearly demonstrated on the River Front, which is illustrated in survey drawings from the 1860s: 'Ventilation. Longitudinal section through portion of river front showing the heating and ventilation arrangements', 1867 (National Archives: Work 29/2989); 'Ventilation. Section through Committee Rooms showing warming and ventilation arrangements' 1867 (National Archives: Work 29/ 2990).

82 Letter from D.B. Reid to C. Barry, 23 April 1845 (Work 11/12, no. 128–29).
83 Letter from D.B. Reid to C. Barry, 26 April 1845 (Work 11/12, no. 131–32).
84 Letter from C. Barry to D.B. Reid, 24 April 1845 (Work 11/12, no. 129–30).
85 Joint Select Committee on Restoration and Renewal of the Palace of Westminster (HL Paper 41/ HC 659 2016), pp. 15, 29, 34. The challenge then and now is to reconcile the use of physical compartmentation with the requirement for an extensive web of continuous voids.
86 'Fire Investigation report, Mackintosh Building 167 Renfrew Street, Glasgow, Scottish Fire and Rescue Prevention', May 2014. (Copy of Report is held in the Archives of the Palace of Westminster Restoration and Renewal Delivery Authority).
87 Surveys undertaken as part of the author's own research show that some of the proposed measures, such as the use of iron valves and fire-resistant materials for the construction of air chambers and passages, were adopted in parts of the building.
88 Letter from Reid to Office of Woods, 10 February 1845 (National Archives: Work 11/12, no. 147); Letter from Charles Barry to Reid, 24 April 1845 (National Archives: Work 11/12, no. 129–30); Letter from Charles Barry to Reid, 10 May 1845 (National Archives: Work 11/12, no. 140–41).
89 Oral Statement by D.B. Reid, 17 March 1846, in HC 574 1846, *Third Report*, Q729.
90 Oral Statement by D.B. Reid, 14 August 1846 (HL 719 1846,) *Select Committee on progress, fourth report*, Q24–26.
91 Letter from D.B. Reid to C. Barry, 6 May 1845 (National Archives: Work 11/12, no. 137).
92 Letter from D.B. Reid to C. Barry, 12 May 1845 (National Archives: Work 11/12, no. 141); Letter from D.B. Reid to Office of Woods, 12 May 1845 (National Archives: Work 11/12, no. 142).
93 Letter from D.B. Reid to C. Barry, 7 May 1845 (National Archives: Work 11/12, no. 138).
94 'Report of Mr Gwilt to the Commissioners of Woods and Forests', 29 September 1845, in HL 719 1846, *Select Committee on progress*, pp. 19–20.
95 Ibid.
96 Answer of Dr Reid to a document issued by the Commissioners of Woods on 2 December 1845; and minutes of evidence of Dr Reid before a Committee of the House of Lords, for inquiring into the progress of the building of the Houses of Parliament (HC 750 1847).
97 Letter from C. Barry to Office of Woods, 4 December 1845, in HC 574 1846, *Third Report*, p. 186.
98 Letter from Office of Woods to Gwilt, 6 December 1845, in HC 574 1846, *Third Report*, p. 44
99 Letter from Gwilt to T. Philipps, 11 December 1845, in HC 574 1846, *Third Report*, p. 47.
100 Oral Statement by A. Milne, 17 March 1846, in HC 574 1846, *Third Report*, Q687-88.
101 Maynooth College Bill, House of Lords, *Hansard's Parliamentary Debates: The Official Report (*3 June 1845 vol 80 cols 1298); Maynooth College Bill, House of Lords, *Hansard's Parliamentary Debates: The Official Report (*2 June 1845 vol 80 cols. 11160–231); Maynooth College Bill - Ajourned debate (Third night), House of Lords, *Hansard's Parliamentary Debates: The Official Report (*4 June 1845 vol 81 cols 6–120).
102 The New Houses of Parliament, House of Lords, Hansard's Parliamentary Debates: The Official Report (05 June 1845 vol 81 cols. 120–2)

103 Address to Her Majesty – The New Houses of Parliament, House of Lords, *Hansard's Parliamentary Debates: The Official Report (*24 July 1845 vol 82 cols. 1033–4).
104 Report from the Select Committee of the House of Lords on the progress of the building, HL SC 1846, 719).
105 Progress of the New Houses of Parliament, House of Lords, *Hansard's Parliamentary Debates: The Official Report (*24 April 1846, vol 85 cols. 970–6); Ventilating the Houses of Parliament, House of Lords, *Hansard's Parliamentary Debates: The Official Report (*7 May 1846, vol 86 cols. 170–1); The New Houses of Parliament House of Lords, *Hansard's Parliamentary Debates: The Official Report (*26 June 1846, vol 87 cols 1033–5).
106 First Report, in HL 719 1846, *Select Committee on progress*, p. 3.
107 Oral Statement by D.B. Reid, 14 August 1846, in HL 719 1846, *Select Committee on progress*, fourth report, Q36–41.
108 Detail of ceiling construction, 10 July 1846 (Work 29/2896); Plans and Elevation of air chambers over House of Peers (Work 29/232).
109 Preamble, House of Commons. *Hansard's Parliamentary Debates: The Official Report (*21 April 1846 vol 85 cols. 788).
110 Minutes, House of Lords, *Hansard's Parliamentary Debates: The Official Report (*28 April 1846 vol 85 cols. 1142).
111 Oral Statement by G. Gurney, 26 May 1846, in HL 719 1846, *Third Report*, Q1418–1581.
112 HL 719 1846, *Third Report*, p. xxxiii.
113 George Stephenson, Philip Hardwick & Thomas Graham, Report upon the system proposed by Dr D.B. Reid, for warming and ventilating the new Houses of Parliament, 25 June 1846 (HC 1846, 447).
114 Ibid.
115 Oral Statement by A. Milne, 17 March 1846, in HC 574 1846, *Third report*, Q672.
116 Oral Statement by the Earl of Lincoln, 19 March 1846, in HC 574 1846, *Third report*, Q852.
117 Ibid, Q873.
118 Oral Statement by C. Barry, 3 March 1846, in HC 574 1846, *Third report*, Q150.
119 The New Houses of Parliament House of Lords, *Hansard's Parliamentary Debates: The Official Report (*26 June 1846, vol 87 cols 1033–5).
120 Robert Brucemann and Donald Prowler, 19th Century Mechanical System Designs Journal of Architectural Education, 30 (3) Energy and Architecture (Feb., 1977), pp. 11–15.
121 Moritz Gleich, Architect and Service Architect: The quarrel between Charles Barry and David Boswell Reid, Interdisciplinary Science Review, 34:4 December 2012, p. 340.
122 'Third Report', 5 August 1846, in HC 574 1846, *Third report*, p. xvii.

PART III

Realisation within new borders: 1847–54

My Lords, I understand that the Prime Minister's letter proposing that both Houses of Parliament might relocate to York had already been sent when the Minister answered questions on this on 15 July, yet he made no reference to it in any of his replies. Was that because he was not aware of the letter or because he chose not to inform this House that the Commons might also be moved?

Lord Wallace of Saltaire, House of Lords, 22 July 2020[1]

Note

1 Parliament: Restoration and Renewal Project, House of Lords, *Hansard's Parliamentary Debates: The Official Report* (22 July 2020, Vol 804. Col. 2213.

6 Barry's system in the House of Lords: 1847–54

A palimpsest of two schemes

The House of Lords was completed and occupied by April 1847, only nine months after Charles Barry was granted sole responsibility for completing the ventilation system. Although he adopted a different approach to warming and ventilation than David Boswell Reid, Barry reutilised most of the physical infrastructure established by him over the previous four years. Having worked closely with Reid for six years, Barry felt confident that his office had acquired sufficient knowledge to complete the system without the help of an external consultant. On 4 December 1846 he wrote to the Commissioners of Woods that he would continue to follow the 'governing principles upon which he [Reid] proposes to apply his system to the New Houses of Parliament, of which I am fully congnizant'.[1]

Barry collaborated with Alfred Meeson, a civil engineer who worked in his practice, and also drew on the expertise of a heating contractor, William Jeake. James Faraday, a gasfitter, assisted his team in the design of the gas lighting, whilst his uncle Michael Faraday, the renowned physicist, was also engaged to conduct a series of experiments in the chamber. Charles Barry Jr, the architect's oldest son, who was also involved in these experiments, reports in his diary that the scientist conducted various tests of the ventilation system and gas lighting between September 1846 and February 1847.[2]

Studies of the historic drawings and new building surveys undertaken as part of the author's own research between 2017 and 2019 have revealed that the main features of Reid's original scheme for the House of Lords were already completed before 1846 and their subsequent integration into Barry's final system, which was largely an adaption of Reid's original plans. It is therefore not surprising that Reid later challenged Barry when Barry claimed sole authorship of the system.[3]

As this primary infrastructure was an integral part of the structural fabric of the new Palace, it was difficult and expensive to alter. The most significant changes made to Reid's planned system were operational rather than physical. The central air supply and a temporary up-cast shaft were retained, whilst the network of air chambers and vertical flues from Reid's own scheme were incorporated. The architectural carcass of the House of Lords, which includes the vertical air flues buried within the structural walls, the ceiling and the cast iron roof were already completed by August 1844.[4] The perforated cast iron floor, which rested on a structure of cast iron beams and columns, was also largely completed. However, Reid's complex scheme of floor inlets was abandoned as Barry was to adopt a fundamentally different approach to handling the air supply and addressing questions of thermal comfort.[5] As such, the ventilation system for the House of Lords, which is the subject of this chapter, has to be understood as a palimpsest of Reid's original design, which provided the foundational infrastructure and Barry's subsequent scheme.

The involvement of Barry's office in the design of the House of Lords ventilation system, however, did not end after its completion. Over the next eight years, Barry and his staff were to be involved in overseeing its operation, studying and improving its performance in use and also participating in formal reviews undertaken by Select Committees in 1852 and again in 1854.[6] As will be seen below, Barry was confronted with significant practical challenges during this post-occupancy phase. Whilst he aimed to create a hermetically-sealed chamber that had an optimal climate, he never did so in practice. Barry's ventilation system failed to deliver a comfortable environment for the Lords during Parliamentary sittings, and they decided to decommission and substantially remodel it after only seven years.

The anatomy of Barry's system: 1847–54

Barry was now responsible for the design and delivery of the ventilation system for the House of Lords. Despite Reid's departure, he retained a central air supply. The House of Lords remained integrated into the fresh air network of the basement level, which, although no longer connected to the House of Commons, still served 80% of the Palace.

Figure 6.1 is a reconstruction of the original air supply network that was completed under Barry's direction. The basement plan shows that the fresh air was still introduced, as Reid had intended, through large shafts located inside two corner turrets of the Victoria Tower (**1**)[7] and distributed through a network of passages inside the basement. The shafts had a height of 250 ft (~76 m) and a diameter of 6 ft (~1.8 m).[8]

A cowl, acting as a wind catcher directing the wind downwards, was installed at the top of the shaft. A weather vane kept the inlet of the cowl facing the wind and reportedly produced a strong downward current within the shaft during windy weather.[9] Inside the basement level the fresh air first entered a filtration chamber (**2**), where it could be purified using water sprays, steam and multiple layers of wire screens.[10] Driven by a large centrifugal fan (**5**) the air was then conveyed through a long, vaulted passage (**3**) into the 'Central Air Chamber' (**6**). The fan, which was driven by a 20-horsepower steam engine, was placed in a separate chamber in front

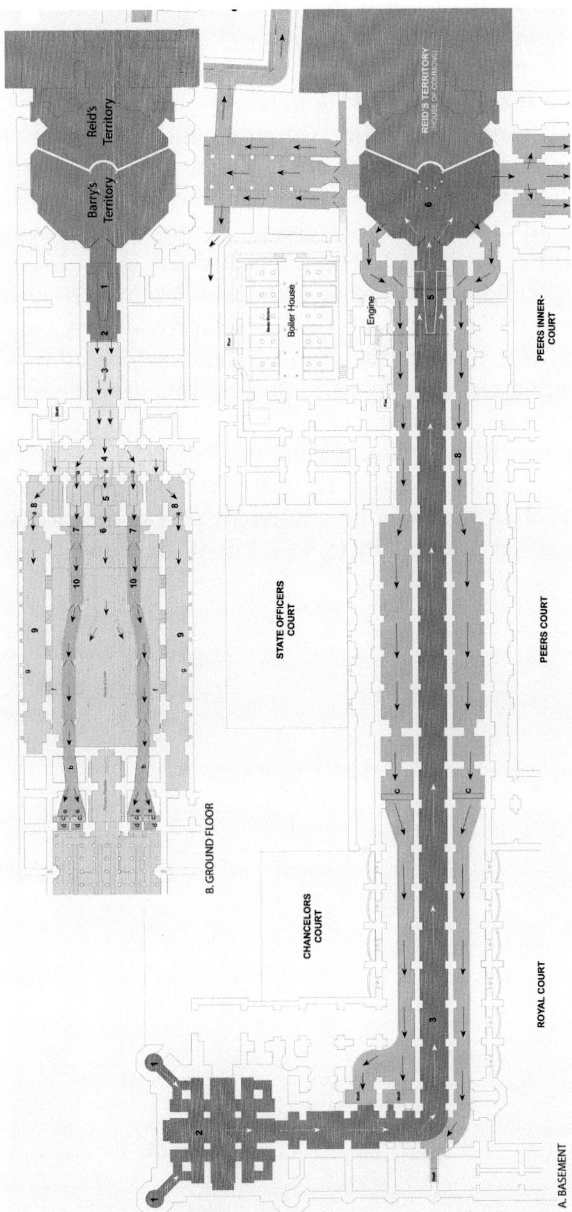

Figure 6.1 Basement and ground floor plans showing the central air supply for the House of Lords (author's own drawing).

of the Central Chamber.[11] The Central Chamber was equipped with air valves, and a heating and humidification plant to pre-condition the supply air and distribute it across four air mains, which led to the different parts of the Palace. The air for the House of Lords itself did not pass through the Central Chamber. Rather, unheated air was supplied directly through a separate main at the south end of the fan chamber,[12] which was equipped with a large valve to switch it to the central supply during sittings.[13] It was operated remotely from the 'ventilator's office' through pulleys.[14] As shown in the ground floor plan in Figure 6.1, the main (**3**) for the House of Lords terminated in a square chamber underneath the Peers' Lobby (**4**), which contained an early form of air-conditioning system, providing heating, cooling and humidification (**5**) for the Lords' debating chamber.[15] The air was heated with 'steam cockles'[16] and high-pressure steam was deployed to raise the relative humidity of the air.[17] Additional cooling was provided by passing cool water from bore holes through the steam cockles, which allowed the supply air for the debating chamber to be cooled, and the iron floor to be transformed into a radiant cooling surface.[18]

As can be seen in the ground floor plan in Figure 6.2, once conditioned the air passed through five circular openings (**a**) that divided the incoming air into separate streams, each of which led to a different area of the House. The openings, which measured 1.8–2.4 m (5.9 -7.8ft) in diameter, were

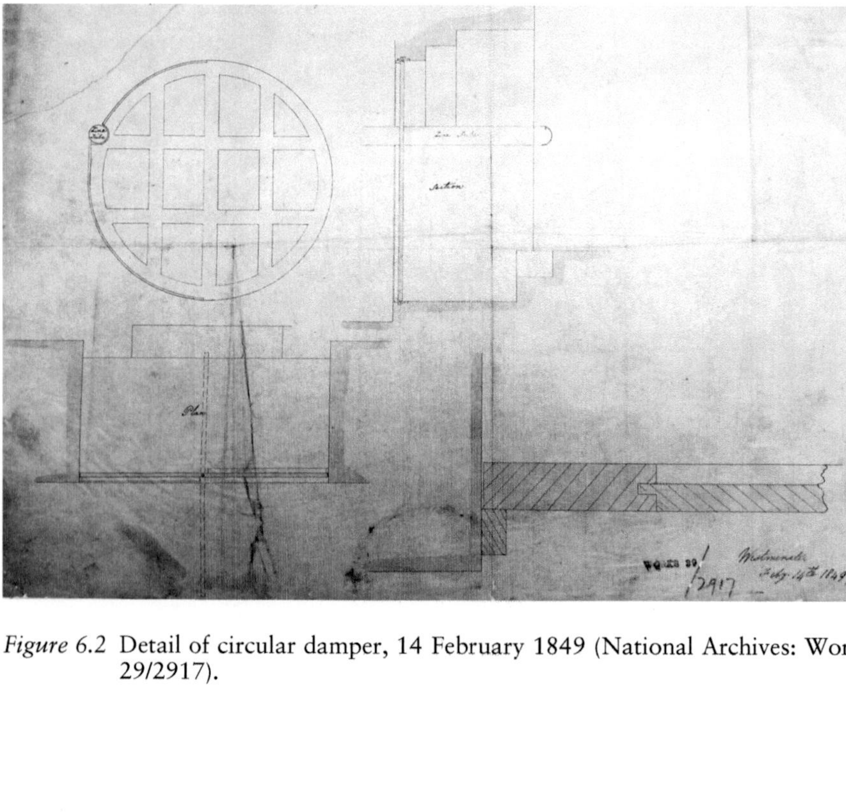

Figure 6.2 Detail of circular damper, 14 February 1849 (National Archives: Work 29/2917).

equipped with rotating timber valves (see Figure 6.2).[19] Attendants operated these valves with the aid of ropes and pulleys (remains of both of which were discovered during the recent site surveys). As shown in Figure 6.1 (ground floor plan), the central valve (**6**) was used to convey warmed or cooled air into the chamber below the main floor. The two smaller apertures on either side of the central valve (**7**) were used to control the air supply to the ceiling, whilst the valves at the east and west end (**8**) led to separate plena under the Division Lobbies (**9**). The air of each stream could also be conditioned separately. The supply air for the ceiling, for instance, was kept at a temperature of 60–62°F (~15–17°C), whilst the air used to warm the floor surface was kept at a higher temperature.[20]

The plenum was in the space above the groin vault of two cloistered passages facing the Peers' and State Officers Courts. The fresh air collected inside the plenum could be conveyed to the Division Lobbies and the corridor at gallery level. It entered the lobbies through perforated cast iron floors, which covered the entire floor area. Each corridor was linked to the plenum through 12 vertical shafts, which were embedded within the external wall. Air entered these shafts through rectangular apertures, which were equipped with individual valves.

'Airs from heaven' vs 'blasts from hell'

Reid's original ventilation arrangements above the ceiling were implemented and incorporated into Barry's ventilation schemes, which was first shown in a series of sketches, produced in 1845,[21] and, in more detail, in Barry's own plans produced after the split from Reid.[22] Indeed, surveys of the existing fabric inside the House of Lords, undertaken by the author between 2016 and 2018, reveal that the arrangements were implemented as outlined in Reid's original drawings.

According to the plans the space above the ceiling was subdivided into three parallel chambers, comprising one 'vitiated air chamber', located above the central row of ceiling panels, flanked by a fresh air chamber on each side of it. The vitiated air chamber was connected to a small up-cast shaft, known as the 'Peers' Shaft', through brick passages above the Peers' Lobby.

The air chamber above the ceiling was divided into three compartments, comprising two compartments for the collection of vitiated air – one situated on the east side of the roof and one on the west side – and one compartment in the centre of the roof for the supply of fresh air. These partitions were equipped with large valves, which Meeson, the civil engineer, referred to as 'inlet doors' and 'exit doors'.[23] According to Meeson, the ceiling outlets also had adjustable valves to control the quantity of air admitted into the chamber below.[24] This was to ensure the air entered the chamber at a uniform velocity across the ceiling.[25]

Figure 6.1 (ground floor plan) shows that the air for the ceiling was blown through a pair of horizontal ducts below the debating chamber (**10**). These ducts, measuring 12 ft × 12 ft in cross section, were composed of iron plates coated with shellac to prevent the air from being tainted with the smell of exposed iron. Its floor was paved with slate.[26] Fragments of this channel can still be found below the floor of the Prince's Chamber. The ducts terminated in six vertical shafts at the south end of the Prince's Chamber, through which the air was carried to two fresh air chambers above the ceiling (see **[d]** in Figures 6.1 and 6.3). The quantity of air entering these chambers was manually adjusted through damper valves at the base of each shaft (see **[c]** and **[e]** in Figures 6.1 and 6.3).[27] These valves could also be used to switch between a warm and cool air supply. For example, when the chamber became too hot due to overcrowding, the warm air supply stream could be closed and cool air introduced directly from the basement, which functioned as a cool air reservoir. According to the *Illustrated London News* the air inside the basement remained between 45°F and 50°F (~5°C and 10°C) during the summer months.[28] The humidity of the air could also be adjusted with the aid of moist canvas screens, which were situated both inside the horizontal ducts below the chamber and inside the fresh air chamber above the ceiling.[29]

Air entered the debating chamber through both the carved, gilded lettering of the ornamental ceiling beams and the gaps between the beams and ceiling panels.[30] Detailed by Augustus Pugin, the carving in the hollow beams allowed the ventilation apertures to be introduced in an inconspicuous way. The apertures were provided with valves that enabled the downward supply to be regulated.[31]

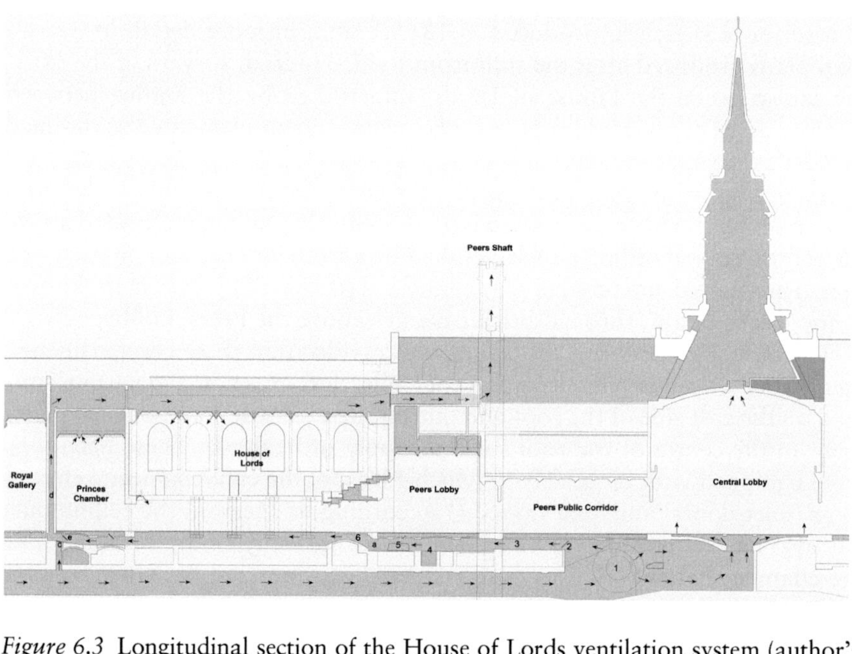

Figure 6.3 Longitudinal section of the House of Lords ventilation system (author's own drawing).

In Barry's scheme the fresh air was supplied and extracted almost entirely through the ceiling, with supplementary outlets provided at mid-level and floor level. Some fresh air was supplied through the coving under the two side galleries, using Pugin's carved quatrefoils as outlets (see Figure 6.5, top-right image). Each gallery was linked to the plenum below the House through four sets of vertical shafts, each equipped with valves to regulate the supply. On rare occasions small quantities of air were admitted through outlets at floor level. These were only intended as a back-up arrangement for 'emergencies'.[32] These outlets were located within the raked floor, within the side panels at the south end of the two blocks of benches[33] and in the centre of the table.[34] These outlets were provided with separate valves, which are shown in a drawing of the floor dated 12 December 1852 (Figure 6.4). The risers of the floor were

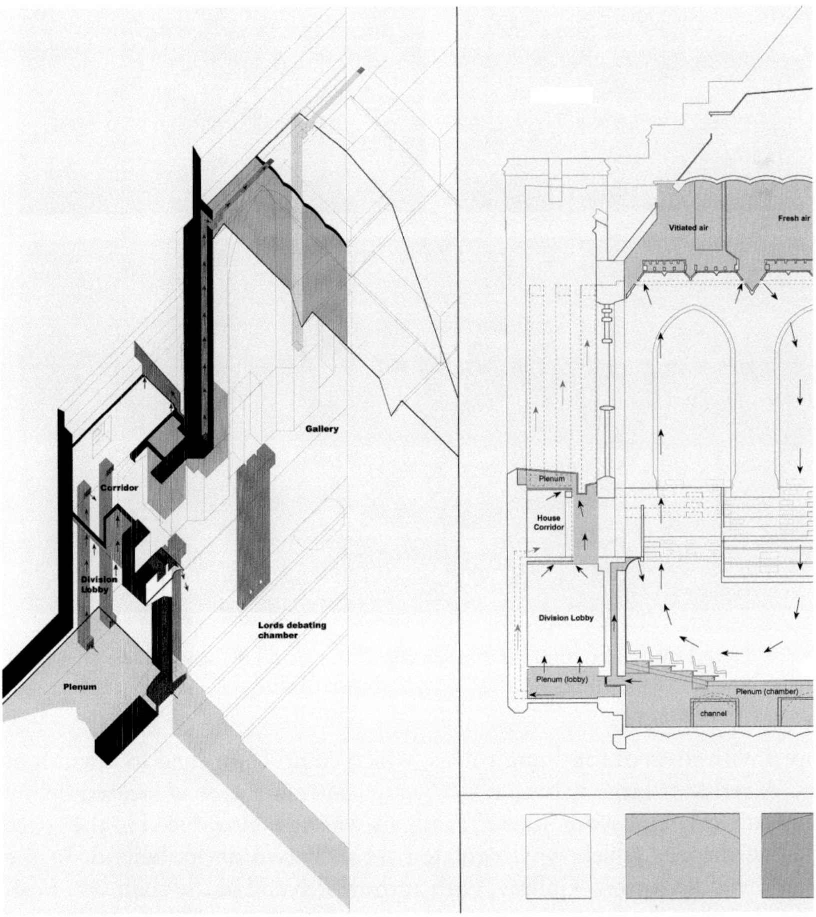

Figure 6.4 Axonometric cutaway (left) and section (right) showing network of air flues and chambers for Division Lobbies (author's own drawings).

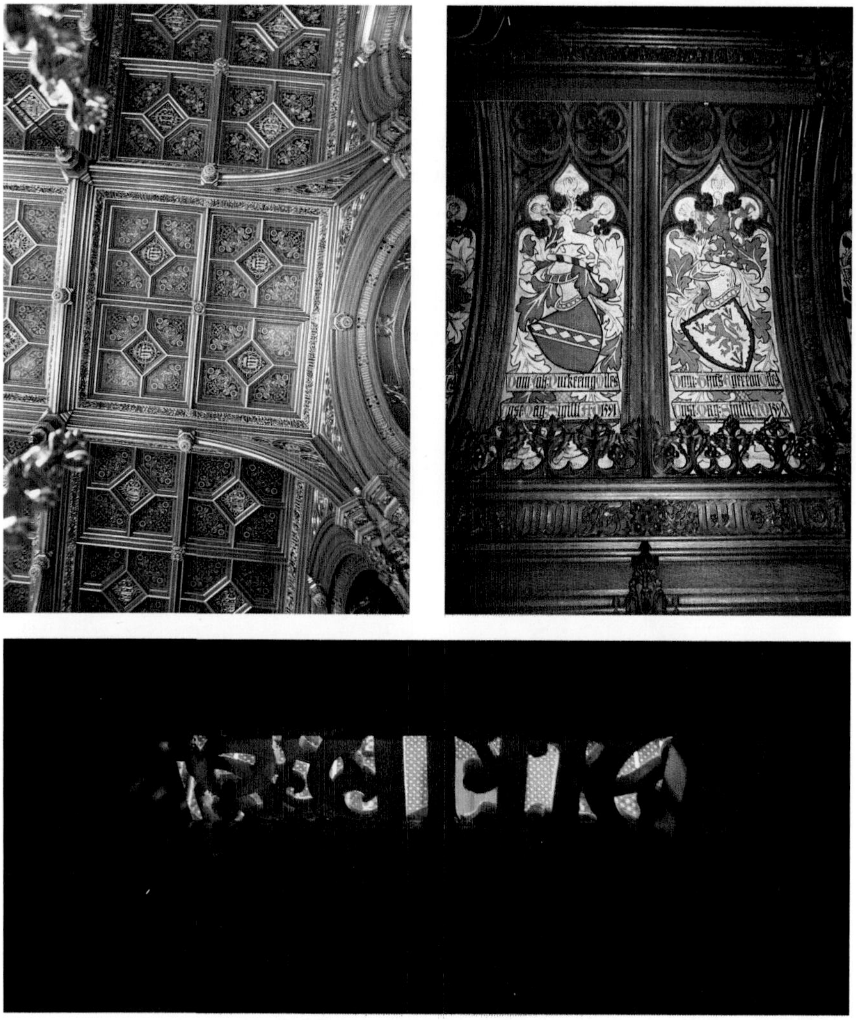

Figure 6.5 Detail of ornamental air outlets in ceiling (top left) and inside coving of galleries (top right) (author's own photograph).

equipped with rows of top-hung valves, which could be opened to a width of 2–3 in. A series of large valves, which were centre-hung, was located below the floor at the back row of seats.[35] Fresh air was also introduced in the space behind the throne, which was regulated through two doors behind. In the Strangers' and Reporters' Gallery, both at the north end of the chamber, fresh air was introduced through grills inside the raked floors. These grills were connected to the central air network through a series of large timbers ducts, each of which was equipped with a wooden shutter for the control of air flow.

This network can be seen in some sketches, dated 14 November 1849,[36] and parts of the original ducts and controls have even survived.[37]

Barry and James Faraday, the gasfitter, argued that introducing the air from above overcame the problems caused by air currents entering close to the body or carrying up dust from the floor.[38] Barry noted that it was 'impossible to avoid the inconvenience of partial currents when air is admitted from the floor, or near the person'.[39] It is evident that Barry's objective was to re-address the fundamental physiological effects that had been the central concern of Reid's earlier inquiries, using a different approach. His scheme took into account the interrelationship between the effects of currents, radiant heat and air temperature on people's perceptions. The temperature of the supply air itself introduced through the ceiling was to be kept relatively low at a temperature of between 61°F and 62°F (~16°C). The Lords, who were seated on benches, were kept warm through the radiant heat of the heated floor, which was kept at 75°F (~24°C) (see Figure 6.6). This reportedly achieved a surface temperature of 66–67°F (~19°C).[40] In a paper presented at the Royal Institution on 26 March 1847, Faraday grandly claimed

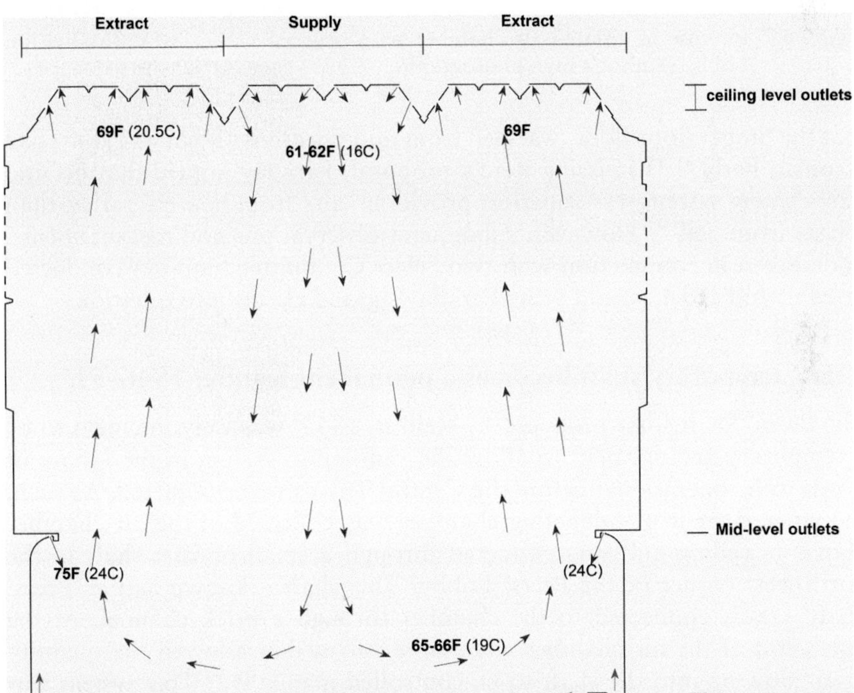

Figure 6.6 Diagram section showing air movement and temperature (author's own drawing).

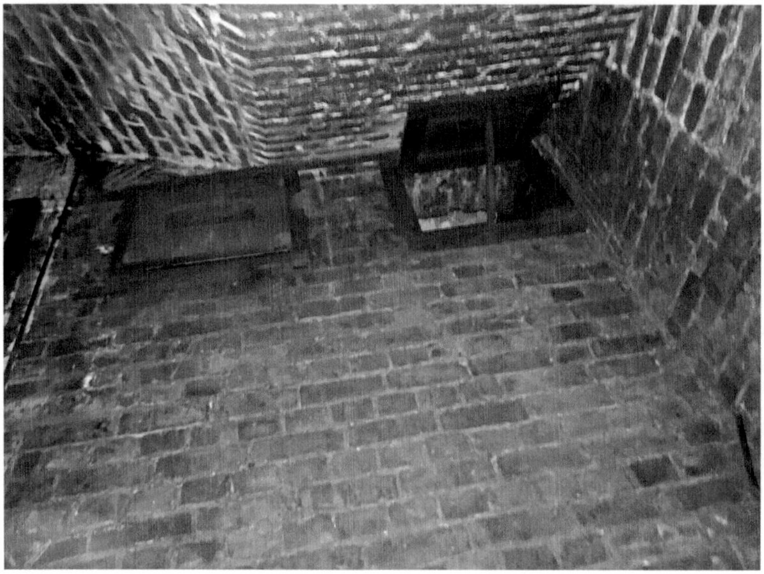

Figure 6.7 Interior of vitiated air chamber with original valves, western Division Lobby (author's own photograph).

that the 'impervious floor, warmed from underneath with hot air, [was] like a roman bath'.[41] Dismissing Reid's proposal, Faraday argued that his and Barry's new system was superior, providing 'airs from heaven' rather than 'blasts from hell'.[42] However, subsequent observations and measurements, undertaken in conjunction with two Select Committee inquiries conducted in 1852 and 1854, would bring Faraday's grand claims into question.

Reid's temporary shaft becomes a permanent feature: 1846–51

The Peers' Shaft, first proposed by Reid in 1845, was only intended to be a temporary arrangement to allow the ventilation system in the House of Lords to be operational before the Central Tower was completed. As such, the vitiated air in the debating chamber was collected in the air chamber above its ceiling and was extracted through a small up-cast shaft in the north-west corner of the Peers' Lobby. This shaft – known as the 'Peers' Shaft' – was connected to the chamber through a brick channel. At the north-end of the air chamber were large valves that allowed the quantity of air flowing into the shaft to be controlled manually.[43] This system was physically separated from the air passage leading to the Central Tower.

The Peers' Shaft could be linked to five separate extraction networks, each of which served a different area (see Figure 6.8). Manually operated valves were used to direct the pull of the shaft to (**1**) the debating chamber,

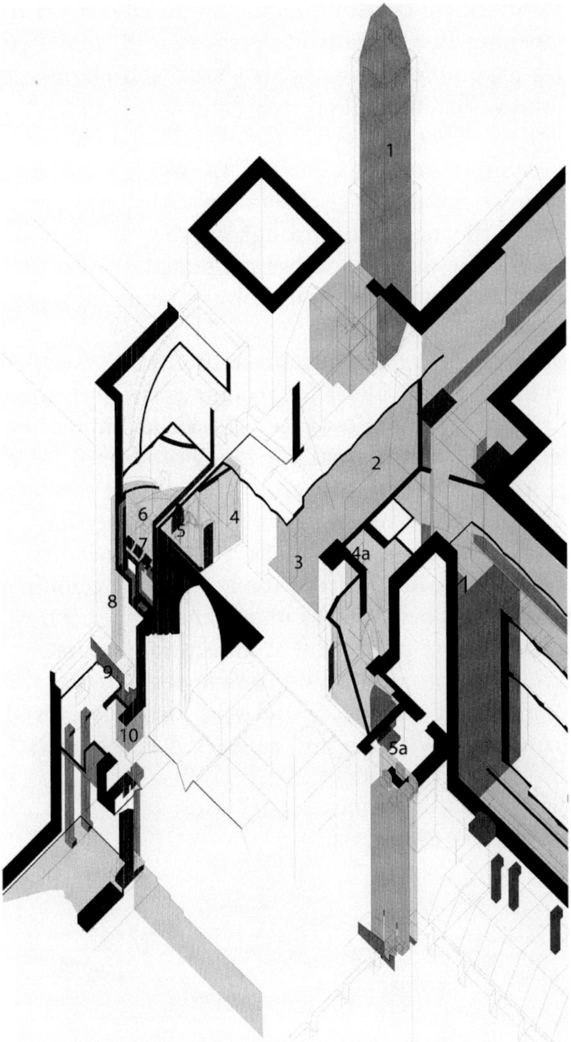

Figure 6.8 Axonometric perspective showing Peers' Shaft and networks (author's own drawing). *Key: 1. Peers' Shaft, 2. Air passage linking shaft to debating chamber and Division Lobbies, 3. Main valve for debating chamber. Western Division Lobby network: 4. Apertures linking air passage to division lobby network, 5. Main valve used to connected/disconnect lobby network, 6. Vitiated air chamber, 7. Valve of vertical flues terminating in vitiated air chamber, 8. Vertical flues connecting air chamber to plenum above corridor, 9. Plenum in roof above Division Lobbies, 10. Vitiated air flues of Division Lobbies. Western Division Lobby network: 4a. Aperture connecting air passage to network of eastern Division Lobby, 5. Main valves used to connected/disconnect network.*

(2) the Prince's Chamber, (3) the Peers' Lobby and to each of the two (4, 5) Division Lobbies. The network of Division Lobbies was connected to the shaft during votes by opening two large iron valves (see [5] and [5a] in Figure 6.8). It constitutes a complex network of shafts and chambers that was concealed within the walls and ceiling (see Figures 6.4 and 6.8). In each lobby the vitiated air was first collected inside a shallow chamber situated above the panelled ceilings, and then carried up through 20 large brick shafts to another plenum chamber above the gallery corridors. These plenum chambers were linked to the main roof through another set of shafts and another large air chamber. The latter chamber contained an array of different valves (with those of the western Division Lobbies being fully preserved) (see Figure 6.8).

The Peers' Shaft was not large enough to sustain natural stack ventilation. Consequently, Michael Faraday, the physicist, recommended introducing a steam jet to artificially boost the air flow rate. The air ascending the Peers' Shaft was to be driven not by the thermal buoyancy of natural stack ventilation, which was to be used by Reid in his earlier scheme, but through the mechanical force of a high-pressure steam jet injected into the shaft.[44] Reid's original Peers' Shaft, completed in 1846, was a plain stone turret, square in plan, and measuring 8 m in height from the level of the ceiling.[45] In November 1848 Barry extended its height to 15 m, enhancing the air flow and transforming the shaft into a more prominent architectural feature.[46]

Several historical illustrations, photographs and drawings show that the exterior of the extended Peers' Shaft was decorated with ornately-carved panels,[47] and had a cast iron hood with sloping sides (see Figure 6.9).[48]

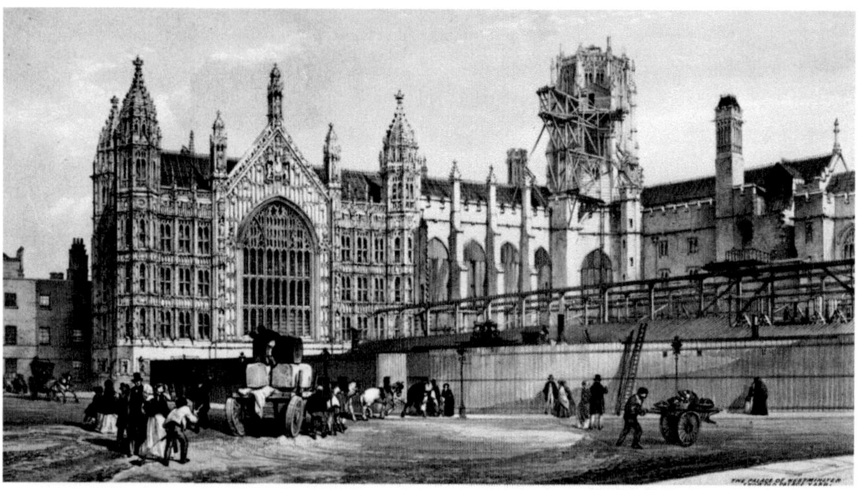

Figure 6.9 Lithograph showing exterior of the Peers' Shaft as seen from the old Palace of Westminster before completion of the west front (undated, Vacher & Son) (Parliamentary Art Collection: WOA 1328).

This hood had a square opening at the top that was surrounded with ornamental cresting. The most dramatic portrait of the Peers' Shaft is provided by a historical lithograph (Figure 6.9) . This lithograph shows the exterior before the western front was completed, whilst the unfinished Central Tower, which was originally intended to replace the Peers' Shaft, is shown in the distance. Unfortunately, the ornamental upper section of the shaft has not survived as it was demolished in 1929 in the context of a stonework repair programme (and following the introduction of the first generation of electrically-powered mechanical ventilation). Only the lower part of the shaft and the original cresting, currently deposited inside the roof space above the Peers' Lobby, are still extant.

The Peers' Shaft did not always provide sufficient ventilation even after it was enlarged in 1848. In June 1854, shortly before the 'new' ventilation system was decommissioned, Barry proposed adding a second turret on the south end of the chamber to relieve the Peers' Shaft and also to assist in the ventilation of the Royal Gallery.[49] Neither this tower nor the plans for connecting the House to the Central Tower were ever realised.

After Reid's and Barry's split in 1846, the idea of using the Central Tower to ventilate the entire Palace was abandoned and what it might be used for became the subject of extensive negotiations between the two men lasting over two years (see Chapter 7). Initially Barry considered the Central Tower to be functionally redundant, and it was only in February 1848 – a year after the House of Lords was opened, during which time the limitations of the small Peers' Shaft had become apparent – that he reconsidered utilising it as the main ventilation shaft for the House of Lords.[50] It was proposed that the height of the Central Tower, which Barry had had reduced to a small spire in 1847, be increased to its full height.[51] Enlarging the Central Tower was to remain an objective up until the mid-1850s although it was never fully implemented. In August 1853 Meeson wrote that the scaffolding was still 're-erected to in such a manner as to admit of the forming a shaft for the discharge of vitiated air'.[52] By 1855, when the ventilation of the House had been rearranged by Goldsworthy Gurney, it still had not been fitted up as a discharge shaft internally.[53] The plan for a taller shaft was finally abandoned after Gurney transformed the even taller shaft inside the Victoria Tower into the principal ventilation tower.[54] Nevertheless, the 'temporary' Peers' Shaft was retained as a second stack for the House of Lords and it remained in active use for 80 years.

As result of these changes, the Central Tower only played a relatively marginal role in the ventilation of the new Palace of Westminster, functioning primarily as an outlet for the vitiated air in the Central Lobby and the area surrounding the Law Lords corridor (which included several offices and the Peers' tearooms and kitchens).[55] The Central Tower was linked to these areas through a separate branch of horizontal and vertical channels. Figure 6.8 shows the branch and the physical walls separating it from the local extraction system of the debating chamber.

Facilities management: the architect remains firmly in control

The House of Lords was completed and occupied in April 1847, but Barry's office was responsible for overseeing its day-to-day operation for the following five years. Barry appointed staff to act as facilities managers, who managed the ventilation, artificial lighting and climate control, on his behalf. Barry and the Office of Works were aware of the fact that a building rarely performs as its designer intends it to and that facilities management is an important factor that can affect a building's performance. In modern building science this phenomenon is referred to as the 'performance gap'. Since the environmental control system of the 19th-century House of Lords was an entirely manual process that relied on the discipline and skill of its attendants, facilities management would have been an even more critical factor in its performance.

During a review of the facilities management arrangements at the Palace of Westminster, undertaken by a Select Committee in 1852, Barry stressed that the effectiveness of the ventilation system was highly reliant on the quality of its management, and warned about the potential negative impact of handing over its operation to new staff.[56] Barry reported that achieving a stable climate was challenging and depended to a 'considerable degree' on the vigilance and attention of his staff. When the Select Committee questioned whether the operational procedures needed to be 'so complicated and so refined', Barry replied that they were necessary for the system to be sufficiently responsive to change, as the climate could be affected by changes in the weather, heat input from the gas lights or fluctuations in attendance. On 30 March 1852 he said:

> *to ventilate rooms under the conditions of the two Houses of Parliament, which are occasionally crowded to excess, and at other times but very thinly occupied, I consider that it is necessary to make special and somewhat complicated arrangements. The same would also apply to the committee rooms, which are more or less crowded, or more or less thinly occupied.*[57]

The continued involvement of Barry's office in the day-to-day operation of the House of Lords ventilation system is significant as it enabled his practice to empirically evaluate its performance in use, undertake modifications and also fine-tune operational procedures before the facilities management responsibilities were formally transferred to the Office of Works in November 1852. This approach has a surprisingly close resemblance to the modern concept of 'extended aftercare' (as outlined, for example, in Stage 7 of the 2020 Royal Institute of British Architects Plan of Work and the Soft Landings Framework).[58]

The Office of Woods had anticipated the need for such extended aftercare from the outset as it was included in Reid's original 1840 employment contract, specifying that he was to supervise its operation for the duration of a whole Parliamentary session after its completion.[59] In January 1851 the Office of Works wrote that the clause was to enable Reid 'to perfect the ventilation and warming arrangements to such an extent as to allow of the future superintendence of those services being entrusted to a person whose duties would entitle him to a much lower scale of renumeration'.[60]

In the case of the House of Lords, this period of aftercare lasted over five years. From April 1847 to November 1852 Barry employed eight members of staff and also appointed Meeson, one his former chief clerks, to act as the superintendent.[61] Some of these staff were even re-employed by the Office of Works after the handover in November, with Meeson continuing to occupy the role of superintendent for another two years until Barry's system was decommissioned and remodelled in the autumn of 1854.[62] Meeson's involvement was also extended to include the supervision of the House of Commons, which up until the autumn of 1852 had been managed independently by a separate team working under Reid's supervision (see Chapter 7). Meeson was responsible for overseeing its operation, maintenance and the process of technical fine-tuning.[63] Day-to-day technical operations were managed by a 'chief engineer' under Meeson's supervision, a 'Mr Scott' under whom worked a team comprising a 'sub-engineer', three firemen, two 'labourers' and a 'boy'.[64] Fulfilling the role of technical operations manager, the chief engineer was responsible for managing the operational staff and as well as monitoring and recording of temperatures.[65] Meeson said, 'Details of ventilating are to his [Scott] charge, subject to reference to myself upon any matters on which he is at a loss or finds any difficulty'.[66] As a result, this Mr Scott had a much more detailed knowledge of its day-to-day performance in use than either Meeson or Barry himself. Meeson noted that he 'can tell you the particular state of the valves of any one room, I could not pretend to do it'.[67]

The chief engineer's team managed the building services for the whole of Barry's territory in the Palace, which included other important spaces, such as the Committee rooms on the River Front.[68] Four attendants, also carefully selected by Barry, were employed by the Department of the Black Rod to manage the system inside the Lords chamber during sittings. During sittings these attendants worked under the supervision of the Usher of the Black Rod, a senior official responsible for the administration of the House of Lords.[69]

Monitoring and control procedures

The ventilation system was closely monitored using an array of scientific instruments, which provided Scott's staff with data on the air velocity, temperature and humidity. The measurements were logged manually in paper registers, which were kept in different locations.[70] Thermometers and hygrometers were deployed to monitor the temperature and humidity of the air in key positions along its path, from the basement to the air chambers above the ceiling.[71] An anemometer was also installed near the fan to measure the quantity of air entering through the fresh air channel. The technical staff followed an operational regime, which set out the range of air flow rates, temperature and humidity to be maintained within the chamber.

Air supply:

- Adjusted according to attendance levels and indoor temperature. The supply is typically 4,000–6,000 ft^3 per minute, but can be raised it to 2,0000 ft^3 per minute.
- Main supply downward through outlets in ceiling, some through coving below the galleries. Auxiliary outlets at floor level were provided as back-up.

Temperature:

- Lord Chancellor has instructed the indoor air temperature to be kept between 65°F and 66°F.
- Supply air only warmed to between 60°F and 62°F to prevent Peers from breathing hot air. Instead, Peers seated on the main floor were provided with radiant warmth through the heated floor.
- Maximum temperature to which the supply air was raised was between 68°F and 70°F.
- If Peers make requests for change in temperature, adjustments were made slowly, typically over a period of 10–15 minutes, to avoid discomfort of a sudden change in temperature.

Humidity:

- The order was to keep the temperature difference between the dry- and wet-bulb thermometers (hygrometers), which was used to determine the humidity of the air, below 6°F.

Control regime of Barry's system: 1847–54

The attendants were instructed to keep the fresh air at a relatively low temperature of 60–62°F (~16–17°C). To prevent it from becoming too dry, they were to keep the temperature difference between the dry- and wet-bulb thermometers of the hygrometers (which were used to determine the humidity of the air) under 6°F. The psychometric chart shows that this equated to a relative humidity between 68% and 69%. Within the interior of the debating chamber itself, the temperature was closely monitored using an array of fixed thermometers in different locations and was to be kept within a narrow range of 65–66°F (~18°C). Thermometers were located on the main floor near the table and throne as well as inside the galleries. The air supply was adjusted based on the indoor temperature and the number of Lords present, which could be subject to large sudden fluctuations.[72]

As in the temporary Houses of Parliament, the operational process also incorporated formal procedures for the collection of user-feedback from the Lords and other users. During sittings, the Usher of the Black Rod and his deputy, the Yeoman Usher of the Black Rod, acted as the formal liaisons between the Lords and the technical staff and responded to complaints or requests from individual Lords who were dissatisfied with the internal climate.[73] Sir Augustus Clifford, who served as the Black Rod from 1832 until 1877, testified that he and his deputy, James Pulman, reviewed complaints made to them by individual Lords, and if necessary, gave instructions to the attendants to make adjustments to the ventilation rate or climate.[74] The Usher's department employed four members of staff, who had been selected by Barry but were under his own management.[75] The Lord Chancellor was also authorised to give instructions for ad hoc adjustments to be made.[76] When instructions were received, the required adjustments were made slowly so as to protect the Lords from a sudden change in temperature. According to Meeson, making adjustments to the system took between 10 and 15 minutes.[77]

Post-occupancy history: 1847–54

In addition to overseeing the routine operation and maintenance of the ventilation system, Meeson also coordinated observational studies of its performance in use, collaborating with Barry and the Office of Woods to conduct a series of technical alterations and trials that focused mainly on improving the air quality and thermal comfort (which had been the focus of the Lords' complaints). Between 1848 and 1852 the ventilation system was gradually improved as Meeson and the operational staff came to understand how the House of Lords performed in real-life conditions. The post-occupancy history of the House of Lords provides a critical insight into the difficulties with realising Barry's objectives in practice. As these difficulties were never successfully resolved the ventilation system was substantially remodelled after only seven years of operation.

A first series of technical trials– convened by Barry Jr and Meeson in collaboration with the scientist Michael Faraday – were undertaken inside the debating chamber prior to the formal inauguration of the new House of Lords in April 1847. The aims of these tests were to assess the performance of the steam jet, which enhanced the rate of air flow in the Peers' Shaft and to carefully study the behaviour of the internal air currents inside the debating chamber.[78] According to Barry, these tests were necessary to prove that handling the ventilation through the ceiling of the debating chamber was technically feasible. Barry noted that they were 'to prove whether they [the incoming and outgoing currents] did or not act simultaneously'.[79] In one of these tests, which was attended by several Lords and journalists, the air movement was traced using an olfactory method in which 'pastilles' – pellets of paste – were burned 'in the chamber below'.[80] As Barry recounts it, observers were occupying:

> *certain positions in the House to discover how far the odour of the pastilles affected the air of the House in certain places, and at certain times; it was invariably found that the first place where the odour of the pastille was experienced, was upon the floor, when it was not felt in the gallery, and eventually in the discharge shaft.*[81]

Reporting on the tests, the *Illustrated London News* wrote that 'the air, laden with the fragrant odours, as a tell-tale, had passed through the flues, and onwards, into the House, and had afterwards passed through the vitiated air chamber exactly in the direction intended'.[82]

Technical adaptations: 1847–51

The ventilation system for the House of Lords went operational in April 1847. Over the next four years, it was gradually refined through modifications to the supply and discharge arrangements. Observations at the time revealed problems with both the air quality on the main floor and sustaining an effective downward current from the ceiling.[83] Up until 1851, the supply air was driven by an air pump. This original system, illustrated in drawings dated 21 October 1849, was composed of three pumps, one that delivered air into the House of Lords and two that moved fresh air into the Central Chamber.[84] Unfortunately, these pumps proved to be insufficient to inject the required amount of fresh air through the narrow air flues that lead to the ceiling.

In a report from 13 April 1852 Barry writes that 'mechanical and other motive power' was needed 'owing to the want of capacity in the flues by which air is transmitted from one place to the other'.[85] The pumps were replaced with a large centrifugal fan,[86] measuring 4 m (13 ft) in diameter (see Figure 6.10).[87] It was installed over the winter recess of Christmas 1851 and went operational in February 1852. The installation of the fan reportedly improved the ventilation of the House of Lords. Barry claimed that it afforded 'an entire and instantaneous command over the movement of the atmosphere' and temperature.[88]

To prevent vitiated air from accumulating on the main floor, Meeson introduced a second extraction system at floor level (see Figure 6.11).[89] Drawings, dated 12 December 1851, show that he introduced apertures inside the first risers of the raked floor.[90] These apertures were connected with the bottom of the Peers' Shaft through a separate set of ducts located underneath the floor.

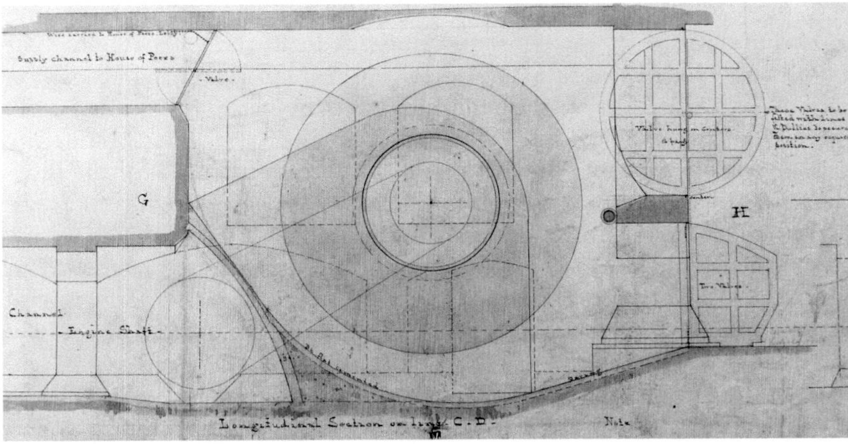

Figure 6.10 Cross-section of centrifugal fan, August 1851 (National Archives: Work 29/2952).

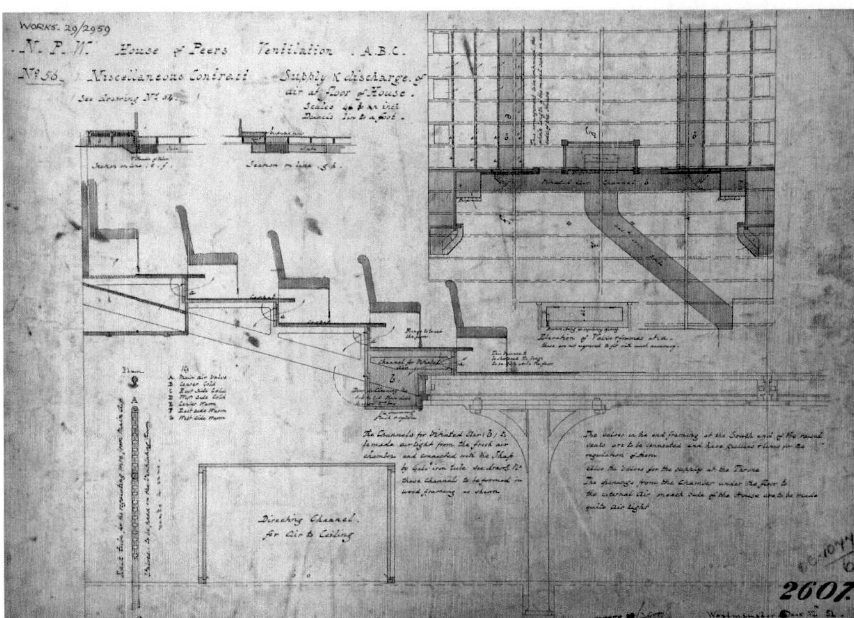

Figure 6.11 Cross-section and floor plan showing floor level outlets, 12 December 1851 (National Archives: Work 29/2959).

Barry's system had clearly deviated from one of the key objectives behind Reid's original scheme: to minimise the use of fossil fuel and mechanically-driven solutions. Barry exploited the energy-intensive technologies that are intimately linked with the first industrial revolution, relying heavily on steam power created through the burning of fossil fuels (i.e. coke and coal) to supply and exhaust the air. Natural principles of ventilation could only be used in his system to a limited extent. The steam required for the heating and ventilation of Barry's part of the Palace was produced from a central boiler and engine house suited inside St Stephen's Court.[91] The engine house was a highly-engineered glass and iron structure, which stood in contrast to the Gothic style of the courtyard elevations surrounding it, and was only one of the few areas in which the mechanical aspects of the new Palace were clearly exposed to public view. It contained two high-pressure steam boilers, which powered both the steam jets inside the shaft and the steam engine driving the centrifugal fan.[92] The engine of this fan was typically run at 12 horsepower although it could provide up to 20 horsepower if needed.[93]

The alterations to Barry's ventilation system for the House of Lords illustrate how Meeson and the Office of Woods adapted the system based on observations about its performance in use. However, the performance of the system was called into question within two years of these alterations being made and it became subject to two formal evaluations by Select Committees, first in 1852 and then again in 1854. The primary focus of these inquiries was the quality of the internal environment and how the Lords themselves experienced it.

The Lords' perception of the internal environment

The first formal investigation into the performance of the House of Lords ventilation system was undertaken between March and May 1852, less than two months after the alterations detailed above had been made, in conjunction with another inquiry, which was evaluating the performance of Reid's own ventilation system in the House of Commons. Unusually, this investigation was led by a House of Commons Select Committee, who was to compare the House of Lords system to Reid's system in the House of Commons, which had encountered serious problems itself. The Committee conducted long and detailed interviews with both Meeson and Barry, examining the technical arrangements and operational procedures in place. Interviews with several Lords were also conducted in order to gain insight into their personal experience of the internal environment. The alterations made to Barry's system had reportedly improved the air quality, whilst smoke tests, undertaken by Meeson to visually trace the air movement, showed that the centrifugal fan enabled more fresh air to reach the floor from the ceiling than previously.[94] Despite this, the Lords themselves were dissatisfied with the quality of the thermal environment as the interior was subject to sudden changes in temperature (caused by the injection of hot or cold air into the system). The Lords were also exposed to internal currents.

During the sitting on the 29 March 1852, one witness for the inquiry, Sir Augustus Clifford, Usher of the Black Rod – who, having managed the Lords' feedback about the internal climate, was in a unique position with respect to how the Lords perceived it[95] – reported that the climate and atmosphere was generally in a good state and that complaints from the Lords were rare. He said that the 'ventilation is very equal and very agreeable' and that 'the general temperature varies very little ... usually 63 degrees to 64'.[96] Sir Clifford noted that the Lords' complaints were primarily about the currents of cold air that rushed into the house every time the doors inside the galleries were opened. The temperature of the House and the air quality were rarely the subjects of complaints. Sir Clifford's views were also shared by a reporter, who stressed that the atmosphere in the gallery was very comfortable except for the draughts.[97]

Despite this, the testimony of other witnesses was more critical in influencing the course of the inquiry. Lord De Ros reported that many Lords perceived the atmosphere as dead and artificial, which he attributed to the uniformity of temperature and the absence any 'sensible' air movement.[98] In contrast to this, Earl Lonsdale perceived the thermal environment as highly changeable noting that it was 'sometimes very hot, other times very cold, with cold draughts'.[99] During the sitting itself on the 29 March, he remarked that the air temperature was 66°F at 5pm but later rose to 70°F, leading to complaints from several Lords. The clerk at the table was sent to ask the engineers to make adjustments and the temperature was reduced back to 66°F. Lonsdale felt it 'too hot and oppressive', observing that there was still a 'deadness and fustiness in the air'.[100] Lord Redesdale, who had sat in the House for five years, also agreed that the quality of the ventilation 'varied a good deal', reporting an 'unpleasant change in temperature' and that the atmosphere felt 'too artificial'. Lord Redesdale also noted that the ventilation system struggled when the House was crowded, in particular during State Openings. He remarked that the chamber had been in a 'very heated state, and in a very offensive state' at the end of the State Opening ceremony on 3 February 1852. Earl Lonsdale, Lord Lansdowne and Charles Grey, 2nd Earl Grey ('The Earl Grey'), all mentioned occasional problems with cold draughts around the legs. The Earl Grey reported that the chamber 'was at one time very hot, and then came a current of very cold air to one's feet'.[101] He stressed that the interior 'gets very oppressive indeed', particularly during summer afternoon sittings when intense sunlight entered the chamber through the large windows, and that the temporary House of Common had felt significantly cooler.[102]

Whilst the experience of the witnesses to the inquiry varied, their accounts provide clear evidence that there were genuine issues of thermal discomfort with the chamber. They were the result of problems with the control of the internal temperature and air movement. During the inquiry Barry defended the system, arguing that it was too early to assess the full impact of the recent changes, and expressed his confidence that it would succeed in preventing the atmosphere from becoming 'oppressive'.[103] After two years had passed without the promised improvements materialising, Ulick de Burgh, 1st Marquess of Clanricarde, raised the issue in the House, criticising the fact that the defects revealed by the 1852 inquiry had not yet been addressed by the Office of Woods.[104]

System diagnosis

On 13 March 1854, the 1st Marquess of Clanricarde tabled a successful motion in the House of Lords to appoint an ad hoc Select Committee to be charged with conducting a formal inquiry 'into the possibility of improving the ventilation and the lighting of the House'.[105] This new inquiry again took place in parallel to another Select Committee, which was appointed in 1854 to examine problems with the ventilation and lighting in the House of Commons. This was significant as the results of this parallel investigation also influenced the outcome of the inquiry into the House of Lords.

The House of Lords Select Committee comprised 15 Lords and was chaired by the 1st Marquess of Clanricarde. Collaborating with the Office of the Works, Meeson and Barry, the Committee examined the ventilation system and conducted technical trials of alternative lighting arrangements. The Committee also engaged four external advisors to provide independent appraisals of its performance: two engineers, John Leslie and Nathan Defries; and two physicians experienced in the field of ventilation and artificial lighting, Neil Arnott and Goldsworthy Gurney. These advisers appraised the system from a technical and physiological perspective. The fact that the inquiry was led by Lords is significant as it allowed their experience as the primary users of the chamber to inform their questions.

The 1854 House of Lords Select Committee also questioned Meeson and Barry whether it was feasible to handle the ventilation of the House entirely through the ceiling. To prove its feasibility, Meeson conducted another smoke test on 26 March 1854 to visualise the air flow inside the chamber. According to Meeson, who described the test in a technical report and oral statement given to the Select Committee on 4 May 1854, the air in the debating chamber behaved as intended. During the experiment, gunpowder was fired inside the main air supply channel at basement level; smoke then passed through the fan and to the air chambers below and above the House; the stream of fresh air could be seen to form a column of cool air in the centre of the House, whilst the warm air could be seen to ascend in two columns on opposite sides of the chamber towards the exit holes in the side panels.[106] The smoke reportedly reached the ceiling inlets in ten minutes, from which it gradually ascended towards the floor. Fifteen minutes later the smoke was 'thick from the table to the back benches'[107] and the atmosphere of the whole House had been filled with smoke.[108] Within another 15 minutes, however, the atmosphere was clear 'except a slight haze at the sides of the ceiling'.[109] This suggests that although the ventilation was apparently sufficient to renew the atmosphere of the House, it could not provide sufficient thermal comfort for its users as the ventilation rate was too weak to mitigate occasional issues of overheating.

The Lords were also still reporting frequent problems with internal draughts. According to the 1st Marquess of Clanricarde, they occasionally complained 'of the strong draughts and currents of air that prevail in the House of Lords'.[110] During a sitting on 23 March 1854, he himself had 'complained to Meeson of the current of air that played about … some apertures were closed, and we felt a very perceptible change in a very short time'.[111] This latter abrupt change was largely caused by the intensity of the fresh air, which was injected through the ceiling. When the stream of fresh air reached the floor, it moved sideways, sweeping across the benches and causing a chill around the Lords' feet and legs. Lord Stanley also echoed the view that there was 'a great draught to the legs'.[112] Gurney, who had undertaken field observations inside the chamber, confirmed[113] in his first report of 10 April 1854 that the 'momentum of the incoming air produces secondary and retrograde currents in the body of the house, which was often traced extending to the floor'.[114]

These problems indicate that the use of the ceiling to provide fresh air to the House failed in its original design objectives, which, according to Meeson, were to prevent 'as much as possible' problems with 'local draughts and local eddies of air being brought against the person sittings in the House'.[115] In reality, however, the Lords inhabited a highly unsteady atmosphere. Meeson testifies that the control of the ceiling currents had been a longstanding issue and that he had made several alterations to rectify it. One of the problems was the fact that the velocity of the descending currents was uneven: those entering through ceiling apertures at the bar end of the chamber were much stronger than those entering near the throne end (which was closest to the supply shafts). Meeson made several modifications to the ceiling, introducing valves to individually control 'the descent of the air' in different sections of the ceiling and also installing a series of 'gauge curtains' above the ceiling to slow the currents.[116]

The ceiling supply of air, however, was not the only cause of problems. Draughts were also caused by the introduction of fresh air through other outlets, such as the openings in the floor of the Strangers' Gallery at the north end of the chamber, and those behind the throne at the opposite end. This produced horizontal currents that swept across the floor and affected the central area occupied by the table and the Woolsack. Lord Ashburton also reported of a 'very strong draught from the throne towards the bar'.[117] Speaking about these same currents, Earl Graham said, 'I have felt it over and over again, I have held up my handkerchief and even that was slightly affected'.[118]

Body heat and gas light

The level of discomfort felt by the Lords due to the chill of the draughts was further accentuated by the heat of the atmosphere. The ventilation rate was not sufficient to counteract the heat produced by a large number of people and multiple gas lights, even after the more powerful centrifugal fan and enlarged shaft were introduced.[119] In March 1854, the Earl of Malmesbury reported that a common complaint was about an 'unwholesome feeling about the air we breathe, and that it very often creates drowsiness and heaviness'.[120] Earl Graham reported that it was most intense inside the gallery, saying that 'any night, you will find that the heat and smell in the gallery is much stronger than in the other parts'.[121]

In an oral statement given to the House of Lords Select Committee on 17 March 1854, Meeson confirmed that the heat in the chamber got 'very oppressive' when there was a sudden influx of people into the chamber before an important sitting.[122] However, he also testified that during sittings at night, the system struggled to mitigate the impact of the gas lights, which emitted large quantities of heat. The gas lights used in the House of Lords had been designed by Barry's team in collaboration with the gas engineer James Faraday (who had in 1843 invented the 'self-ventilating gas burner'): glass cylinders contained and concealed gas flames and the combustion fumes were extracted through a separate network of flues. In the House of Lords, these gas,light lamps were attached to the piers above the galleries (see Figure 6.12); the fumes from each pair of lights were carried to the roof space through vertical flues inside the external buttresses (see **Figure 6.4**).[123] These flues were connected to the Peers' Shaft through a horizontal duct above the ceiling.[124] Technical details of the gas lights were published in the *Illustrated London News* in May 1847.[125] Although this arrangement was successful in protecting the atmosphere from noxious fumes, the shades and brass brackets of the lamps were found to emit large quantities of heat. According to Meeson, the ornamental brass of the arms and burners got very hot, whilst Barry stressed that there was a 'very large quantity of tubing' that was 'radiating heat' into the space.[126] This not only raised the air temperature, it also exposed the Lords who were sitting inside the galleries to a strong radiant heat.[127] Indeed, Meeson reported that there was a strong 'radiation on the forehead and person from the heat of the lamps'.[128] Attendants generally maintained an atmospheric temperature of between 65°F and 66°F. However, the atmospheric temperature rose by a further 7°F within 60 minutes of the gas lights being switched on, something which could not be prevented.[129] It should be noted that the number of gas lights in the chamber was high as it was illuminated using no fewer than 32 large lamps. Historical illustrations show that ten pairs of lights were attached to the piers between the windows on the east and west side and six lights at either end of the chamber.[130] Measurements taken by Meeson when the chamber was unoccupied show that the gas lights alone were capable of raising the air temperature from 62°F to 68°F within three hours.

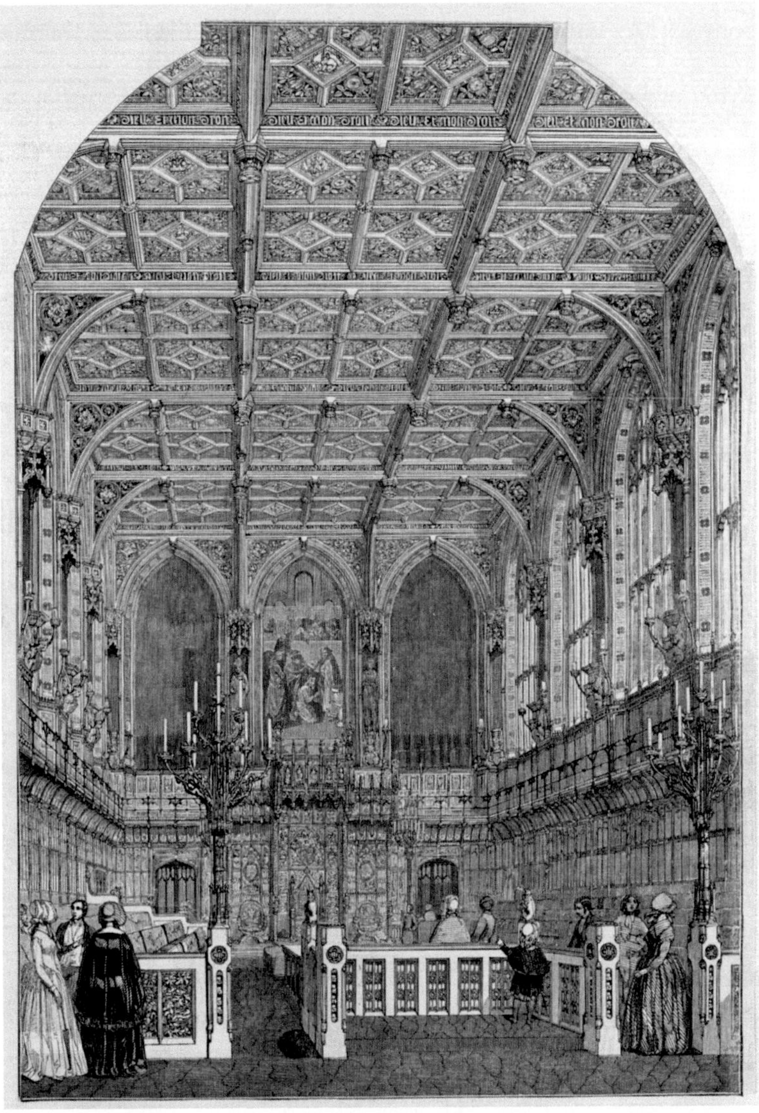

Figure 6.12 Historical interior of the House of Lords showing original bracket lights, April 1847 (*Illustrated London News*, 17 April 1847, p. 249).

As it was impossible to mitigate the heat, the doors inside the galleries were frequently opened in order to exhaust the heat naturally through windows inside the corridors. However, this often led to new complaints about draughts as 'cold air rushes down on the Peers heads'.[131] Unfortunately, there was no means of natural ventilation as all the windows inside the chamber were completely sealed.

Inquiries into treatments

The testimony of Meeson and the Lords show how difficult it was for the former to

realise a climate-controlled space in practice. The Lords on the 1854 Select Committee had, throughout the inquiries, discussed their experience of the House of Lords debating chamber with Barry and Meeson. They also examined the causes of their own discomfort and explored measures to address them. Some of the suggested measures were for relatively simple interventions, such as the introduction of operable windows for natural ventilation (proposed by Earl Graham on 17 March 1854).[132] However, the main focus of their suggestions was on reducing the heat load of the lamps, for example, by using more efficient burners or alternative fuels.

In order to lower the demand for artificial lighting, Barry also proposed replacing the stained glass, which was relatively dark, with a much 'lighter coloured glass'. Although the windows were large, much of the daylight was obstructed by the dark colour of the stained glass, and at times gas lighting was required during both the day and the night. Barry installed samples of the new glass in one window to demonstrate its impact, and claimed that it could increase the amount of daylight in the chamber by between 30% and 50%.[133]

In May 1854, another more radical intervention – decommissioning the existing lighting and installing new lighting – was considered. Admitting that it might not be possible to 'reduce the temperature of the side lamps to the degree of coolness which would be desirable',[134] Meeson conducted a trial with a mock-up of an alternative lighting arrangement: pendant lights were attached to the ten ornamental ceiling bosses; each pendant was composed of several rings of open burners; the fumes produced by the gas burners were extracted upwards through iron tubes, which were linked to the ventilation shafts through a horizontal flue above the ceiling. The Lords approved of the pendant lights. The existing lighting arrangements were thus removed and the pendant lights were permanently adopted in the House of Lords chamber.

The interventions to reduce the heat load of the artificial lighting in the chamber are significant as the lighting – with its bold and ornamental brass fittings – had been designed as an integral part of the interior architecture. The installation of the pendant lighting – shown in a 1857 watercolour painting of the interior by Joseph Nash – resulted in a significant change in the architecture of the chamber. Despite the fact that the pendant lighting was more utilitarian in character than the original gas lighting scheme, Barry was surprisingly supportive of the new lighting arrangements, arguing that it improved both thermal and visual comfort and enhanced the appearance of the interior details. Indeed, 'the architectural character of the House is more strikingly developed,' and the light was more 'equally diffused', but Barry also believe that it would be more comfortable as it was 'more out of the eyes of Members'.[135]

The Lords were beginning to lose faith that the existing ventilation and lighting problems with the House could be resolved by simply adapting or refining the existing arrangements. The 1854 House of Lords Select Committee thus considered an even more radical option: remodelling the existing system entirely. From March to July 1854, Gurney had tested an alternative ventilation scheme in the House of Commons for a parallel Select Committee inquiry. The House of Lords Select Committee evaluated Gurney's scheme, interviewed MPs about their experiences of it and commissioned Gurney to produce a proposal for the House of Lords chamber.

Gurney summarised his scheme in two reports, published on 10 April and 17 June 1854. In Gurney's system, as with Barry's, the air was to be introduced from above. To reduce the intensity of the internal currents, Gurney proposed to utilise the entire ceiling for the supply and to extract the hot air through the floor. He argued that with the existing arrangement, it was impossible to achieve thermal comfort due to the high velocities that were necessary to sustain the upward and downward flow.[136] It was therefore necessary to 'break down the momentum' of the current and enable air to 'pass quietly into the house'.[137]

Gurney also recommended returning to a pure stack-driven system rather than a combined fan and stack system. Similar to Reid's ventilation system in the temporary House of Commons, this system did not (unlike the current House of Lords system) rely on a separate system of mechanical fans to produce the air supply. The existing supply strategy was inefficient, he claimed, as the air was mechanically forced through an expansive and convoluted network of ducts. Gurney proposed to disconnect the central air supply from the debating chamber and replace the former with a series of local intakes immediately above or below the latter. He therefore suggested reinstating the perforated cast iron floor, connecting it to a new up-cast shaft inside the Elizabeth Tower, using the existing fresh air channels inside the basement as a conduit. The ceiling inlets were also to be connected to new intakes inside the roof directly above the House. Gurney believed that if the fresh air could more easily enter, then the convection inside the Elizabeth Tower would be sufficient to draw fresh air in. He reasoned that the openings for these new intakes would have to be large and in close proximity to the chamber in order to maintain 'self-acting and perfect pneumatic balance between the open air and the House, under every possible condition'.[138]

To allow for some natural ventilation inside the chamber Gurney proposed introducing four operable windows on each side of it. Up to this point, the stained glass of the windows had been fixed as the chamber had been designed as a hermetically-sealed space. The new windows were only intended to be opened temporarily for 'refreshening' the atmosphere before and after sittings and at 'other times when desirable'.[139] To protect the atmosphere from the heat produced by the artificial lights, he proposed to move the lamps into the roof space and to conceal them behind a layer of glazed ceiling, which would replace the original wood panels.

In contrast to his House of Commons proposal, Gurney's proposal for the House of Lords was not tested through trials. Despite the differences in the technology installed in the two chambers, the House of Lords Select Committee deemed the trials conducted in the other House sufficient to demonstrate its merits. The Lords were confident in Gurney's technology and that it would be able to overcome the problems associated with Barry's original ventilation system. In its final report, dated 7 July 1854, the Select Committee recommended adopting Gurney's plans except from the lighting. It wrote that 'it appears, from the evidence of Member of the House of Commons who usually sit in different parts of it, as well as from general report, that the ventilation of the House has been much improved'. [140]

This was a radical step as it departed from Reid's original concept of a centralised air supply and a permanently hermetically-sealed environment. The remodelling of the House of Lords was completed in 1855 and it was to be the last major alteration of the House for the next 60 years.

These interventions marked the end of an experimental phase in the post-occupancy history of the House of Lords. Gurney's stack system, examined in detail in Chapter 9, continued to be used until the introduction of mechanical fans in 1911 The House of Lords replaced its 19th-century ventilation system with modern mechanical air conditioning in 1966, a full 16 years after this had happened in the House of Commons.

Conclusion: the end of an evolutionary path

Joseph Nash's first painting of the House of Lords from 1851 portrays the splendour of the Gothic-revival interior. It shows that its technical features – such as the gas lighting, and the ornamental outlets for the mechanical ventilation – were an integral part of the architecture. However, this intimate relationship between technology and interior architecture was only short-lived due to the significant technical challenges raised by the problems that the Lords themselves experienced with the chamber.

Between 1847 and 1854 the interior was assessed on the basis of both its formal architectural qualities and its environmental qualities. This latter is a dimension of architecture that can only be understood from a phenomenal perspective, taking into people's experience. The personal accounts of the Lords provided crucial insights into the performance of the ventilation system with respect to, for example, the effects of the atmospheric temperature, the internal air currents and the radiant heat from surfaces and gas lights.

Despite the fact that the inquiry had called on expert opinion and carefully examined the results of numerous studies, the ultimate decision about the future of Barry's ventilation system for the House of the Lords was made by the Lords themselves, largely on the basis of their own experiences. Acting as both commissioners and users of the chamber, the Lords assessed its performance based mainly on the perceived quality of the thermal environment. The chamber was too variable: either too hot, too cold, too changeable or too draughty. These were not marginal issues in determining the future of Barry's system and whilst these experiences were relatively ephemeral and transient, the failure to address them had a lasting and decisive impact on the design of the debating chamber. The changes made to rectify the deficiencies in the chamber can be clearly seen in Joseph Nash's second painting of 1857.

The development of the House of Lords chamber was also significantly influenced by political decisions, first in 1846 when Reid's involvement in the Lords chamber was ended, and again in 1854 when Barry's system was decommissioned after only seven years in operation. In both cases, the Select Committee was used as an instrument for coordinating the process of asking questions about the technical aspects of the ventilation system. The 1854 inquiry was significant as it marked the end of a phase in which the responsibility for the improvement and evaluation of the ventilation system was mainly that of the architect, Barry, and the engineers, led by Meeson.

For the first seven years of its operation, the system was subject to a process of continual refinement, led by Meeson, Barry, the operational staff and the Office of Works, and involved the gradual adaption of technologies and existing operational procedures. However, the 1854 Select Committee inquiry resulted in a loss of confidence in Barry's system and its eventual decommission. For better or worse, Barry's ventilation system was no more than a short-lived experiment in the technological history of the House of Lords. The fate of the ventilation system in the House of Lords was not, however, unique. As will be seen in the next two chapters, Reid's system in the House of Commons was subject to a similar, yet even more, intense process of political scrutiny, which would also lead to its decommission and replacement in 1854.

Notes

1 Letter from C. Barry to Office of Woods, 4 December 1845, in HC 574 1846, *Third Report*, p. 186.
2 Diary of Charles Barry, Jr., 22 September 1846-15 April 1847 (Parliamentary Archives: Bar 27).
3 House of Commons (HC) 402 1852, *Second report from the Select Committee on Ventilation and Lighting of the House*, pp. 551–52.
4 Oral statement by C. Barry on 23 February, House of Lords (HL) 719 1846, *Report from the Select Committee of House of Lords on Progress of Building*, Q6.
5 Reid implemented a similar system in the House of Commons. Its design is discussed in Henrik Schoenefeldt, 'The Lost (First) Chamber of the House of Commons', *AA Files*, 72 (2016), 161–73.
6 House of Commons. 1852 Select Committee on Ventilation and Lighting of the House; House of Lords. 1854 Select Committee of the House of Lords appointed to inquire into the possibility of improving the Ventilation and Lighting of the House; House of Commons. 1854 Select Committee on improving the Ventilation of the House of Commons.
7 In the following paragraphs, numbers or letters in **BOLD** parentheses refer to the labels depicted in the relevant figures.

8 Meeson, A. Letter to J. Thornborrow, 13 August 1855 (National Archives: Work 11/14, no. 881–884).
9 'Victoria Tower: detail of wind cowl to air turrets',13 January 1852. (National Archives: Work 29 2962).
10 Oral statement by C. Barry on 5 April, HC 402 1852, Q1382–88; 'Schematic plan of air supply', 1852 (National Archives: Work 29/2928).
11 'Enclosure to large fan', August 1851 (National Archives: Work 29/2952).
12 Oral statements by A. Meeson on 23 April, HC 402 1852, Q2584, Q2589.
13 In his report of 10 April 1854, Gurney described the supply as a network of 'tortuous passages' through which the air had to be forced mechanically.
14 'Plan of principles floor', 1852 (Strategic Estates Archive). Surveys undertaken by the author have revealed that small parts of the original pulleys and valves have survived.
15 Charles Barry, 'STATEMENT required by the Committee relative to the Structural arrangements made by Dr *Reid*, with reference to the WARMING and VENTILATING of the HOUSE of LORDS, prior to his Removal from the Works in the Year 1846, and the use made of such Arrangements by the Architect, in the System now in use for Warming and Ventilating that Chamber', Appendix, No. 3, HC 402 1852, pp. 550–51.
16 Ibid.
17 Gurney, G. Letter to the House of Lords Select Committee, 10 April 1854.
18 Ibid.; 'Ventilation of the New House of Peers', *Daily News*, 30 March 1847, p. 3.
19 'Detail of valve', 14 February 1849 (National Archives: 29/2917).
20 Oral statement by A. Meeson on 4 May, HL 384 1854, Q594–7.
21 D.B. Reid, undated, 1845 (National Archives: Work 29/2888).
22 C. Barry, 'Plan of roof', December 1846 (National Archives: Work 29/2900) and 20 October 1849 (National Archives: Work 29/2918).
23 Oral statements by A, Meeson on 23 April, HC 402 1852, Q2578–80.
24 Oral statement by A. Meeson on 4 May, HL 384 1854, Q588.
25 Oral statement by A. Meeson on 17 March, HL 384 1854, Q85.
26 Oral statements by A. Meeson on 17 March, HL 384 1854, Q102, Q125–26.
27 These valves can be seen in historical drawings. Recent surveys have shown that their frames and hinges have survived to this day. See 'Plan and sections of chambers under House of Lords showing ventilation ducts', 9 December 1851 (National Archives: Work 29/2958).
28 'The New House of Lords', *Illustrated London News*, 15 May 1847, p. 309.
29 Oral statement by A. Meeson on 17 March, HL 384 1854, Q125.
30 Drawings of the ceiling.
31 Oral statement by A. Meeson on 4 May, HL 384 1854, Q588; Oral statements by A. Meeson on 12 May, HL 384 1854, Q829–32.
32 Oral statement by C. Barry on 5 April, HC 402 1852, Q1422.
33 Oral statements by A. Meeson on 23 April, HC 402 1852, Q2638–46; Oral statements by A. Meeson on 17 March, HL 384 1854, Q70–74.
34 Air was admitted through an 18 in^2 aperture, which was covered with a perforated zinc plate.
35 Oral statements by A. Meeson on 23 April, HC 402 1852, Q2650–52.
36 The sketches are on the back of the drawing 'Elevation of north gallery, with sketch of timber ducts at back', 14 November 1849 (National Archives: Work 29/247).
37 Henrik Schoenefeldt, *The Victorian ventilation system of the House of Lords – A reconstruction: Volume 2: Survey, October 2018*. (Archives of the Palace of Westminster Restoration and Renewal Ltd [Restricted Access, Security]).
38 'The New Houses Parliament', *The Times*, 5 April 1847, p. 3.
39 Oral statement by C. Barry on 4 May, HL 384 1854, Q505.

40 Oral statement by A. Meeson on 4 May, HL 384 1854, Q594–7.
41 'Ventilation of the House of Lords', *Builder*, 1847 (5–217), p. 156; 'The New Houses Parliament', *The Times*, 5 April 1847, p. 3.
42 Ibid.
43 'Details of valves in air chamber above ceiling', 1846 (National Archives: Work 29/232).
44 'Ventilation of the House of Lords', *Builder*, 1847 (5–217), p. 156; 'The New Houses Parliament', *The Times*, 5 April 1847, p. 3; Oral statements by D.B. Reid, HC 402 1852, Q1417–21.
45 The early ventilation shaft was much plainer and shorter than the extended shaft. See 'Cross-section of early ventilation shaft', December 1846 (National Archives: Work 29/2900).
46 'House of Peers. Detail of stone ventilation shaft', 22 November 1848 (National Archives: Work 29/244).
47 See for example: *Illustrated London News*, 18 January 1860; *Exterior of Peers' Shaft, as seen from old Palace before the West front was completed*, undated, lithograph, Vacher & Son (Parliamentary Art Collection: WOA 1328); Unknown, *Peers' Shaft*, 1929, photograph (Historic England Archives: Folder AL 1016).
48 'Elevation and cross-section of extended shaft with steam-jet', 1851 (National Archives: Work 29/244). The drawing shows the ornamental elevation of the upper section of the shaft.
49 Oral statement by A. Meeson on 17 March, HL 384 1854, Q124.
50 Barry, C. Letter to A. Milne, 3 February 1848 (National Archives: Work 11/13, no. 370); Commissioners of Woods. Letter to D.B. Reid, 9 February 1848 (National Archives: Work 11/13, no. 374).
51 Barry, C. Letter to A. Milne, 3 February 1848 (National Archives: Work 11/13, no. 370).
52 Alfred Meeson, Second Report on the ventilation (National Archives: Work 11/14, no. 791–812).
53 Meeson, A. Letter to J. Thornborrow, 13 August 1855 (National Archives: Work 11/14, no. 881–4).
54 Gurney, G. Letter to , 13 August 1855 (National Archives: Work 11/14, no. 881–4).
55 The plans held by the Office of Works show that by 1881 the air passage linking the Central Tower to these areas was abandoned to create space for offices. See 'Plan of the Second Story', September 1881 (PED, no. 101688).
56 Oral statement by C. Barry on 30 March, HC 402 1852, Q789.
57 Oral statement by C. Barry on 30 March, HC 402 1852, Q790.
58 The Soft Landings Framework was introduced to extend the involvement of architects beyond the design and construction stages of a project by engaging them in post-occupancy evaluations and the process of operational fine-tuning and inducting facilities management staff. See Henrik Schoenefeldt, 'The House of Commons: a precedent for post-occupancy evaluation', *Building Research & Information*, 47 (2019), pp. 635–65.
59 Milne, A. Letter to D.B. Reid, 24 January 1840 (National Archives: Work 11/12, no. 30–32).
60 Office of Woods (signed by the Earl of Carlisle and A. Milne). Letter to HM Treasury ('Dr Reid's Salary'), January 1851 (National Archives: Work 11/16, no. 28–33).
61 Oral statement by C. Barry on 30 March, HC 402 1852, Q788.
62 Commissioners of Works. Letter to Lords of HM Treasury, February 1854 (National Archives: Works 11/26/3, no. 23);
Oral statement by A. Meeson on 17 March, in HL SC 384 1854, Q1–5.

63 Oral statement by C. Barry on 30 March 1852, in Second Report, HC 402 1852, Q791–92.
64 Transcript of oral statement by C. Barry on 30 March, HC 402 1852, Q785–6.
65 Transcript of oral statement by A. Meeson on 23 April, HC 402 1852, Q2679–85.
66 Ibid., Q2684.
67 Ibid., Q2685.
68 Oral statement by A. Meeson, 23 April, HC 402 1852, Q2678.
69 Oral statement by A.W. Clifford on 29 March, HC 402 1852, Q544–47.
70 Oral statement by A. Meeson on 23 April, HC 402 1852, Q2584.
71 Ibid., Q2584–99.
72 Oral statement by A. Meeson on 23 April, HC 402 1852, Q2556; Oral statement by A. Meeson on 17 March, HL 384 1854, Q75.
73 Oral statement by A.W. Clifford on 29 March, HC 402 1852, Q544–46.
74 Ibid.
75 Ibid., Q544–46.
76 Questions. Observations, House of Lords, *Hansard's Parliamentary Debates: The Official Report (22 July 1879, vol. 241, cols. 2018–21)* (London: Hansard); Oral statement by A. Meeson on 17 March, HL 384 1854, Q55.
77 Oral statements by A. Meeson on 17 March, HL 384 1854 384, Q96–97.
78 Notes on these experiments can be found in *Diary of Charles Barry Junior, 1 January 1846 and 12 April 1847* (Parliamentary Archives: BAR/27).
79 Oral statement by C. Barry on 5 April, HC 402 1852, Q1405.
80 Ibid., Q1412–20.
81 Ibid., Q1412–20.
82 'New House of Lords', *Illustrated London News*, 15 May 1847, p. 309.
83 Oral statement by C. Barry on 5 April, HC 402 1852, Q1405–1637.
84 'Cross-section of Boiler and Engine House', 21 October 1849 (National Archives: Work 29/2922).
85 Oral statements by C. Barry on 5 April 1852, HC 402 1852, Q1643.
86 Ibid, Q1669–74; Charles Barry, 'STATEMENT…', Appendix 3, HC 402 1852.
87 Oral statements by C. Barry on 5 April, HC 402 1852, Q1639–40; Q1664–73.
88 Oral statement by C. Barry on 30 March, HC 1852, 402, Q780.
89 Oral statement by C. Barry on 5 April, HC 402 1852, Q1580.
90 Ibid.
91 'Cross-section of Boiler and Engine House', 21 October 1849 (National Archives: Works 29/2922); 'Longitudinal Section of Boiler and Engine House', August 1852 (National Archives: Works 29/2964); 'Details of Boiler House', 1852 (National Archives: Work 29/2966).
92 Oral statements by C. Barry on 5 April, HC 402 1852, Q1555–67, Q1676.
93 Ibid., Q1555–56.
94 Oral statements by A. Meeson on 23 April, HC 402 1852, Q2606, Q2743–45.
95 Oral statements by A.W. Clifford, HC 402 1852, Q544–46.
96 Ibid., Q541.
97 Ibid., Q560–3.
98 Oral statements by Lord De Ros on 30 March 1852, HC 402 1852, Q689–94.
99 Oral statement by the Earl Lonsdale on 30 March 1852, HC 402 1852, Q679.
100 Ibid.
101 Oral statement by Earl Grey, 30 March 1852, HC 402 1852, Q702.
102 Ibid., Q703.
103 Oral statement by C. Barry, 30 March 1852, HC 402 1852, Q780.
104 'Parliamentary business for the week', *Morning Post*, 13 March 1854, p. 2.
105 'Imperial Parliament', *Morning Chronicle*, 15 March 1854, p. 2.
106 Oral statement by A. Meeson on 17 March, HL 384 1854, Q114.

107 See 'Notes made by Mr Meeson during the "Experiment on the ventilation of the Houses of Peers, and General Building, made on Sunday, 26 March 1854"', HL SC 1854, p. 57.
108 Oral statement by A. Meeson on 4 May, HL 384 1854, Q590.
109 See 'Notes made by Mr Meeson during the "Experiment on the ventilation of the Houses of Peers, and General Building, made on Sunday, 26 March 1854"', in HL 384 1854, p. 57.
110 Oral statement by Earl of Clanricard on 24 March, HL 384 1854, Q429.
111 Ibid, Q324.
112 Oral statement by Lord Stanley, 8 May 1854, HL 384 1854, Q719.
113 Oral statement by G. Gurney on 8 May, HL 384 1854, Q732.
114 HC 105 1854-55, 'Copy of Mr Gurney's report to the office of works, on his experiment for withdrawing and decomposing the noxious effluvia from the sewers in the neighbourhood of the Houses of Parliament'.
115 Oral statement by A. Meeson on 17 March, HL 384 1854, Q89.
116 Oral statements by A. Meeson on 12 May, HL 384 1854, Q826–29.
117 Oral statement by Lord Ashburton on 12 May, HL 384 1854, Q830.
118 Oral statement by Earl Graham on 17 March, HL 384 1854, Q85.
119 As the chamber was 180,000 ft^3, the air change rate per hour (ACPH) would have been between 1.3 and 2. This is a low ventilation rate compared to Victorian or modern standards. Meeson, however, claimed that the system had the capacity to provide 6.5 air changes, but this was rarely used.
120 Oral statement, HL 384 1854, Q89.
121 Oral statement by Earl Graham on 24 March, HL 384 1854, Q441.
122 Oral statement by A. Meeson on 17 March, HL 384 1854, Q108.
123 'Longitudinal cross-section' (National Archives: Work 29/2986); 'Floor plan of ground floor mezzanine' (National Archives: Work 29/2980).
124 For a detailed survey of the remains of the historical gas lighting infrastructure, see Henrik Schoenefeldt, *The Victorian ventilation systems of the House of Lords – A reconstruction* (London: Archives of the Palace of Westminster Restoration and Renewal Delivery Authority [Restricted Access, Security], March 2018).
125 'New House of Lords', *Illustrated London News*, 15 May 1847, p. 309.
126 Oral statement by C. Barry on 4 May, HL 384 1854, Q510.
127 Oral statements by A. Meeson on 17 March, HL 384 1854, Q55–6.
128 Ibid., Q89.
129 Ibid., Q55–6.
130 The lights are depicted in, for example: 'Interior of the House of Lords from the south', *Illustrated London News*, 17 April 1847, p. 248; 'Interior of the House of Lords from the Bar', *Illustrated London News*, 17 April 1847, p. 249; Joseph Nash, *House of Lords*, 1851, body colour, Parliamentary Art Collection: WOA 2940.
131 Oral statement by A. Meeson on 17 March, HL 384 1854, Q110–111.
132 Oral statement by Earl Graham on 17 March, HL 384 1854, Q110.
133 Oral statements by C. Barry on 4 May, HL 384 1854, Q79–80.
134 Oral statement by A. Meeson on 4 May, HL 384 1854, Q553.
135 Oral statement by C. Barry on 4 May, HL SC 1854, Q501.
136 Oral statement by C. Barry on 4 May, HL SC 1854, Q501.
137 Letter from G. Gurney to Office of Works, 7 April 1854 (National Archives: Work 11/14 no. 842)
138 Gurney, G. Letter to the Lords of the Select Committee, 10 April 1854, HL 384 1854. pp. 110–11.
139 Gurney, G. Letter to Francis Stone, 17 June 1854, HL 384 1854, pp. 118–19.
140 Report, 7 July 1854, in HL 384 1854, *Select Committee on improving*, p. iii

7 Towards an independent system for the House of Commons

Introduction: implementing Reid's principles within a confined habitat

The study of the House of Lords in the previous chapter examined Barry's contribution to the history of environmental design in the Palace of Westminster after the collapse of his working relationship with David Boswell Reid in 1846. The vote by the Houses to award responsibility for the House of Lords to Barry and to confine Reid's involvement to the ventilation of the House of Commons on 26 June 1846, however, had even further-reaching consequences as it resulted in fundamental changes to the ventilation system that affected the entire Palace of Westminster. The idea of having a central outlet for the Palace was discarded and replaced with a new decentralised approach to stack ventilation. In this scheme, there was to be an array of local turrets, each of which would serve a particular area of the Palace. This change in the ventilation plans caused another architectural transformation and the new ventilation turrets, which were added gradually between 1847 and 1855, would be responsible for creating the now iconic silhouette of the Palace of Westminster.

As a consequence of these changes, the ventilation of the House of Commons also became physically separated from the infrastructure within Barry's area of responsibility, the House of Lords. This forced Reid to develop a new arrangement that could be operated as an independent system, which he designed and constructed between 1847 and 1852. Reid continued to act as Barry's ventilation consultant but only for the House of Commons. The remaining parts of the Palace were Barry's responsibility.

Although it was only operational for a mere two years before it was decommissioned, Reid's ventilation system of the House of Commons fundamentally affected the final design of the Palace of Westminster. The debating chamber of the House of Commons became the only area of the Palace in which Reid's original ideas regarding ventilation and climate control were ever fully implemented and demonstrated under real-life conditions.

Whilst its short operational history from 1852 to 1854 will be examined in Chapter 8, this chapter reconstructs Reid's ventilation system for the House of Commons, and retraces its development from 1847 to 1852. The history of its development shows how the restrictions placed upon Reid affected both the final design of the Palace and his working relationship with Charles Barry. Despite the restriction of Reid's involvement to a small area of the Palace, problems in the collaboration between Reid and Barry did not cease. If Reid was to realise his principles within the more confined area of the House of Commons, he would have to still collaborate closely with Barry in order to resolve various design problems.

A new beginning

The breakdown of Reid and Barry's working relationship in 1846 was a major incision, both physically and operationally. In addition to reducing his level of involvement to the House of Commons, MPs considered operational changes to address previous problems of collaboration between Reid and Barry. On 1 April 1854, three months before the Lords had voted to end Reid's involvement, a House of Commons Select Committee had endorsed his continued engagement in designing the ventilation of the House debating chamber. There was, they proclaimed, a 'great improvement' in the atmosphere of the temporary House of Commons. However, the Committee acknowledged that changes in the project governance were required in order to effectively manage the collaboration between Reid and Barry.[1] In a resolution, the Committee requested that the Office of Woods establish a new body to take on the responsibility for mediating between the two parties in the case of difficulties. Indeed, it should intervene

> *in the event of any objection being made by Mr Barry to the plans of ventilation and warming submitted by Dr Reid, or, in the event of any difference arising between Mr Barry and Dr Reid, either as to the amount of information requisite for the preparation of these plans, or in their execution, or otherwise.*[2]

The Select Committee emphasised that the body should be granted the authority to make and enforce a final decision regarding the ventilation system, and that the architect would be expected to comply with their instructions. The proposed body was to be composed of two people, one recommended by Reid, the other by Barry. A third person, appointed by the Office of Woods, was to act as an 'umpire' in case the first two were unable to resolve a problem. The records show that such a body was never appointed.

The Office of Woods continued to oversee the development of Reid's scheme for a further two years. In 1848, oversight of the project was transferred to a Royal Commission, which was appointed to manage it on their behalf. This was due to the fact that for the first year after Reid's and Barry's split in 1846, Reid developed his plans for the House of Commons ventilation system without involving Barry and his office.

Unlike previously, Barry and his staff were not involved in the creation of the ventilation drawings. This was significant as it was the outcome of a new model of practice in which Reid's office worked more independently of Barry's office. The first set of drawings and the majority of the later ventilation drawings were produced by Reid's office,[3] bearing Reid's signature as the lead designer. On 5 April 1847 Reid submitted a large set of over 40 drawings to the Office of Woods, providing a comprehensive and detailed outline of the proposed ventilation system for the House of Commons.[4] This set of drawings was produced independently by draughtsmen in Reid's own office under his direct supervision, including William Patrickson (who had been an assistant in Reid's office since 1842).[5]

It was only after he had submitted this first set of plans that Reid was to look at potential design problems jointly with Barry. During this period of collaboration, however, Reid and Barry (therefore their offices) continued to act as two independent practices, each focusing on their own areas of responsibility. Most of the communications between the two men was through the medium of drawings and letters, suggesting that the greater focus resulting from the restrictions in Reid's responsibilities, combined with the experience that he had gained in the six previous years, had enabled him to become more competent in his role.

The correspondence between Reid's and Barry's 'practices' was closely monitored by the Office of Woods, which initially received any drawings before they were passed on to the relevant party. As before, the information contained in the specialist ventilation drawings was integrated into the working architectural drawings by Barry's team, who also supervised construction. Barry imposed tight restrictions on Reid's and his office's involvement in this process: they were to supply Barry's team with specialist ventilation drawings and written specifications, which was to guide them in developing the full construction details; they were not permitted to take original drawings from Barry's office; once the working drawings were completed, they could only visit Barry's office to view the original drawings or takes copies of them; and they were to return any copies to Barry's office complete with written annotations and amendments. Two clerks, Alfred Meeson and a 'Mr Quarm', were also appointed to act as Reid's liaisons in Barry's office. They were authorised to discuss proposals on Barry's behalf and give Reid access to any needed drawings and minutes.[6] The communication regime introduced by Barry in 1845 continued.

Reid repeatedly voiced his frustration with the working arrangements as they limited his access to the construction work undertaken by Barry's team. With little opportunity to examine the technical arrangements, Reid felt unable to guarantee the ventilation system would perform as intended. [7]

A new scheme built on old foundations

The drawings produced by Reid and his office show that the proposed ventilation system for the House of Commons debating chamber was substantially similar to his proposed system for the House of Lords and that the primary physical infrastructure, which had already been constructed based on his original drawings, were repurposed to accommodate the now independent ventilation system of the House of Lords. The physical infrastructure, composed of features that were constructed as an integral part of the masonry structure (such as the air chambers and vertical shafts) had been completed in the spring of 1847 and had been outlined in two earlier sets of drawings.[8] The first set of drawings, completed between 1843 and 1844, outlined the arrangements of passages and shafts for the fresh-air supply at basement and ground-floor level.[9] These included the equalising chamber, supply ducts for the East and West Galleries,[10] and four large shafts that supplied the galleries at the north end.[11] The second series of drawings, completed between July and October 1845, outlined the ventilation network in the upper part of the chamber.[12]

As the primary network within the structural carcass of the Palace was already completed, Reid and his assistant focused on developing the technical details for the interior features and adapting the existing supply and extraction network below and above the chamber.

The first set of drawings, dated 5 April 1847, includes drawings of many mechanical engineering details, showing the heating, boilers, fan and intricate mechanism of the air valves. Other drawings in the set show layout and construction details for an array of outlets integrated within the floors, ceilings and benches. Some parts of the original arrangements were remodelled to enable the ventilation system to work as an independent system as the House of Commons was no longer part of the central supply and discharge systems of the Palace. For example, arrangements of the air channels and fresh-air intakes at roof level are shown, as well as a first proposal for a local up-cast shaft to enable the ventilation to become independent from the Central Tower.

Following the same principles as Reid's original scheme for the House of Lords (see Chapter 5), the House of Commons ventilation system introduced fresh air through the ceiling and the floor. However, this was designed as, and achieved using, two separate systems, each equipped with its own set of fresh-air inlets, up-cast shafts, fan and air-conditioning.

A return to a difficult collaboration: 1847–52

Reid's ventilation system for the House of Commons was further developed and implemented between May 1847 and February 1852. During this period, the plans for the system underwent several modifications and the design of some of its important features, such as the air supply for the ceiling inlets, was to be the outcome of intense negotiations between Reid, Barry and the Office of Woods.

Although Reid was no longer involved in the design of the rest of the new Palace, Barry's approval was still needed for critical features of Reid's own system.

In Spring 1848 a Royal Commission was appointed to take over the project management of the Palace of Westminster. This new body, known as 'Her Majesty's Commission for Superintending the Completion of the Houses of Parliament' (henceforth referred to as 'the HM Commission')[13] was also responsible for monitoring the communication between Barry and Reid.[14] According to minutes of the HM Treasury, dated 17 March, one of its roles was to 'control and finally decide upon all arrangements with reference to the warming, ventilating and lighting of the building, when such arrangements are at present undecided, subject as above to the approval of the board'.[15]

The HM Commission was composed of three members: Thomas Philip, 2nd Earl de Grey of Wrest, who was the first President of the newly-founded Royal Institute of British Architects (RIBA); John Fox Burgoyne, an engineer and senior army officer; and Thomas Greene, MP for Lancaster.[16] All correspondence between Reid and Barry – including drawings and letters – were sent via the office of the HM Commission until December 1851, when it was dissolved and the responsibilities were transferred to the newly-created Office of Works (which replaced the Office of Woods).[17]

From a project management perspective, the appointment of the HM Commission is significant as it reflects a recognition that the Office of Woods was not able to manage a project of the scale and complexity of the House of Commons. As discussed in Chapter 5, insufficient strategic oversight of the project resulted in the Office of Woods taking a reactive approach to expanding the brief and building programme. The recommendations of the 1846 House of Commons Select Committee had recognised that the collaboration between Reid and Barry needed to be more tightly managed if the previous complications and delays were to be avoided. As will be shown in this and the next chapter, the HM Commission was to become more active in the day-to-day project management of the ventilation system than the Office of Woods ever was. It took a more active part in design decisions, and at times, also had to be proactive in resolving design problems or acting as a moderator between Reid and Barry when collaborative decisions were needed.

Repurposing the Central Tower: 1847–48

Reid's adoption of a decentralised approach to stack ventilation made the Central Tower obsolete, and its design and function was to be the subject of long discussions lasting over 12 months. During the 1846 Joint Select Committee inquiry, Barry had alluded to the fact that the tall tower would

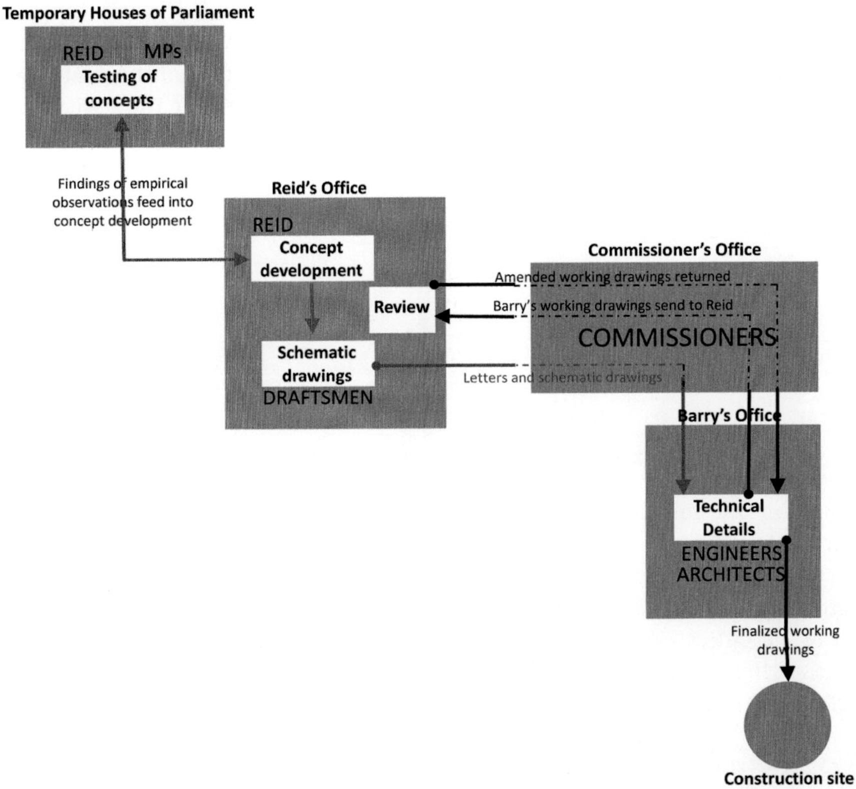

Figure 7.1 Two independent practices: work flow between Reid's and Barry's offices in the design of the House of Commons (author's own drawing).

become functionally obsolete unless Reid was to reuse it for the ventilation system in the House of Commons.[18] However, Barry advocated for its retention on architectural grounds even if it was not needed for the ventilation system.[19] As discussed in Chapter 6, Barry had produced a new design for a much smaller tower in May 1847,[20] arguing that the great height of earlier tower proposals was not required as it was no longer to be the main air outlet for the entire Palace.[21] Barry thus contemplated converting it into a local up-cast shaft for the House of Lords to replace its small temporary up-cast shafts.

This idea, however, was not compatible with Reid's new scheme for the House of Commons. In his drawings from April 1847, Reid proposed to retain the Central Tower and adapt it to be a fresh-air intake shaft, which could be linked to the ceiling of the debating chamber. Some fresh air was to be supplied downwards through the central row of ceiling panels,[22] whilst hot air was to be extracted through the side panels using a new up-cast shaft at the north end of the House. The drawings show the position and schematic outline of this new up-cast shaft (see Figure 7.2).[23] Barry himself produced a first architectural design for the shaft on 1 May 1847.[24]

Only two weeks later, on 14 May, Reid informed Barry that the shaft was not tall enough to effectively discharge vitiated air without the assistance

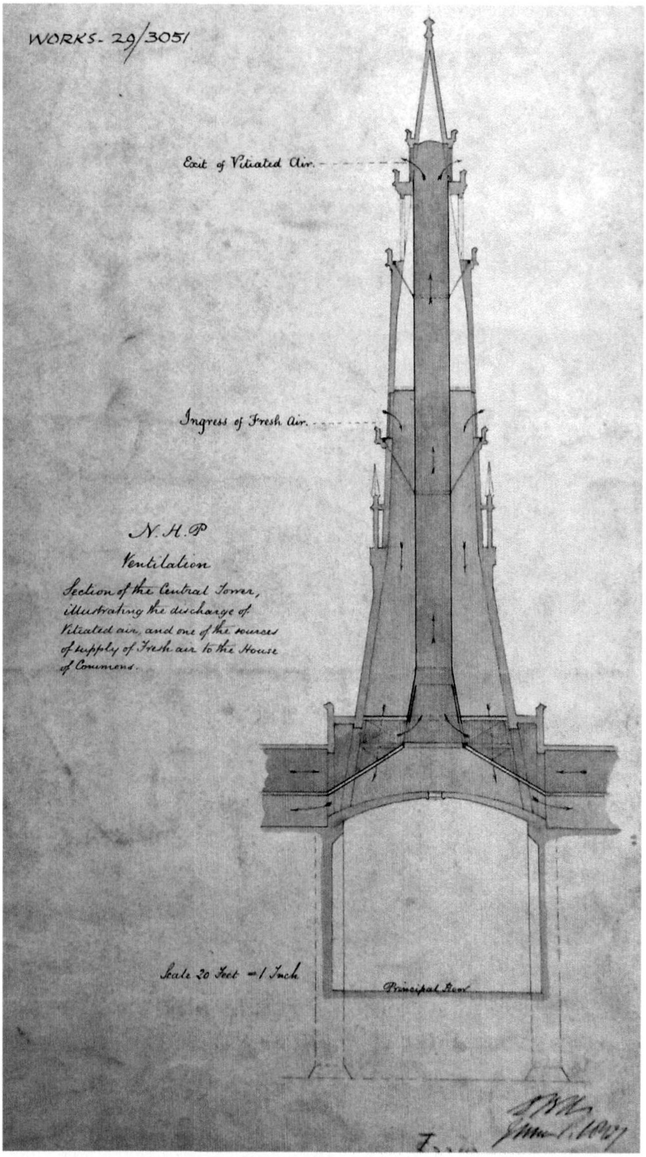

Figure 7.2 Reid's proposal for integrating the intake and up-cast shafts into the Central Tower (National Archives: Work 29/3051).

of a powerful ratification system.[25] Barry proposed introducing a second discharge shaft to complement the northern shaft and requested that Reid provide details about the height and design requirements.

Reid failed to supply this information, instead submitting an alternative proposal to use the Central Tower as a principal up-cast shaft for the House of Commons. Depicted in three drawings from 31 May to 1 June 1847, the height of the Central Tower was significantly increased back to its original height to enable the vitiated air to be discharged at a high altitude;[26] use of the lower part of the shaft for the House of Commons' second air supply was continued.[27]

To facilitate this without compromising Barry's plans for the House of Lords, Reid proposed to transform the Central Tower into a shared up-cast shaft for both debating chambers: a large iron flue in the middle of the Central Tower could be constructed to exhaust the hot air of both chambers at the top of the spire, whilst the new intake for the ceiling supply of the House of Commons would be located half way up the shaft. The fresh air would be propelled from the Central Tower to the ceiling by a steam-driven fan,[28] while the inlets in the floor of the House would continue to be served with fresh air from the Elizabeth (Clock) Tower.[29]

Barry opposed the idea of using the Central Tower for the air supply on architectural grounds, namely that that the tower was too high. In a letter to the Office of Woods dated 4 June 1847, Barry claimed that he did not approve of a tower of such a height as it appeared to him 'very extravagant notions on his [Reid's] part, as to altitude, which at one time he [Reid] wished to exceed that of St. Paul's'.[30] He was to later claim that he had only previously accepted a tall shaft due to pressure from the Office of Woods.[31] Indeed, after the rejection of Reid's scheme in 1846, Barry had proposed to reduce the height of the Central Tower by more than half, arguing that a tall tower was neither required for architectural effect nor ventilation purposes.[32]

Reid disagreed with Barry.[33] In drawings returned to Barry on 15 June 1847, Reid commented that the Central Tower had to be taller than the surrounding smoke and air turrets: first, to avoid the fresh-air inlets being exposed to their emissions, and second, to gain access to the 'atmosphere free from fog and suspended dew, as well as from the smoke from numerous manufactory chimneys'.[34] He urged Barry to specify the greatest height he would be willing to accept.

In his plans Reid also indicated the position of the local discharge towers relative to the Central Tower.[35] Barry had introduced two turrets on the River Front, which were connected to the chimney and air flues of the various libraries and Committee Rooms.[36] The number of discharge towers was gradually increased and their position relative to the air intakes was the subject of lengthy discussions. In early July 1847 Barry wrote that six local discharge shafts '110 feet above Trinity Standard High water' were introduced, with two smoke shafts also planned on the west side of the Central

Tower. Barry confessed that the ultimate number of turrets could only be determined 'upon the success or failure of the shafts already enumerated'.[37]

Although Reid continued to argue that the Central Tower would be more effective as a second discharge shaft,[38] he submitted five drawings on 19 July 1847 for the second up-cast shaft that Barry had suggested.[39] These drawings show that the new up-shaft, unlike the northern shaft, was to be a rectangular stone shaft without a turret.[40] The top of the shaft was to be open but was to have a damper in the shape of a hipped roof, which was operated from the base of the tower by pulleys. At the bottom of the shaft was to be a fire, 'to be used more especially in sultry weather or when the movement of machinery is not in operation'.[41] This up-cast shaft was agreed upon by Barry and became the main up-cast shaft for the House of Commons (see [2] in Figure 7.3).[42]

Three typologies of shaft, shown in Figure 7.3, were developed for the Palace after 1846 with their different functions articulated by distinct architectural forms. First, up-cast shafts for vitiated air (of which the up-cast shaft for the House of Commons is an example, see [2][43]). These were rectangular masonry towers with crenelated parapets. This shaft type was also used for the Peers' Shaft in the House of Lords (see Chapter 6). Second, smoke stacks ([3]). Third, combined air and smoke turrets ([1], [4] and [5]). These were tall shafts crowned by stepped turrets, which allowed openings for the discharge of air and smoke fumes at different altitudes

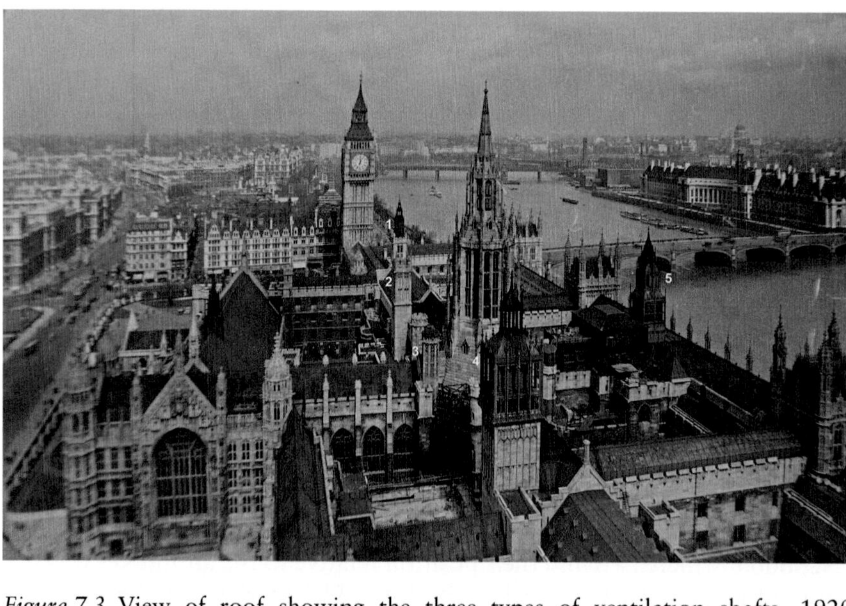

Figure 7.3 View of roof showing the three types of ventilation shafts, 1920s (Historic England Archive: 'View from Victoria Tower', Album 1017, number H16505).

to be accommodated. The form of church spires or Medieval font covers thereby provided a suitable Gothic precedent for the design of the modern ventilating towers. Examples of this shaft typology are the three turrets above the River Front. Although they had the appearance of ornate Gothic spires, they incorporated technical features of the ventilation system as they were designed to protect the vitiated air and smoke outlets from the winds through the use of a system of iron louvres, which was responsive to wind pressure: The louvre blades were kept open by counterweights, but closed themselves if they were exposed to higher wind pressures. According to the *Illustrated London News*:

> *The louvres are ingeniously contrived to open or shut, as the wind may play against- that is, when the wind is beating against one side, or at the angle-for though quadrangular at the base, it assumes the form of an octagon, just at the commencement of the louvres, thus be fanning any one side, or angle, the opposite apertures remain open; thus the foul air readily finds egress.*[44]

The unresolved question of the fresh-air supply

In December 1847, HM Treasury approved the budget required to cover the additional cost, estimated at £4,567, of adapting the ventilation system in the House of Commons.[45] The design for the fresh-air intake shaft, however, was not finalised for another several months. In early February 1848 Barry finally accepted that the much reduced height of the Central Tower that he had proposed was unsuitable as an upper-air supply. However, he continued to refuse to increase it to the full height .[46] Instead Barry proposed moving the second inlet inside the courtyards and using the Central Tower only as a discharge shaft for the House of Lords.

Disagreeing with Barry's plan, Reid complained to the Office of Woods that Barry was interfering with his current plans as his proposal would deprive him of the Central Tower as a potential air supply.[47] However, although Reid favoured using the Central Tower as an air intake, he agreed that parts of it could be used to function as an air discharge for the House of Lords.

In mid-February 1848 Barry tentatively accepted Reid's proposal to use the Central Tower as a combined supply and discharge shaft for the House of Commons and as a discharge shaft for the House of Lords. Asked to furnish Barry with detailed drawings of the arrangements inside the Central Tower, Reid refused and claimed that sufficient explanation of his plans had already been provided.[48] On 26 February 1848 Barry informed the Office of Woods that Reid's reply did not contain details of any 'arrangements required by him to be made in the Central Tower'.[49] He noted that he had already sufficiently explained his intentions, which was to use the Central Tower as a discharge shaft for vitiated air only, and that he had provided the required details to ascertain how far Reid's plans could be reconciled with his own.[50]

The Office of Woods urged Reid and Barry to cooperate as they were becoming increasingly impatient with the two men's lack of collaboration.[51] Unfortunately, both Reid and Barry remained stubborn and the design process again became obstructed by severe communication problems.[52] In March 1848 Reid produced a questionnaire asking Barry to clarify his intentions.[53] In his reply, Barry explained that the tower was solely deployed for the supply of fresh air to the House of Commons, and that it would only be necessary to apply 'valves or shutters to all apertures communicating with the open air, so that fresh air could be conveyed to the House of Commons by the channels indicated from any of the apertures, according to the state of the external atmosphere'.

Knowing that Barry 'desired to use a portion of the tower for the discharge of vitiated air', Reid asked him to supply detailed sections and plans of the arrangements inside the tower.[54] To ensure that the fresh air supply was protected from air pollution, Reid asked that the drawings indicate the 'precise position and altitude of every intended discharge of vitiated air from the roofs'.[55]

In his reply to Reid's questionnaire Barry confirmed that he still intended to use the entire Central Tower as an air discharge, but indicated that the smoke flues from the fireplaces were to be connected to separate smoke shafts on its west side. Barry alleged to the Office of Woods that Reid – through his continued refusal to 'explain the mode by which he wishes to make use of it as a means of supply for fresh air' – was responsible for the delays in resolving these design questions.[56]

On 21 March, Reid sent a second questionnaire to Barry asking for the specifications of the position of the smoke shafts and the height at which vitiated air would be emitted by the various discharge turrets.[57] Barry refused Reid's requests again.[58]

Despite this, Reid decided to amalgamate the existing information on the arrangement of the various ventilating towers on a drawing. Dated 5 April 1848, it indicates the position and height of the various discharge towers introduced since 1846. A total of seven vitiated air shafts and two smoke flues, including shafts whose exact position Barry had not specified, are shown. Reid again urged Barry to finalise the number and position of the towers.[59] In reply, Barry stressed he was unable to determine the ultimate number of discharge shafts until the performance of the existing shafts had been tested.[60] The final arrangement of discharge shafts was not completely settled and in the end it was allowed to evolve gradually.

The Office of Woods also submitted an alternative proposal: they suggested that the architectural towers on the north end of the River Front's central curtain could be converted into a new fresh air intake, which could then be connected to the House of Commons through ducts under the roof.[61] Reid himself objected to this proposal on the grounds that the northern tower was unsuitable as it was located in the path of the prevailing westerly winds, which carried foul air from the exhaust shafts.[62] In response, he submitted annotated block plans showing various alternative arrangements for the upper-air supply that had been considered since June 1847,[63] and maintained his insistence that the most adequate position for the second inlet would be the Victoria Tower or, if it was increased in height, the Central Tower.[64]

Following pressure from the newly-appointed HM Commission, Barry submitted tracings of the original drawings that outlined four alternative schemes for the supply of fresh air to the House of Commons.[65] In schemes 1 and 2 the Elizabeth Tower would function as the only fresh air supply. In scheme 3 the smoke from the boiler flues would be emitted through apertures on the top of the Central Tower; new turrets above the Upper Waiting Hall, and in the south-east corner of Westminster Hall, would function as the principal smoke shafts for the fireplaces. In scheme 4, the Central Tower would be used as a combined inlet and air discharge shaft. To protect the inlets, Barry proposed moving all of the smoke discharges located within a 120 ft radius of the Central Tower to a shaft in the south-east corner of Westminster Hall.

On 15 May 1848 Reid warned that even if the fresh-air apertures were positioned at the bottom of the Central Tower, they would still be exposed to emissions from the adjacent smoke shafts.[66] He now recommended relocating the air intakes from the Central Tower into two new inlets: one facing the River Thames and one inside the turret above St Margaret's Porch, near Westminster Hall.[67] The plans illustrate that the air from both the Thames and Westminster Hall would converge in the air chamber above the Central Hall before it was directed to the ceiling in the House of Commons.

Two alternative solutions for the fresh-air intakes on the River Front were considered. First, similar to the earlier proposal by the Office of Woods, air intakes could be located to the southern tower of the River Front, where it would be less exposed to emissions from other turrets.[68] Second, the inlets could be introduced in the roof plane facing the River Thames.[69]

The new second intake for the House of Commons: 1848

On 22 June 1848, following 12 months of intense negotiations, Reid and Barry finally agreed on a new arrangement for the ceiling-level supply involving two diagonal partitions inside the Central Tower. This was outlined in a drawing that was signed by all parties (see Figure 7.4). The additional cost for its construction, estimated at £3,789, was approved by HM Treasury in March 1849.[70] HM Commission wrote that the additional expense was necessary as Reid was:

> *excluded from a second supply of fresh air from an upper louvre in the addition to that desired from the clock tower without which he did not*

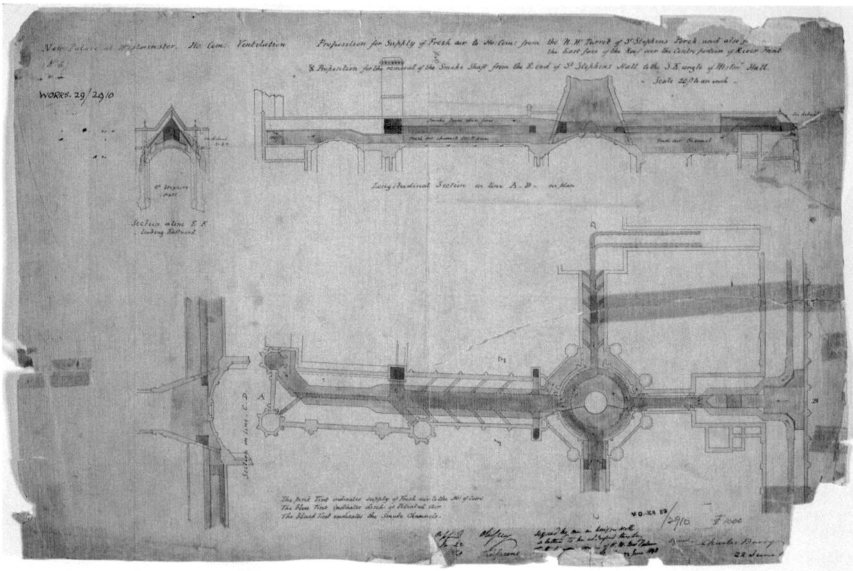

Figure 7.4 Plan and sections outlining agreed arrangement of fresh-air intakes and channels for ceiling supply of the House of Commons, 22 June 1848 (National Archives: Work, 29/2910).

> *consider that the ventilation of the New House of commons could be satisfactorily carried out with effect as it was found to be impracticable to meet Dr. Reid's wishes for deriving that supply from the Victoria tower and central tower.*[71]

Accordingly, the only function of the Central Tower was now as the principal air discharge shaft for the House of Lords. Later drawings and sketches, which are held at the RIBA drawing collection at the Victoria and Albert Museum in London, show that Barry increased its height and developed it into an ornate Gothic spire, which was completed in the mid-1950s.[72]

The primary inlet for the ceiling supply of the House of Commons was now to be embedded inside the cast iron roof facing the river.[73] Unlike the architecturally-prominent discharge towers, these inlets were made inconspicuous by integrating the openings, which were covered by adjustable louvres, into the existing roof plane.[74] The supply was switched to a second inlet, which was located inside a turret in the north-west corner of St Margaret's Porch, when the inlets were exposed to pollution.[75] Fresh air was conveyed to the House of Commons through a large iron duct located under the roof. Remains of the original duct can be found above the ceiling vaults of St Stephen's Porch and the Central Lobby (see Figure 7.5). The ceiling supply also had its own fan and heating.[76] The fan was located on

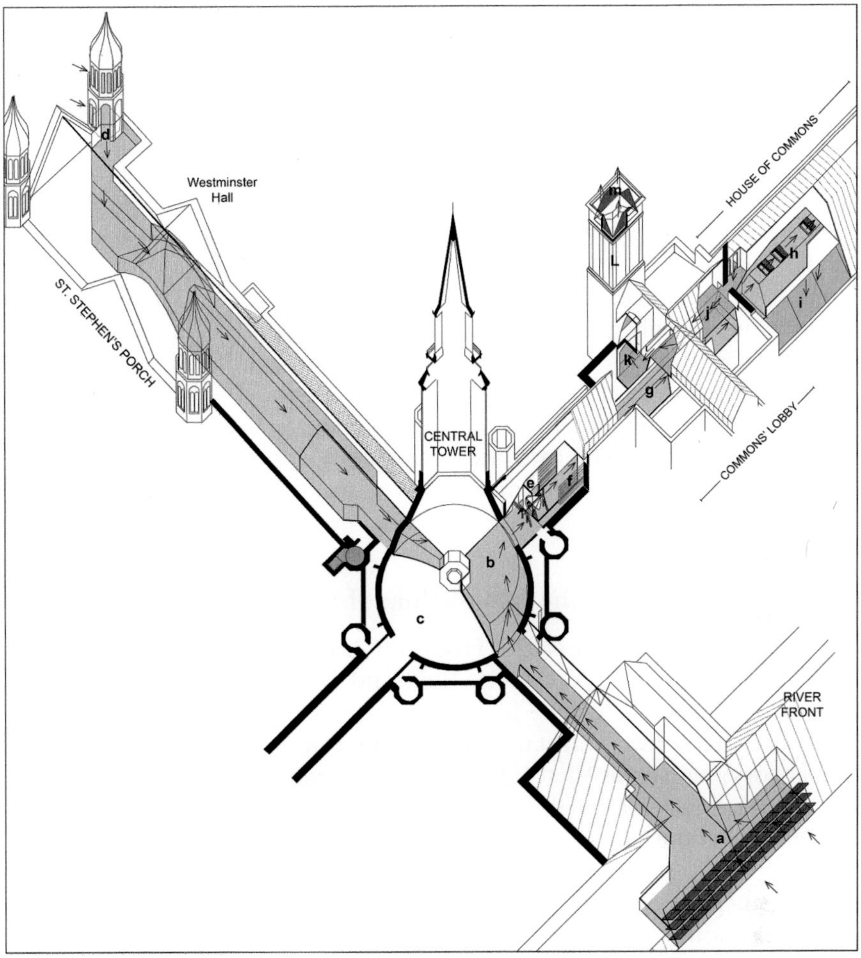

Figure 7.5 Axonometric showing ventilation and warm air central heating of ceiling system (author's own illustration). *Key: A. Principal air inlet of the ceiling system with adjustable cast iron louvres, B. fresh air channel passing through Central Tower, with diagonal wall marking boundary between Barry's and Reid's territories, C. Area connected to the House of Lords (Barry), D. Back-up inlet for ceiling system, inside turret, E. Fan, F. Steam pipes, G. Supply passage leading to House of Commons, H. Fresh air chamber above central ceiling panels of House of Commons; I. Vitiated air chamber above sloping side panels of ceiling, J. Passage connecting vitiated air chamber with up-cast shaft, K. Base of up-cast shaft with coke fire; L: Up-cast shaft, M. Louvres valves at outlet of up-cast shaft.*

the North side of the Central Tower, whilst the air was warmed inside a passage lined with steam pipes. This passage ended in the fresh-air chamber above the central ceiling panels. Fresh air was admitted into the debating chambers through gaps between the panels and through openings inside the ornamental beams, which were hollow.[77] The size of the gaps could be adjusted manually by means of sliding valves.[78]

The vitiated air chambers, which were situated above the side panels, were connected to the new up-cast shaft. Air entered at the base of the shaft and was exhausted through cast iron valves on the top, which could be adjusted using pulleys.[79] The pull produced by the rising hot air, which at times was enhanced by a coke-fuelled fire, drove the vitiated air out of the debating chamber. As the pull was not strong enough to simultaneously ventilate both the debating chamber and surrounding lobbies (e.g. the East and West Division Lobbies and Commons Lobby) valves were used to connect the shaft to individual spaces, including the Commons Lobby, and the Ladies' and Stranger's (now known as the Visitors' Gallery) Galleries.[80] During votes the valves were adjusted to redirect the pull from the House to the Division Lobbies.[81]

The Clock Tower and the remodelling of the lower-air supply

Shortly after agreeing how the House of Lords and Commons systems would be separated at roof level, Reid and Barry agreed on which passages would be used for the main air supply to the floor outlets within the House. Barry was to retain parts of the centralised supply located in his territory, using the Victoria Tower as the main intake, but the ventilation system was to be physically separated from that of the House of Commons. As the House of Commons only had access to the high-level inlet in the Elizabeth Tower, Reid adopted the Central Chamber as a new back-up for periods when the Tower could not be used due to high levels of air pollution.[82] Reid wanted to make the building more responsive to air pollution levels and wanted to use switchable inlets to allow air to enter, or prevent it from entering, the ventilation system, just as he had at roof-level. His initial proposal was to use the entire Central Chamber in order to gain access to its four different intakes, each of which was connected to a different courtyard. This would have enabled the supply air to be switched to a less polluted source if needed.[83] However, as discussed in Chapter 6, Barry had in July 1848 partitioned the Central Chamber into two spaces using diagonal walls (as he did at roof level), with the south half containing the heating and humidification arrangement for Barry's system. As a result, there were only two apertures in the North part of the Chamber that Reid could use to provide fresh air from the inlets located in the Cloister Court and the Common Inner Court.[84] Figure 7.6 shows two sketches from July 1848 of how the Central Chamber was ultimately partitioned using the diagonal walls.

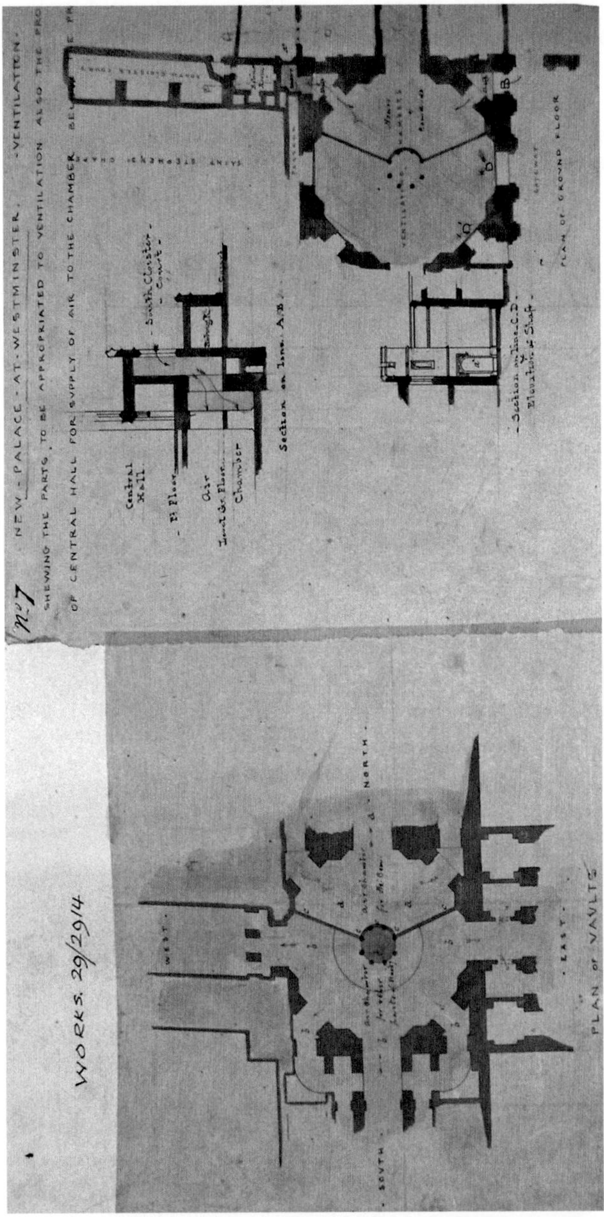

Figure 7.6 Sketches showing how the Central Chamber was partitioned by Barry at basement level, 6 July 1848 (National Archives: Work 29/2914).

In the part of the Palace that was Reid's area of responsibility the supply air was tempered using the 'air-conditioning' principles that Reid had tested in the temporary House of Commons. This allowed the supply air for the new debating chamber to be cooled, heated, humidified or dehumidified as needed.[85] In Reid's system, air was admitted through the Central Chamber and the Elizabeth Tower and conveyed to the House of Commons through basement passages ([a] in Figure 7.7). It could then be directed to either a heating chamber (which was filled with pipes) or a cool air chamber on the ground floor through the use of 12 circular and three rectangular valves ([b] and [c] respectively) located in the ceiling vaults of the basement.[86] The supply air temperature was controlled by varying the temperature of the water inside the pipes. However, the air temperature was cooled not using ice (which had been briefly tested in the temporary House of Commons) but using a passive, non-mechanical method of passing cold water through the water pipes (or, if the air was sufficiently dry, by evaporating water).

A process of mixing heated and unheated air was used to lower the temperature in response to sudden changes in attendance. The valves above the heating chamber could be closed, and heated air redirected into the surrounding cool air chamber in order to lower the air temperature more rapidly ([e] in Figure 7.7).[87] When the temperature of the House needed to be cooled quickly, the heating chamber was closed completely and cool air was admitted into it directly from the basement.

The humidity of the supply air could be raised using steam or by evaporating water, and lowered using an "absorbent of moisture" (although Reid does not specify what this substance was) in the air chamber on the ground floor. The air temperature was monitored using 20 thermometers whilst its relative humidity was monitored using a hygrometer.[88] Once tempered, the air was passed through separate rectangular and circular valves into the Equalising Chamber ([f] and [g], respectively, in Figure 7.7).

In addition to lowering the air temperature by non-mechanical means, Reid exploited the fact that people's perception of the air temperature is affected by air currents passing over the skin, which can produce a cooling sensation.

The floor of the debating chamber was covered with perforated cast iron plates. However, in contrast to the temporary House of Commons where fresh air was admitted across the entire surface of the floor, the new permanent system had outlets to extract vitiated air downwards as well as inlets to supply fresh air upwards. The configuration of the air supply followed the same principles that Reid had attempted to implement inside the House of Lords in the mid-1840s. Inlets were restricted to areas where MPs would not be exposed to currents. Air was supplied through the floor between the table and bar, through risers in the gangways,[89] and along the back of the benches. The chairs for the Speaker of the House and the Sergeant-at-Arms had their own individual supplies. The air supply could be adjusted individually by attendants inside the Equalising Chamber, using over 60 sliding valves. Along

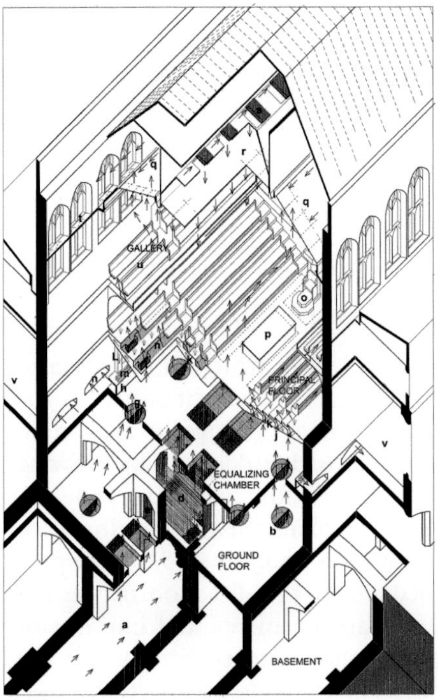

Figure 7.7 Reconstruction of Reid's environmental system in the House of Commons showing underfloor ventilation and air-conditioning arrangement with ducts leading to individual benches (author's own drawing). *Key: A. Fresh air passage linking inlet shafts to Central (air) Chamber, B. Valves for conveying air from basement into cool-air chamber (circular valves with shutters shown in open position), C. Valves for conveying air from basement to central heating chamber (rectangular valves with adjustable curtains below grating shown partially opened), D. Pipes of hot water apparatus in heating chamber, E. Vertical door valves for conveying hot air into cool-air chamber, F. Rectangular curtain valve through which heated air is admitted into Equalising Chamber, G. Circular shutter valves to admit unheated air from cool air chamber into Equalising Chamber, H. Horizontal duct in which vitiated air extracted through perforated floor was collected before it was exhausted via boiler in north-west turrets of the Central Tower, I. Sliding valves for supply of individual benches, J. Sliding valves for supply through treads inside the gangways, K. Vitiated air chamber under perforated iron floor (extract), L. Vertical ducts connecting vitiated air chamber with horizontal ducts, M. Steam and hot water pipes (heating and humidification), N. Valves conveying air to fresh-air chamber under the perforated floor of the Division Lobbies, O. Speaker's chair, P. Table, Q. Vitiated air chamber above sloping side panels, extract of ceiling system, R. Fresh air chamber used to supply tempered fresh air through central ceiling panels (ceiling system), S. Sliding valves for regulating air supply to ceiling, T. Line of acoustic ceiling retrofitted in 1851, covering half of Barry's original window, U. Gallery with air supply through floor, V. Division Lobbies.*

the back of every bench was an inlet, which had ducts with individual valves.[90] Air was continuously extracted downwards through the floor, immediately in front of the benches,[91] which then entered a vitiated air chamber below the floor. The air was then discharged via the boiler chimney, which terminated in the turrets in the north-west corner of the Central Tower.[92]

Special provisions were made to enhance the thermal comfort of front benchers, the Speaker of the House and the Sergeant-at-Arms, such as the use of hot-water plate radiators attached to the underside of the iron floor to warm their feet.[93]

Finalising the interior design: March–August 1850

Although parts of the floor were already under construction in December 1848, the ventilation system could not be completed until the seating layout and design for the benches was finalised.[94] However, changes to the interior design of the House of Commons introduced further delays. The drawings that Reid had produced of the proposed ventilation system in 1847 and 1848 were originally based on Barry's earlier plans for the interior of the debating chamber. However, between May 1849 and July 1851, these plans were still in the process of being revised under the direction of the HM Commission to accommodate new seating arrangements and improve the acoustics of the space. A life-size mock-up of the proposed interior was thus erected in 1850 to enable the new plans to be evaluated and refined and a series of trial debates for this purpose involving MPs was undertaken.[95]

Concerned that the debating chamber did not provide sufficient space for seating the required number of MPs, the Office of Woods, the Speaker of the House and several MPs all worked with Barry to explore potential seating arrangements and details for the benches.[96] Thomas Greene, one of the three members of the HM Commission, provided Barry with the required dimensions for the benches, and, acknowledging Reid's requirements, also specified space for fresh-air inlets.[97] Apart from sketches that outlined the details for the ventilation arrangements in the floor, however, the issue of ventilation did not receive any further attention during these inquiries. The mock interior had not been equipped with a ventilation or heating system,[98] and the trial debates could thus only be held during mild weather.[99]

The first mock-up trial, completed in March 1850, revealed that the main floor was too tight to comfortably seat 462 MPs, which required 12 rows of benches. In April it was reduced to 10 rows and a temporary gallery was also erected behind the Speaker of the House's chair to make up for the lost seats.[100] Another trial with the MPs seated in the Strangers' Gallery at the Bar end of the debating chamber was also conducted but was rejected.

Four more trial debates were held between May and June 1850[101] to enable MPs to assess the seating arrangements and evaluate the audibility of speech in different parts of the chamber.[102] However, the MPs found the acoustics to be unsuitable for debates. *The Times* reported that MPs on the main floor were unable to hear the Speaker of the House, and that those sitting in the gallery at the bar end complained that they could not follow the debates on the floor below.[103]

Another House of Commons Select Committee, chaired by Greene, was appointed to resolve the acoustics problem.[104] Several scientists, which included Reid himself, John Scott Russell, Charles Wheatstone and Michael Faraday, were consulted. However, the Select Committee failed to agree on a solution and as a result, The Office of Woods subsequently asked Barry to improve the acoustics through trial and error.[105] There was precedent for this: as the concept of room acoustics, and accurate methods for calculating and modelling the sound properties of spaces, were not developed until the 20th century,[106] the Victorians relied on undertaking experiments using full-scale structures to evaluate and improve the acoustic performance of enclosed spaces. Indeed, the history of large concert halls, such as the Royal Albert Hall in South Kensington, and Crystal Palace (after its relocation to Sydenham Hill, now in the London Borough of Bromley) also show that the use of experimental adaptations was not uncommon.

Rectifying the poor acoustics incurred additional costs: £500 for the experiments themselves and £6,200 for the construction of the new ceiling.[107] The Chancellor of the Exchequer argued that the cost of such experiments was unavoidable due to the lack of reliable guidelines on acoustic design.[108]

A first new ceiling in the permanent House of Commons was installed in July 1850.[109] It had a horizontal panel in the centre, which was 8 ft (~2.5 m) below the original ceiling, and was surrounded by sloping side panels, which were intended to reflect the sound back to the main floor and into the galleries.[110] This fundamentally changed the spatial proportions and light quality of the interior.[111] The new debating chamber, which mirrored the interior of the chamber in the House of Lords, originally had a high, completely horizontal ceiling. Designed by Barry and Pugin, it was completed in June 1847.[112] According to the *Morning Chronicle* it was 'divided by carved beams into eighteen panels, each of which is subdivided into sixteen panels, both beams and panelling being of richly carved oak'.[113] The chamber also had tall windows above the side galleries. Together, the original ceiling and windows gave the space the light and airy quality of an Elizabethan hall.

This first 'new' ceiling was tested during a trail debate on 31 July but the feedback from MPs was inconclusive.[114] A second trial with a second different ceiling was conducted on 7 August 1850: in this case, the ceiling was the same shape as the first one but the edge of the sloping panels, which had covered more than half of the window area, was raised to increase the amount of natural light.[115] Barry writes that the acoustics achieved through this second ceiling received 'general approval on the part of the Members of the House, as being efficient for that purpose'.[116]

The second 'new' ceiling was replaced with a permanent oak ceiling, which was completed in July 1851 (see Figure 7.8).[117] Reid adapted the

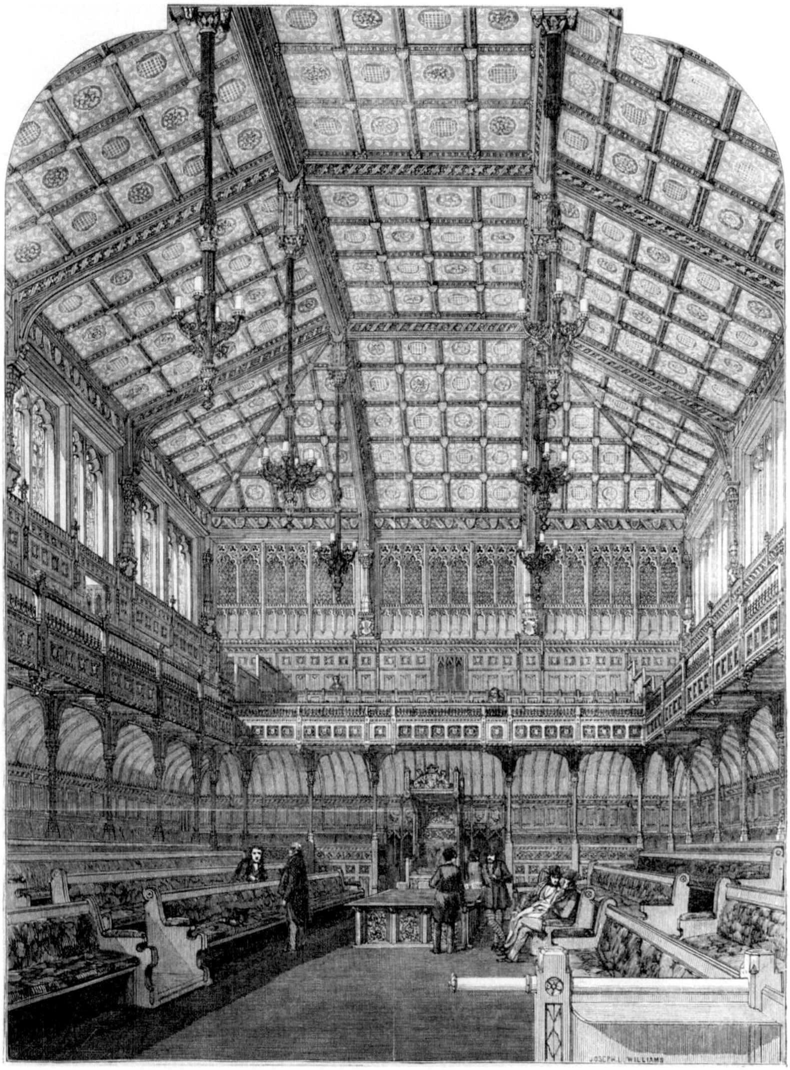

Figure 7.8 Interior of chamber in 1852 after alterations (*Illustrated London News*, 7 February 1852, p.121).

fresh-air and vitiated-air chambers above the new ceiling to fit it,[118] and advised Barry on the details of the apertures needed for the extraction and supply of air.[119] Similar to the original ceiling, the air passed through gaps between the panels and carved beams.[120] There was a one inch gap between the panels and the ornamental beams to enable air flow.[121] In addition hot air was conveyed into the vitiated-air chamber through openings inside the hollow ornamental beams of the ceiling.[122]

Assisted by Benjamin Hall, MP for Marylebone (who was a civil engineer and member of the Select Committee), Barry finalised the seating layout in July 1850. The main floor could now seat 318 MPs; an additional 150 MPs could be accommodated in the side galleries, which had been widened to hold two rows of benches.[123] Barry's team developed the final working drawings for the floor, which incorporated Reid's details, based on this new layout.[124] Tracings of Barry's working drawings were taken by Reid for review, who returned them replete with written annotations and sketches explaining any necessary revisions. Copies of a first set of drawings were forwarded to Reid between October and December 1850.[125] In January 1851 Reid returned revised details for the floor and benches,[126] and further refinements were made between June and November.[127]

Conclusion: a physical manifestation of a political decision

Resolving the details for the interior of the House of Commons debating chamber would prove to be an important step in the development of Reid's final ventilation scheme as seven years had already elapsed since it had been separated from the central ventilation network of the new Palace. The physical consequences of political decisions made in 1846 by the various Select Committees can be clearly seen in the final ventilation scheme.

Some features, such as the diagonal partition walls in the basement and inside the Central Tower at roof level, were introduced to create physical boundaries between the ventilation systems of the two Houses, thereby reflecting the political border drawn between Reid's and Barry's areas of responsibility in 1846. This border was the outcome of long and intense negotiations, not the least as new features, such as the fresh-air supply, occupied areas in the roof that were in Barry's territory. The Central Tower, once a central feature of Reid's original masterplan, and how it was to be repurposed, was the subject of negotiations between Reid and Barry for over 12 months and as a result, it became physically isolated from the House of Commons.

In order to realise a decentralised approach to stack ventilation, a second new ventilation shaft was added (only one of numerous new shafts that were added to the roof after 1846). Although the scale of Reid's involvement in the design of the new Palace of Westminster was significantly reduced, the collaboration between Reid and Barry continued to be beset by problems. As has been shown in this chapter, the design of the ventilation system was significantly delayed due to these. The House of Commons system is significant as it was the only part of Reid's original plans that was ever fully realised.

Reid was subsequently employed for a Parliamentary term to oversee the House of Commons ventilation system's day-to-day operation, during which time he gained invaluable experience of its performance in use. In the next chapter, the post-occupancy history will show that the ventilation system did not work optimally and, similar to Barry's system in the House of Lords, it thus became subject to extensive scrutiny from its occupants – Members of Parliament – as well as external scientific advisors.

The prevailing issue of institutional oversight through the Office of woods was readdressed through the appointment of a second management body, the HM Commission. This resulted in a constellation of two executive bodies not dissimilar to the Sponsor Board/Delivery Authority model used in contemporary infrastructure projects, such as Crossrail, the 2012 London Olympics and the current Palace of Westminster Restoration and Renewal Programme.[128] In the current context, the Office of Woods would be the sponsor board – a single client body acting on behalf of Parliament – whilst the HM Commission would be the delivery authority (although, unlike its modern counterpart, it was not an arms-length organisation). Whilst the latter body dealt with the practicalities of project management, the former body was responsible for the review and approval of any major alterations and budget extensions that the HM Commission or Barry himself were considering.

Notes

1 House of Commons (HC) 177 1846, *Report from the Select Committee on Westminster Bridge and new palace*, p. 3.
2 Ibid.
3 Example of later drawings produced within office are: Reid's 29 June 1848, Work 29/3068; 17 June 1849, Work 29/3068; 22 January 1850, Work 29/3078).
4 Set of Drawings, 5 April 1847 (National Archives: Works 29/2905 to 29/2959).
5 Oral statement by D.B. Reid on 14 August 1846, in House of Lords (HL) 719 1846, *Third Report from the Select Committee appointed to inquire into the progress of the building of the Houses of Parliament*, Q29.
6 Barry, C. Letter to Henry Cole, 23 April 1849 (National Archives: Work 11/16, no. 16); Letter to HM Commissioners for the completion of the New Palace of Westminster, 3 May 1849 (National Archives: Work 11/16, no. 18).
7 Greene, T. Letter to Sir Burgogne, 5 August 1850 (National Archives: Work 11/16, no. 34).
8 'Architect's Report on the present state of works', 30 June 1847, HC 684 1847.
9 The set of original drawings, included, amongst others: 'Sections of ground and vault floors, north end, showing back movement of supply', 1843 (National Archives: Work 29/2841); 'Plan of the ground floor at the north end of the House, showing situations of air chambers for back movement of supply', March 1843 (National Archives: Work 29/2853); 'Plan of ground and principal floors and sections below principal floor level of central hall and commons' corridor, showing provision for ventilation', 29 January 1844 (National Archives: Work 29/1656); 'North end. Elevations of walls under principal floor, shewing provision for ventilation', 16 July 1844 (National Archives: Work 29/2867); 'South end. Plans and elevations of walls under principal floor line, shewing manner of setting arch flues to supply the galleries above', 20 September 1844 (National Archives: Work 29/2865).
10 Reid, D.B., 'Plan and elevation of risers in east and west wall', 19 February 1844 (PRO: Work 29/2864).

11 Reid, D.B., 'House of Commons plan and section showing ventilation at north end', 11 October 1845 (PRO: Work 29/2892).
12 'No 129. Plan and section shewing air chamber of commons public corridor', 1 July 1845 (National Archives: Work 29/2890; 'no. 133. Longitudinal section of House of Commons chamber', 11 October 1845 (National Archives: Work 29/2891); 'No 136. Ventilation. House of Commons: plan and section showing ventilation at north end', 11 October 1845 (National Archives: Work 29/2892); 'No 136. House of Commons: transverse section at north end, showing passage of smoke and air into House of Commons roof', 11 October 1845 (National Archives: Work 29/2893).
13 Barry, C. Letter to H. Cole, 5 March 1849 (National Archives: Work 11/13, no. 551).
14 Barry, C. Letter to HM Commissioners for the completion of the New Palace of Westminster, 2 June 1848 (National Archives: Work 11/13, no. 456); Commissioners of Works. Letter to Reid, 19 December 1851, (National Archives: Work 11/14, no. 672).
15 'Treasury Minute', 17 March 1848, reproduced in 'Copies of Correspondence relative to the New Houses of Parliament', HC 205 1848, p. 1.
16 Ibid.
17 House of ?, *Hansard's Parliamentary Debates: The Official Report (30 May 1852, vol. ?, cols. ?)* (London: Hansard); Commissioners of Works. Letter to Reid, 19 December 1851.
18 Oral statements by C. Barry on 23 June, HL 719 1846, Q29, Q38–41.
19 Ibid., Q41–2.
20 'Tracing of Central Tower', 15 June 1847 (National Archives: Work 29/3050).
21 Gore, C. Letter to C. Barry, 27 July 1841 (PRO: Work 11/12, no. 68).
22 Drawing, 5 April 1847 (National Archives: Works 29/3009, 29/3027).
23 Drawing no. 29, 5 April 1847, signed by D.B. Reid (National Archives: Work 29/3046).
24 'Ventilating tower at north end of the House of Commons', signed by C. Barry, 1 May 1847 (Work 11/12, no. 298).
25 Reid, D.B. Letter to Commissioners of Woods and Forests, 14 May 1847, including answers to Barry questionnaire of 10 May 1847 (National Archives: Work 11/12, no. 297).
26 Drawing dated 31 May 1847 (National Archives: Work 29/3051), Drawing dated 1 June 1847 (National Archives: Work 29/3047).
27 Two drawings dated 31 May 1847 (National Archives: Works 29/3047, 29/3051).
28 Drawings, dated 5 April 1847, signed by D.B. Reid (National Archives: Work 29/3009, 29/3046).
29 'List drawings with brief explanatory remarks', 5 April 1847, submitted by D.B. Reid to A. Milne (National Archives: Work 11/12, no. 271).
30 Barry, C. Letter to A. Milne, 4 June 1847, reproduced in 'Copies of correspondence…', HC 205 1848, pp. 1f.
31 Oral statement by Jacob Prim, 20 July 1891, HC 371 1890–91, *Report from the Select Committee on House of Commons (ventilation)*, Q949.
32 'Drawing no. 48', discussed in Dr Reid's letter of 15 June 1847 (National Archives: Works 29/3050).
33 Barry, C. Barry, C. Letter to A. Milne, 10 June 1847, reproduced in 'Copies of correspondence…', HC 205 1848, 205, p. 2.
34 Ibid., pp. 2f.
35 'Drawing no. 49' with Reid's commentary, 15 June 1847 (National Archives: Work 29/3048).

36 Reid, D.B. Letter to C. Barry, 9 May 1845 (National Archives: Work 11/12, no. 139-40); Barry, C. Letter to D.B. Reid, 10 May 1845 (National Archives: Work 11/12, no. 140–41).

37 Barry, C. Letter to Commissioners of Woods and Forests, 6 July 1847 (Work 11/12, no. 308).

38 Reid, D.B. Letter to Commissioners of Woods and Forests, 19 July 1847 (National Archives: Work 11/12. no. 302–5).

39 'Drawing 53', 19 July 1847 and signed Reid (National Archives: Work 29/3053).

40 Drawings from Reid's office, 19 July 1847 (National Archives: Works 29/3054, 29/3055).

41 Notes on drawing from Reid's office, 19 July 1847 (National Archives: Work 29/3055).

42 Note that the northern shaft does not appear in these drawings because Barry had reappropriated it to use as a combined smoke and ventilation tower for the Speaker's Court.

43 In the following paragraphs, numbers or letters in **BOLD** parentheses refer to the labels depicted in the relevant figures.

44 'The Ventilating Shaft', *Illustrated London News*, 16 December 1854, p. 621.

45 Commissioners of Woods and Forests. Letter to D.B. Reid, 9 December 1847 (National Archives: Work 11/13, no. 356).

46 Barry, C. Letter to A. Milne, 3 February 1848 (National Archives: Work 11/13, no. 370).

47 Commissioners of Woods and Forests. Letter to D.B. Reid, 9 February 1848 (National Archives: Work 11/13, no. 374).

48 Barry, C. Letter to Commissioners of Woods and Forests, 16 February 1848 (National Archives: Work 11/13, no. 376); Reid, D.B. Letter to Commissioners of Woods and Forests, 22 February 1848 (National Archives: Work 11/13, no. 378).

49 Milne, A. Letter to D.B. Reid, 26 February 1848 (National Archives: Work 11/13, no. 383).

50 Ibid.

51 Commissioners of Woods and Forests. Letter to D.B. Reid, 24 February 1848 (National Archives: Work 11/13, no. 379) Reid, D.B. Letter to Commissioners of Woods and Forests, 25 February 1848 (National Archives: Work 11/13, no. 381); Reid, D.B. Letter to A. Milne, 2 March 1848 (National Archives: Work 11/13, no. 384).

52 Milne, A. Letter to Reid, 26 February 1848 (National Archives: Work 11/13, no. 383); Reid, D.B. Letter to A. Milne, 2 March 1848 (National Archives: Work 11/13, no. 384); Barry, C. Letter to A. Milne, 7 March 1848 (National Archives: Work 11/13, no. 386).

53 Barry's replies accompanying his letter to Commissioners of Woods and Forests, 16 March 1848 (National Archives: Work 11/13, no. 395); Queries to Mr Barry as the appropriation of the upper portion of the central tower, accompanying Dr Reid's letter to the Commissioners of Woods and Forests, 13 March 1848 (National Archives: Work 11/13, no. 395).

54 'Barry's replies ...', 16 March 1848 (National Archives: Work 11/13, no. 395).

55 Reid, D.B. Letter to the Commissioners of Woods and Forests, 13 March 1848 (National Archives: Work 11/13, no. 394).

56 Barry, C. Letter to Commissioners of Woods and Forests, 16 March 1848 (National Archives: Work 11/13, no. 400); Barry, C. Letter to Department of Woods and Forests, 18 March 1848 (National Archives: Work 11/13, no. 402).

57 Reid, D.B. Letter to the Commissioners of Woods and Forests to Barry, 20 March 1848 (National Archives: Work 11/13, no. 404); Queries to Mr

Barry, accompanying Dr Reid's letter addressed to the Commissioners of HM Woods, 21 March 1848 (National Archives: Work 11/13, no. 406).

58 Reid, D.B. Letter to Woods, 24 March 1848, 2 PM (National Archives: Work 11/13, no. 414).

59 Diagram accompanying Dr Reid's letter to the Commissioners of Woods, 5 April 1848 (National Archives: Works 29/3061); Reid, D.B. Letter to Alexander Milne, 5 April 1848 (National Archives: Work 11/13, no. 426).

60 Barry, C. Letter to the Office of Woods, 6 July 1847 (National Archives: Work 11/12, no. 308).

61 Letter from Department of Woods and Forests, 12 April 1848 (National Archives: Work 11/13, no. 439); Letter from the Department of Woods to Reid, 20 April 1848 (National Archives: Work 11/13, no. 440).

62 Reid, D.B. Letter to H. Cole, 26 April 1848 (National Archives: Work 11/13, no. 441–2); 'Drawing no. 58', 26 April 1848 (National Archives: Works 29/3064).

63 Ibid.

64 'Drawing no. 47' (National Archives: Works 29/3047) and 'Drawings 46' (National Archives: Works 29/3051), accompanying Dr Reid's Letter to the Commissioners of H. M. Woods, 31 May 1847.

65 Four tracings enclosed in a letter from C. Barry to the Commissioners of New Palace, 5 May 1848 (National Archives: Work 11/13, no. 441–2).

66 Reid, D.B. Letter to the Office of Woods, 15 May 1848 (National Archives: Work 11/13, no. 446).

67 Reid, D.B. Letter to Commissioners for the completion of the New Palace of Westminster, 29 June 1848 (National Archives: Work 11/13, no. 462); 'Drawing no. 64', 29 June 1848 (National Archives: Work 29/3068).

68 'Plan and section showing space available for ventilation between cone of central tower and contiguous parapet and proposed arrangements for supply of air to ceiling when obtained from the river front', 22 May 1848 (National Archives: Work 29/3065).

69 Reid, D.B. Letter to the HM Commissioners for the completion of the New Palace of Westminster, 22 May 1848 (National Archives: Work 11/13, no. 453).

70 Barry, C. Letter to H. Cole, 5 March 1849 (National Archives: Work 11/13, no. 551); H.M. Commissioners for the completion of the New Palace of Westminster. Letter to the Treasury, 10 March 1849 (National Archives: Work 11/13, no. 550); HM Treasury. Letter to HM Commissioners for the completion of the New Palace of Westminster, 28 March 1849 (National Archives: Work 11/13, no. 558).

71 HM Commissioners for the completion of the New Palace of Westminster. Letter to HM Treasury, 10 March 1849.

72 Pencil sketches and drawings of Central Tower are kept in the RIBA Drawing Collection and Parliamentary Archives, e.g. undated, Study of central lantern and octagon, undated, (Parliamentary Archives: Arc/RIBA/222); Four pencil sketches of Central Tower (Parliamentary Archives: Arc/RIBA/218–21); Preliminary studies for Central Tower (RIBA Drawing Collection: SA36/6–9).

73 Reid D.B., 'Drawing showing modification proposed for the supply of fresh air', 7 June 1848 (PRO: Work 29/3066).

74 Reid, D.B. Letter to HM Commissioners for the completion of the New Palace of Westminster, 29 June 1848 (National Archives: Work 11/13, no. 462); Drawing no. 64 (National Archives: Work 29/3068).

75 Reid D.B., 'Section through roof over St Stephen's Hall showing construction and arrangement of smoke and air flues', 29 June 1848 (PRO: Work 29/3067); Reid D.B., 'Plan and section over the groins of St Stephen's Hall and St Stephen's Porch showing flues and air channels in roof', 27 December 1851

(PRO: Work 29/2960); Oral statements by D.B. Reid on 26 March, HC 402 1852, Q373–407.

76 Reid D.B., 'Plan and section of roof space between House lobby and central tower', 5 April 1847 (PRO: Work 29/3046); Reid D.B., Details of new apparatus in air chamber in the roof over Commons' corridor, 22 January 1850 (PRO: Work 29/3083).

77 Barry, C., 'Plan of Principal Floor', 10 January 1851 (PRO: Work 29/3090); Oral Statement by C. Barry on 5 April, HC 402 Ev 224 Q1491); 'Dr Reid's Arrangement for Warming and Ventilating the New Houses of Commons', *Civil Engineer and Architect's Journal* (September 1852), 'Dr Reid's Arrangement for Warming and Ventilating the New Houses of Commons', 292–3.

78 These valves are shown in several of Reid's drawings, for example: 'Longitudinal Section', 5 April 1847 (PRO: Work 29/2905); 'Plan of roof', 5 April 1847 (PRO: Work 29/3027).

79 Reid D.B., 'Plan, elevations and sections of vitiated air shaft', 24 February 1848 (PRO: Work 29/2916); 'Plans and sections showing valves at top of vitiated air shaft', 19 July 1847 (PRO: Work 29/3054); 'Cross section through House and divisional lobbies looking south' (PRO: Work 29/2994); 'Longitudinal section through House and Commons lobby looking west' (PRO: Work 29/2993).

80 Reid D.B. *Statement explanatory of the arrangements for warming and ventilating the new House of Commons*, 5 April 1852, *in* GB. Parliament. 1852a, 545–48. HC 650 1850, *First Report.*

81 Oral Statement by D.B. Reid on 30 April, HC 402 1852, Q3573.

82 Barry, C., Letter to HM Commissioners for the completion of the New Palace of Westminster, 6 July 1848 (National Archives: Work 11/13, no. 464); Oral Statement by D.B. Reid on 26 March, HC 402 1852, Q402–3.

83 Reid D.B., 'Plans of basement and ground floor', 5 April 1847 (National Archives: Works 29/3007, 29/3008); 'Plan and section of Central Chamber', 26 April 1848 (PRO: Work 29/3063).

84 Barry, C., Letter to HM Commissioners for the completion of the New Palace of Westminster, 6 July 1848 (National Archives: Work 11/13, no. 464).

85 Reid D.B. *Statement explanatory of the arrangement, in* HC 650 1850, *First Report*, pp. 545–48.

86 Reid D.B., 'Plan of equalizing chamber', 5 April 1847 (National Archives: Work 29/3014); 'Plans and sections of principal heating apparatus', 5 April 1847 (PRO: Work 29/3026); 'Detail of air valves', 5 April 1847 (National Archives: Work 29/3041 to 29/3043); 'Cross-section', 10 January 1851 National Archives: Work 29/3094).

87 Oral statements by S.W. Daukes on 5–6 April, HC 402 1852, Q1812, Q1931.

88 Oral Statement by D.B. Reid on 20 April, HC 402 1852, Q361.

89 Oral statement by D.B. Reid on 30 April, HC 402 1852, Q3545.

90 Barry, C., 'Ceiling above equalizing chamber with supply tubes, valves and flaps', 28 June 1851 (National Archives: Work 29/3100); 'Section through gangway and seats', 10 January 1851 (PRO: Work 29/3093).

91 Oral statements by D.B. Reid on 26 March, HC 402 1852, Q346-362; Oral statements on 17 June, HC 448 1844, Q317–28, Q387.

92 Reid D.B., 'Plan of basement', 5 April 1847 (National Archives: Work 29/3007).

93 Reid D.B., 'Details of underfloor plate radiator, 22 January 1850', (National Archives: Work 29/3081-89); 'The New Houses of Commons', *Builder,* 1851 (9-442), pp. 460–61.

94 Barry, C. Letter to H. Cole, 20 February 1849 (National Archives: Work 11/13, no. 546); Barry, C. Letter to D.B. Reid, 28 February 1849 (National

Archives: Work 11/13, no. 549); Reid, D.B. Letter to New Palace Commission, 22 December 1848 (National Archives: Work 11/13, no. 549).

95 'The New House of Commons', *Daily News*, 31 March 1850, p. 5; 'The New House of Commons', *Hansard's Parliamentary Debates: The Official Report (16 February 1849, vol. 102, cols. 756–8)* (London: Hansard); 'The New Houses of Parliament', Houses of Commons. *Hansard's Parliamentary Debates: The Official Report (25 April 1850, vol. 110, cols. 795–7)* (London: Hansard).

96 'Sections of iron floors and iron framing for seating', 1848, (National Archives: Work 29/1291); 'Details of iron floor to the House of Commons, showing construction of binders, etc. for seating', 1848, (National Archives: Work 29/1290); 'House of Commons seating. Elevations and sections of members benches', 1849, (National Archives: Work 29/1485); 'Sections of seats, as originally intended and as altered' (National Archives: 29/1486).

97 HC 325 1849. New Houses of Parliament. A return, showing how many members the new House of Commons will accommodate, specifying the space allowed for each.

98 'The new House of Commons', Houses of Commons. *Hansard's Parliamentary Debates: The Official Report (10 May 1850, vol. 110, cols. 1314–5)* (London: Hansard).

99 'The New Houses of Parliament', Houses of Commons. *Hansard's Parliamentary Debates: The Official Report (28 May 1850, vol. 110, col. 433)* (London: Hansard).

100 'The New Houses of Parliament', Houses of Commons. *Hansard's Parliamentary Debates: The Official Report (25 April 1850, vol. 110, cols. 795–7)* (London: Hansard).

101 'The House met at 12.00pm in the New House of Commons', Houses of Commons. *Hansard's Parliamentary Debates: The Official Report (11 June 1850, vol. 111, col. 1067)* (London: Hansard). 'Mr Speaker took the chair at Twelve of the clock in the New House of Commons', Houses of Commons. *Hansard (12 June 1850, vol. 111, col. 1119)*. 'Mr Speaker took the Chair at Four of the clock in the New House of Commons', *Hansard (13 June 1850, vol. 111, cols. 1158–9)*.

102 'The new Houses of Parliament', House of Commons. *Hansard;* 'The new House of Commons', House of Commons. *Hansard*.

103 'Opening of the House of Commons', *The Times*, 31 May 1850, p. 5.

104 HC 650 1850, *First Report of the Select Committee on New House of Commons*; 'The New House of Commons', *Hansard (2 August 1850, vol. 113, cols. 726–39);* The New House of Commons, *Hansard (16 May 1850, vol. 111, cols. 102–4);* 'Temporary Accommodation for the Houses of Parliament', *Hansard (24 May 1850, vol. 111, cols. 328–59, 328)*.

105 'House of Commons', *The Times*, 3 August 1850, p. 2.

106 Bill Addis, 'A brief history of design methods for building acoustics', in *Proceedings of the Third International Congress on Construction History, Brandenburg University of Technology Cottbus, Germany*, ed. by Karl-Eugen Kurrer, Wener Lorenz, & Volker Wetzk (May 2009), pp. 1–10.

107 Barry, C. Letter to New Palace Commissioners, 3 September 1850 (Work 11/16/2, no. 1); H.M Commissioners for superintending the completion of the new Palace of Westminster. Letter to G.P. Lewis, 10 September 1850 (Work 11/16/2, no. 3).

108 'Alterations at the Houses of Parliament', Builder, 1850, (8–392), p. 377.

109 'Imperial Parliament', *Standard*, 31 July 1850, p. 2; 'Imperial Parliament', *John Bull*, 26 July 1851, p.471.

110 'Parliamentary Intelligence', *The Times*, 1 August 1850, p. 3.

111 'The New House of Commons from the Bar', *Illustrated London News*, 7 February 1852, p.121.
112 'Architect's Report...', 30 June 1847, HC 684 1847.
113 'Opening of the New House', *Morning Chronicle*, 31 May 1850, p. 2.
114 'House of Commons', *The Times*, 3 August 1850, p. 2.
115 'Parliamentary Intelligence', *The Times,* 8 August 1850, p. 4; 'House of Commons', *John Bull*, 12 August 1850, p. 501.
116 Barry, C. Letter to Commissioners for superintending the completion of the New Palace at Westminster, 3 September 1850 (National Archives: Works 11/16 (2), no. 9).
117 'New House of Commons has been remodelled', *The Times*, 26 July 1851, p. 5; 'House of Commons', *Morning Post*, 4 February 1852, p. 3.
118 (National Archives: Work 29/3094); 'Plan and sections shewing air chamber and site of fanner in roof', 8 February 1851 (National Archives: Work 29/2940).
119 Reid, D.B. Letter to the Commissioners for the New Palace, 31 October 1850 (National Archives: Work 11/16 (2) no. 9).
120 'Reid's statement explanatory of the arrangements for the warming and ventilating of the New House of Commons', 5 April 1852, in HC 402 1852, *Second report from the Select Committee on Ventilation and Lighting of the House*, pp. 545–8).
121 Oral statement by C. Barry on 5 April, HC 402 1852, Q1491; 'Dr Reid's Arrangement for Warming and Ventilating the New Houses of Commons', *Civil Engineer and Architect's Journal*, 1852 (222-15), pp. 292f.
122 'Plan of new House of Commons at principal floor level showing ventilating flues', 10 January 1851 (National Archives: Work 29/3090).
123 HC 650 1850, *First Report of the Select Committee on New House of Commons*; 'Imperial Parliament', *Morning Chronicle*, 8 August 1850, p. 2; 'Imperial Parliament', *Daily News*, 8 August 1850, p. 2.
124 HC 650 1850.
125 Note on drawing (National Archives: Work 29/3098); Barry, C. Letter to H. Cole, 12 October 1850 (National Archives: Work 11/16 (2), no. 8).
126 'Plan and details', 27 June 1851 (National Archives: Work 29/3092); Reid, D.B. Letter to New Palace Commissioners, 10 January 1851 (National Archives: Work 11/13, no. 621); Set of drawings, 10 January 1851 (National Archives: Work 29/3074 to 29/3100).
127 'Section through raised platforms and also through seats showing system of under-seat ventilation', 10 January 1851, (National Archives: Work 29/3093).
128 Joint Select Committee on Restoration and Renewal of the Palace of Westminster (HL Paper 41/ HC 659 2016); Joint Committee on the Draft Parliamentary Buildings Bill – Governance of Restoration and Renewal (HC 1800/HL Paper 317 2017–19).

PART IV

Critical evaluations and adaptation: 1852–1950

May I burnish my credentials with those who say that we have to get on with this work by saying that 25% of the space in the Palace is currently taken up by unseen historical ventilation systems?

This debate has often falsely been portrayed as a debate between pragmatic modernisers such as the hon. Member for Rhondda (Chris Bryant), who by the way I cannot help liking and respecting—I do not know why, but I do—and traditionalists who are putting their own comfort first. Actually, the arguments are far more complex. I have spent quite a lot of time over the past few years talking to architects and mechanical and electrical engineers, and I personally believe that compared with the present plans, there are greener, cheaper, faster and better solutions

Sir Edward Leigh, MP for Gainsborough,
16 July 2020[1]

Note

1 Restoration and Renewal, Hansard, House of Commons, *Hansard's Parliamentary Debates: The Official Report* (Vol 678, col. 1763).

8 Reid's short-lived experiments in the House of Commons: 1852–54

Thermometer tells one tale, and the human body tells another.

John Leslie, 1852 [1]

Introduction: studying Reid's system in use

The post-occupancy history of the House of Lords illustrates that Charles Barry did not succeed in achieving his core design objective in practice, namely, to sustain an optimal thermal environment during sittings of Parliament. The Lords were not comfortable in their debating chamber and, following the recommendations of a Select Committee, Alfred Meeson and Barry's system was decommissioned after only seven years. The quality of the Lords' experience became the primary criteria for assessing the chamber during the investigations. As suggested in Chapter 6, this is a historic example of what contemporary researchers in building science, such as Coleman and Robinson have termed a 'qualitative performance gap': that is, the 'perceived gap between what inhabitants expect and their actual experience of the building'.[2] This performance gap also became an issue for the users of Reid's ventilation system for the House of Commons.

Reid wanted to create a ventilation arrangement that would enhance user-satisfaction, an ambition he pursued over a period of 15 years starting with the temporary House of Commons in 1836.[3] The post-occupancy history of the House of Commons, which is the focus of this chapter, will show that Reid faced even more serious difficulties than those encountered by Barry (see Chapter 6). Indeed, Reid's system was decommissioned after only two years. Throughout its operational life between February 1852 and April 1854, its performance was continually monitored and also became the subject of several formal investigations, including two studies led by Select Committees. The historical data and related studies offers an intimate insight into the actual performance of Reid's system and how it was assessed by MPs, staff and technical advisors, revealing the technical challenges as well as political pressure that ultimately led to its withdrawal from service

Insufficient pre-occupancy trails: 1851–52

From the time he was originally commissioned to work on the Palace of Westminster the Office of Woods intended to employ Reid to manage and fine tune the operation of his ventilation system for a full year.[4] In the summer of 1851, Reid began preparing for this, discussing with the HM Commission the recruitment of operational staff, the development of operational regimes and a series of first technical tests. It is clear from Reid's letters that he was worried that the ventilation system would fail to perform as intended if he was not given the opportunity to test and fine-tune its operational procedures. In a letter to the HM Commission, dated 3 July 1851, Reid argued that it would be necessary to recruit and train engineers, and to test and refine operational procedures before the House was formally occupied in February 1852. He wrote that his objective was to 'examine the working of the ventilation apparatus, draw up such rules and instructions in reference to it as can be determined with most advantage after a full trail'.[5]

These trials were to be conducted when the debating chamber was occupied during a series of mock sittings, which Reid proposed to undertake in the the autumn of 1851. The HM Commission initially rejected this proposal, arguing that conducting the trials in January 1852 would provide sufficient opportunity to test and refine procedures.[6] This only gave Reid one month to ensure the system was operating effectively.

In a letter to the HM Commission, dated 25 July 1851, Reid again requested permission to conduct tests before the end of the year as he wanted to assure himself 'that the apparatus is placed on that footing in which it ought to be before the January trials commence'.[7] With the backing of Lord Charles Russell, the Sergeant-at-Arms who had first-hand experience with the operation of the ventilation system of the temporary House of Commons, Reid gained the support of Lord John Manners, MP for Colchester, who persuaded the Commission to rethink their earlier decision.

On 18 August 1851 the HM Commission finally authorised Reid to conduct the trials as soon as possible.[8] Unfortunately, Reid was only able to conduct a few short technical trials by the end of the year due to delays in the construction of the House of Commons.[9] In January 1852, when Reid was originally supposed to conduct tests, Charles Barry was conducting trials of the gas lighting in the chamber, which were under his and not Reid's jurisdiction.[10] As a result several technical issues could not be successfully resolved in advance. In December 1851 the steam engine that drove the fan for the air supply was found to be too loud, disrupting debates, and was thus deactivated.[11] In early February 1852, staff were employed to manually operate the fans instead.[12] The records show that Reid's concern about the potential failure of his system was not unjustified and in February 1852 he was already facing significant operational challenges. Before discussing these, however, it is important to examine the design of the operational procedures, and the roles of staff that were employed in its day-to-day operation, to systematically monitor its performance in use.

Gathering intelligence on performance in use: 1852–54

From February 1852 until April 1854 the performance of the ventilation system was systematically monitored by staff, who followed operational procedures that Reid had previously tested and refined in the temporary Houses of Parliament. Reid, who acted as the superintending engineer of the ventilation system, and his technical staff were employed by the Office of Works and Public Buildings, who paid an annual sum of £1,280 to cover their wages and the cost of regular maintenance of, and repairs to, the system.[13] This sum, however, did not cover the full operational costs of the system as responsibility for covering these was shared with the Department of the Serjeant at Arms, which was responsible for estate management. For example, Reid was allocated annual budgets of between £400 and £500 for the procurement of coal and coke, and £1,000 for the regular cleaning and repair of the debating chamber's horse-hair carpet. The Sergeant-at-Arms was also in charge of monitoring operations whilst the House was sitting and he employed his own 'attendant on ventilation', who earned an annual salary of £105, for this purpose.[14]

The Superintending Engineer had a small private office in the south-west corner of the Members' Lobby, which ensured that he was in close proximity to MPs and messengers.[15] Staff responsible for the technical operation of the system did not interact with Members as they were stationed inside the air chambers themselves, and, unlike the messengers of the Sergeant-at-Arms, were not permitted to enter the chamber during sittings The main ventilation office, which was used for coordinating these operations, was situated on the ground floor and was linked to Reid's own private office by a staircase.[16]

Reid acted as the superintending engineer for the House of Commons from February 1852 until the Office of Works terminated his employment in October 1852. Meeson – who had worked in Barry's office and had, since 1847, been overseeing the ventilation system in the House of Lords – was appointed as the superintending engineer in Reid's stead.[17]

Feedback about the environmental conditions in the debating chamber was collected and communicated in one of two ways depending on the type of data, quantitative or qualitative (see Figure 8.1). The technical staff for the ventilation system recorded measurements inside the different air chambers below and above the House, whilst both measurements and MPs' qualitative feedback from inside the debating chamber itself were collected and transmitted to the office by a personal messenger of the Sergeant-at-Arms (who was authorised – unlike the technical staff – to be in the debating chamber during Parliamentary sittings).[18] Registers of the subjective feedback and various measurements were then sent to the main ventilator's office, where they were transcribed into a central logbook and analysed.

The original logbooks, which are held in the Parliamentary Archives in Westminster, contain standardised registers for these different types of data. For example, there are separate columns for the number of MPs in

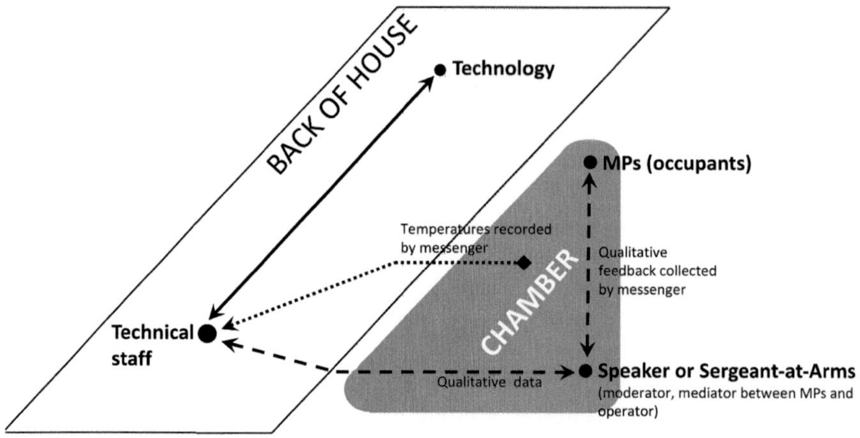

Figure 8.1 Social network enabling the collection, processing and communication of qualitative and quantitative data within the chamber (author's own drawing).

the debating chamber, the external weather conditions (e.g. humidity, wind direction, temperature and air pressure), the internal climate (e.g. temperature, humidity and air speed), and operational data (e.g. data on changes in fuel consumption, the position of valves and heating capacity). Inside the House of Commons, the temperature of the debating chamber itself was measured by eight thermometers: four were fixed to the back wall of the galleries,[19] whilst the other four were on the main floor near the chairs of the Speaker of the House and the Sergeant-at-Arms and behind the benches on the Opposition and Government sides.[20] The internal humidity was only measured by the technical staff in the ventilation chamber below the floor of the debating chamber, before the fresh air was admitted into it.

Each page of the central logbooks also had a wide margin, headed 'Remarks', which was used for writing notes about operational procedures, air quality (which was only monitored through direct observations), feedback and any instructions received from the House itself. Detailed notes on how the air supplies were switched in response to external air pollution conditions were also recorded in this margin. References to orders and feedback can also be found in the logbooks.[21] For example, a ventilation attendant writes that the Speaker of the House was 'too warm' in 31 March 1853, whilst he 'complained of draughts round his head' on 13 April; on 7 April the Sergeant-at-Arms is noted as having 'wished the House [was] a little cooler'. On 6 March 1854, the ventilation attendants noted that the atmosphere was 'very foggy and charged with smoke' and that the supply was switched to the Central Hall as that from the Clock Tower was very smoky. This problem persisted for two weeks, with the attendants remarking that switching the supply made the air 'better but not good'. On one occasion a 'foggy atmosphere loaded with smoke of the neighborhood penetrated the building'.

Any feedback received from MPs was carefully reviewed by the Sergeant-at-Arms, Lord Charles Russell, before any orders were given to the Superintending Engineer (see Feedback path 1 in Figure 8.1). Lord Russell reported that he was the 'medium of communication, as respects the ventilation, between Dr Reid and the Members'[22] and also highlighted the challenge of amalgamating MPs' feedback especially when it conflicted.

Reid hoped to collect routinely a wide variety of different types of environmental data in the new debating chamber. However, the logbooks show that the quantity of recorded data that the attendants collected varied significantly from day to day and the attendants therefore rarely succeeded in collecting the sheer volume of data required to completely fill these registers. This is not surprising as monitoring the ventilation system was a very labour-intensive procedure: every reading was recorded individually by hand without the assistance of automated recording devices or modern digital data loggers. To gain a full set of temperature readings alone, attendants had to take over 50 readings per hour, each of which had to be manually logged in different locations. Reid was clearly aware of this resource issue as he proposed to introduce ropes and pulleys to enable the dampers to be operated remotely, and speaking tubes and bells to improve communication between the ventilation attendants and the Ventilation Office.[23]

An unsuccessful beginning: February–March 1852

On 3 February 1852 the new House of Commons was formally inaugurated and the ventilation system went operational for the first time. It was a difficult first day for Reid and his team of attendants: They could not maintain a comfortable indoor climate and thus received numerous complaints about it from MPs. On the following day, the ventilation system became the subject of a debate during which MPs described their experiences. Joseph Hume, MP for Montrose Burghs, reported that he left the chamber as he could not bear the heat and asked for measures to 'keep the place moderately cool'.[24] Vice-Admiral Robert Fitzroy, an officer in the Royal Navy and a member of the Royal Society, mentioned that MPs were exposed 'to puffs of alternate hot and cold air'. Ralph Bernal Osborne, MP for Middlesex, called for Reid to be questioned at the Bar of the House (the boundary of the chamber beyond which guests or visitors may not pass when the House is sitting). Hume, Vice-Admiral Fitzroy and Lord Russell (the Sergeant at Arms) argued that the ventilation of the House was a complex problem that required a full technical inquiry under the direction of a Select Committee. Lord Seymour, MP for Totnes and First Commissioner of Works and Public Buildings, tried to calm the feelings in the House by stressing that 'the ventilation was not yet brought to full perfection, so that it could not be said to have had a fair trial'.

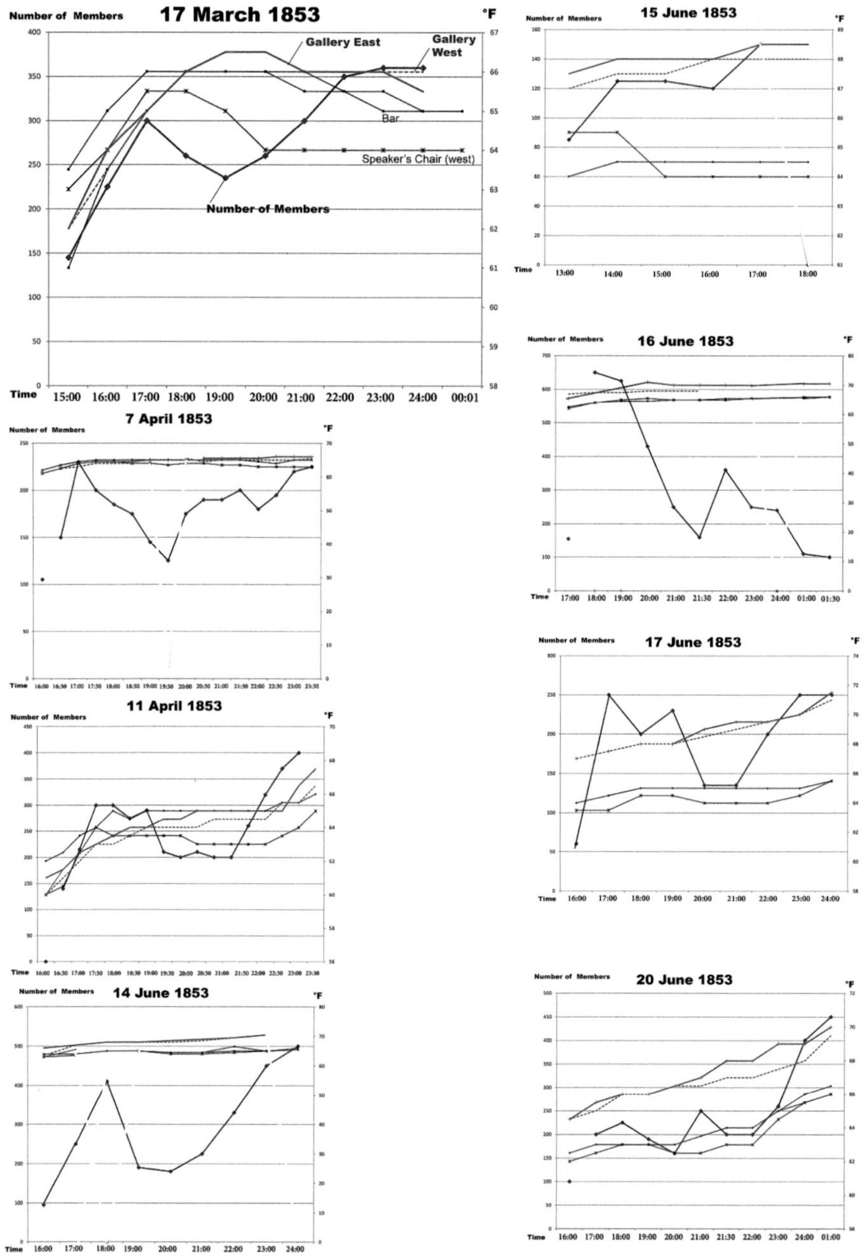

Figure 8.2 Extract from data collected inside the debating chamber, 17 March to 20 June 1853 (author's own graph).

On 6 February the ventilation of the House became the subject of another debate during which seven MPs voiced their dissatisfaction with Reid's system. The Sergeant-at-Arms admitted that the system 'was not perfect', whilst Hume observed that the large financial investment in technology had not yet yielded a solution to the longstanding problems with the internal climate of the debating chamber. George Scobell, MP for Bath, argued that the atmosphere 'had been almost insufferable' and Charles Newdegate, MP for North Warwickshire, reported that he 'suffered severely from the state of the atmosphere'. Other MPs highlighted specific problems with draughts and unstable temperatures, with Bernal Osborne reporting that he suffered from 'hot and cold blasts', whilst William Brown, MP for Lancashire Southern, referred to a 'blast of cold air'. Similarly, Lord Dudley Stuart, MP for Marylebone, claimed that the temperature was unstable and that 'Members were warm, at another time cold', which made it difficult for MPs to concentrate.

The debate on 6 February was only attended by 86 MPs. Nevertheless the motion for the appointment of a Select Committee was approved with a majority of 31 votes (55 ayes vs 24 noes).[25] Bernal Osborne's motion asking for Reid to be questioned at the Bar of the House was also approved and Reid presented himself to Parliament during the same sitting. Standing behind the bar, Reid gave a lengthy oral statement about the problems. In his statement he became very defensive and claimed that the system's problems were caused by factors outside his control.[26] He said:

> *doors were torn off in some passages leading to the House, from which gusts of air came into the house from every side. You might as well ask me to regulate the winds and currents of the Bay of Biscay, as expect me to ventilation the house if the doors and windows of the entrances leading to the house are not placed under my control.*[27]

On 7 February Reid submitted a memorandum to the Office of Works, outlining the problems and offering proposals for remedial measures.[28] In this and another letter dated 14 February,[29] he argued that his system was not working effectively as the two supply fans could not be deployed as intended: the fan for the floor level supply was being operated manually by attendants as the original steam engine was too noisy for MPs during sittings;[30] moreover, the second fan could not be deployed at all as the downward supply from the ceiling carried the hot air produced by the gas-lit chandeliers into the body of the House.[31] Indeed, on 3 February, the ceiling supply had been suspended for this reason and it was only being used in debates during the day when artificial lighting was not required.[32]

Further complaints were made during a debate on 10 February. It was too hot, MPs complained, and there were 'tremendous draughts of cold air'. The heat was also particular intense in the galleries due to the chandeliers, which not only raised the air temperature in the upper part of the chamber but also produced a strong radiant heat.[33] The next day Bernal Osborne persuaded the House to consider Reid's proposals for improvements, which included a new lighting system that the latter claimed was compatible with the downward supply. Reid was asked to produce detailed plans and estimates to be reviewed by the Select Committee in March.[34] Since the heat produced by the gas-lit chandeliers was a major cause of discomfort, Reid was granted permission to improve the lighting immediately. However, Reid's proposed lighting system was not implemented as Barry objected that that the chandeliers were an integral part of the architecture and should thus be retained.[35]

Members of Parliament continued to voice their discontent with Reid's system and on 12 March the House voted in favour of commissioning an independent technical study of it. Lord Manners, who had by this time succeeded Lord Seymour as First Commissioner of Works, recommended the Cornish surgeon and inventor Goldsworthy Gurney.[36] Gurney had acted as an advisor to the Government on several issues (including lighthouse lighting, the sewers and the ventilation of mines) and was very familiar with the issues. He had also conducted empirical assessments of Reid's system in the temporary House of Commons, which had included inquiries into how to improve the artificial lighting in the context of the ventilation system (see Chapter 3).

Gurney – assisted by a member of Select Committee, Sir Denham Jephson-Norreys, MP for Mallow – took initial spot measurements during the sitting on 19 March to examine the conditions inside the galleries. On the main floor the temperatures were as low as 61.5°F (~16°C) but rose to 68°F (20°C) on the gallery floor and 73°F (~23°C) above the seats. Sir Jephson-Norreys reported that the chandeliers produced 'a burning sensation, such as if ... exposed to a red hot iron'.[37] Logbook entries for the period 22 March to 23 April 1852 show that the temperature in the galleries fluctuated from 63°F (~17°C) to 73°F (~23°C) and was typically 2–6°F above those on the main floor, which was in the 62–70°F range.

By modern standards, the peak temperature of 73°F is not exceptionally high although it should be noted that MPs of the time wore heavy clothing and preferred lower temperatures than today's MPs.[38] Records of the set temperature for the permanent House could not be located by the author, but in the temporary House of Commons attendants maintained levels of 60–63°F (~15–17°C) in the winter and prevented temperatures from exceeding 67°F (~19°C) in the summer. Gurney claimed that 64°F (~18°C) was the 'most satisfactory temperature'.[39]

Thermal comfort is affected by a number of environmental factors including radiant temperature, relative humidity and air movement. However, up to this time, only air temperature data was routinely collected, whilst relative humidity only began to be routinely collected in December 1853.[40] The physicians Neil Arnott – who had been commissioned by Thomas Wakley in 1837 to independently evaluate Reid's system for the temporary House of Commons (see Chapter 3) – and John Leslie, who jointly reviewed Reid's monitoring system, argued that air temperature measurements on their own were not sufficient to fully understand MPs' experience, in particular the effect of air currents on their thermal comfort.[41]

A first formal investigation

Reid's system underwent a detailed performance evaluation between March and April 1852. This was coordinated by the Select Committee, which was chaired by Lord Robert Grosvenor, the MP for Middlesex, and involved the engineers Joseph Locke and Robert Stephenson as technical advisors. Interviews and two independent technical examinations were conducted to examine the climate conditions and MPs' perceptions of it. One of the technical studies was conducted by Gurney, whilst the other was jointly conducted by the architect Samuel Whitfield Daukes and the heating engineer Henry Cruger Price. The committee interviewed the Speaker of the House, the Sergeant-at-Arms and five MPs about the environmental factors that affected their experience. The civil servant Thomas Thornton reported that temperatures were unsteady and that strong currents frequently swept over the galleries, noting that in 'some parts of the evening [...] the temperature is very high, and others comparatively low'.[42] The Clerk of the House of Commons also highlighted that the air was too dry, which caused MPs to cough and experience 'considerable irritation in the chest and throat'.[43] The latter experience was also observed by the Sergeant-at-Arms and the Speaker, who reported a particularly severe instance of it on 24 March:

> *I sent once or twice to Dr. Reid to beg that he would make some change in the state of the air, for it was so dry that it caused an irritation in the throat, and I could hear the Members coughing all around.*[44]

The technical study by Daukes and Price confirmed that the climate was unstable and found that the ventilation rate was often too low, at other times it was too high, resulting in uncomfortable currents. Daukes and Price claimed that it was caused by managerial problems, with the former observing both that cold and hot air entered the House as separate currents, creating an uneven temperature across the floor, and that currents could suddenly change between hot and cold when the valves inside the air chambers below the House were adjusted. Cold and hot air were thus not sufficiently mixed to prevent these issues.[45] In their final report, Daukes and Price argued that Reid's control regime was impractical and recommended that it be simplified, for example by abandoning the practice of adjusting climates to the 'continual and conflicting wishes of individual members'.[46]

Gurney also came to similar conclusions. In his first report, dated 6 April, he writes that the simultaneous extraction and supply of air through the floor and ceiling was difficult to coordinate.[47] He further examined this issue with the assistance of three engineers – James Mather, James Hann and John Hutchinson – by conducting tests with differential barometers and anemometers. In his second report published on 13 April, Gurney reports that the ventilation rate was insufficient to counteract overheating.[48] Moreover, the atmospheric pressure inside the debating chamber was lower than it was outdoors, which caused air to enter with great force when the doors to the chamber were opened. He claimed that the quantity of vitiated air extracted through the up-cast shaft was not matched by the fresh air provided by the fan-driven supply. The fans were originally introduced by Reid to maintain this balance artificially: air was to be blown in with 'such a force that the air shall have equal pressure within and without'.[49] However, Gurney claimed that the quantities of air supplied by the fan and extracted by the stack were difficult to synchronise effectively. Instead, he proposed that a purely-stack driven system be adopted in which fresh air was able to enter naturally responding to the pull induced by the up-cast shaft.[50]

After these inquiries, and despite their conclusions, the Select Committee authorised Reid to implement the alterations that he had outlined in his memorandum and also overturned Barry's earlier rejection of Reid's proposed new lighting system.

A second investigation

The original lighting for the House of Commons was designed by Barry and Michael Faraday to harmonise with the Gothic detailing. It comprised six gas chandeliers that hanged around the horizontal section of the ceiling.[51] This lighting system was adopted after Barry had rejected a different system that Reid had designed in 1848 to be compatible with the ceiling supply. Details of Reid's proposal showing how the lighting was to be integrated into the ceiling were submitted to the Office of Woods on 10 March 1848. Aiming to cast the whole chamber in a soft and uniform light, the entire ceiling was to be covered with 336 small lights.[52] Conical light reflectors, which also functioned as hoods through which the gas fumes could be extracted upwards, were to be inserted into the centre of each ceiling panel and would terminate in flues connected to the up-cast shaft.[53] Fresh air was to be supplied downwards through gaps around the edge of each ceiling panel, whilst the up-cast shaft ensured that fumes were instantly evacuated before they could contaminate or overheat the supply air.

Reid's drawings were forwarded to Barry on 22 March 1848.[54] Barry opposed Reid's involvement in the design of lighting in several letters he sent to the Office of Woods and rejected Reid's proposed lighting scheme because it interfered with the architecture of the ceiling.[55] Instead he advocated the use of self-ventilating gas chandeliers that Michael Faraday had developed for the House of Lords,[56] where fresh air was also supplied downwards through the ceiling.[57] The lighting issue was the subject of negotiations for nine months. However, in the end, Barry persuaded the Commissioners of

Woods to adopt Faraday's system.[58] Reid wrote several very emotional letters warning the Commissioners that the lighting and ventilation needed to be designed as an integrated system.[59]

The problems encountered in February 1852 show that these warnings were not unjustified and, indeed, Lord Seymour saw them as evidence that Reid and Barry had not cooperated sufficiently on the design.[60] Reid argued that reconciling the lighting and ventilation could be achieved by returning to his original plans, and that the new lighting system installed during the Easter recess of 1852 neither adhered to his original plans nor facilitated the use of a downward supply through the central panels.[61] Despite these objections, he reorganised the ceiling system to accommodate the new lighting. The fresh air chamber in the centre of the ceiling was converted to extract vitiated air and the supply was moved to the sloping side panels. The new vitiated air chamber was connected to the up-cast shaft and gas lights were installed in 16 of the 64 oak panels.[62] Prints in the *Illustrated London News* show that it was composed of cone-shaped reflectors, in line with Reid's original proposal. However, unlike his proposal, rings with open gas flames were suspended below the reflectors so that their fumes simply rose through the top of the reflectors into the vitiated air chamber instead of being extracted through separate flues as Reid had originally intended.[63]

A first demonstration of the new arrangements was conducted during the sitting of Parliament on 19 April 1852.[64] Reid claimed that the modifications had reduced the temperature difference between the floor and gallery to 2°F (i.e. within ~1°C of each other), a claim that data in the logbooks seem to confirm. The largest recorded difference was 3°F (~1.5°C), compared to 11°F (~5°C) before the alterations.[65] Temperatures were also consistently lower and more stable. The average daily temperature inside the gallery had fallen by 2°F and varied by no more than 4°F (~3°C), compared to 7°F (~4°C) before the Easter recess. In another letter to the Office of Works, dated 29 June, Reid stressed that although the ventilation had further improved after a new steam engine for the fan driving the floor supply was introduced,[66] he was still concerned with the state of the system. He wrote that important features were 'executed promptly and in some cases in a merely temporary manner many arrangements that should now be put on a more systematic and permanent footing'.

Reid received permission to improve the system in September, but he was unable to complete this work before his contract was terminated in October 1852. [67]

Reid's ventilation system under new management

Reid's employment came under review by the Office of Works, prompted by the Select Committee's recommendation that the ventilation systems inside the Palace should be placed under one Superintending Engineer rather than the current two. The ventilation in the House of Lords was supervised by Meeson, an engineer, who had also worked as the Clerk of Works in Barry's office. In several letters Barry warns Lord Manners that appointing Reid for this new post would be a risk and that he was not prepared to tolerate Reid's interference.[68] In response, Reid threatened Lord Manners with legal action.[69] Lord Manners discussed this issue with the Chancellor of the Exchequer Benjamin Disraeli, the Prime Minister Lord Derby, and consulted the engineers Stephenson and Locke. The engineers agreed that Reid – despite his expertise in the field of ventilation – was unsuitable due to his evident inability to cooperate with Barry.[70] Lord Manners terminated Reid's employment in October 1852.[71]

The Office of Works appointed the engineer Meeson as the new superintending engineer in November 1852, and immediately set about attempting to make the chamber usable for the MPs, who were due to resume sitting on 4 November. Meeson writes that alterations to the system had to be completed quickly, which resulted in the work being roughly executed.[72] This is not surprising as Reid was not able to complete his system (let alone optimise its performance) during the nine months that he was in charge of its day-to-day operations.

Meeson undertook a first survey of Reid's system in January 1853 and in his report to the Office of Works he warned that it was in a poor condition, which was preventing it from working effectively.[73] The fan and heating system of the ceiling supply were in disrepair and critical features, such as the valves below the floor, were either not completed or poorly executed, which caused air to rise through parts of the floor unchecked.[74] However, Meeson came to the conclusion that the main problem with Reid's system was the operational procedures that Reid himself had implemented rather than the technology itself. He noted that the control procedures were too complex, as attendants had to perform a large number of tasks in different locations,[75] and that the temperature was difficult to regulate as the hot-water system could not be adjusted quickly enough to respond to the extreme fluctuations in the number of people. According to the logbooks the number of people in the chamber could range between 50 and 800 people during a single sitting, resulting in sudden changes in the internal heat load. Temperature control was thus a major issue. The Speaker of the House reported that Reid had had difficulties managing the heat load during crowded debates,[76] and the Sergeant-at-Arms observed that the temperature and air quality was highly susceptible to changes in attendance. Members of Parliament, he noted:

> *suffered from very high temperatures, which Reid was unable to control; frequently, hour after hour, I have requested him, at the desire of the Members and in accordance with my own feelings, to lower the temperature; he appeared to be unable to do so; it sometimes increased rather than diminished during the progress of the evening.*[77]

Meeson also reported that Reid's lighting system had not resolved the overheating problem inside the galleries, and could only be counteracted by a higher ventilation rate (which were neither required to maintain a good air quality nor desirable from the point of thermal comfort).

In March 1853 the 'Standing Committee on the Ventilating and Lighting of the House of Commons' was appointed to review Meeson's recommendations.[78] Chaired by the First Commissioner of Works, Sir William Molesworth, MP for Southwark, it also included Locke and Stephenson as technical consultants. Gurney was also commissioned to develop a new lighting system, and he presented his proposal to the Office of Works on 10 March:[79] the lights were to be moved into the vitiated air chamber above the ceiling and the wooden panels replaced with panes of painted glass.[80] The fumes from each light were to be conveyed to the up-cast shaft through separate flues, which would protect the firemen and attendants stationed inside the air chamber.[81]

Gurney's lighting system was tested in April 1853 and measurements taken by Gurney's assistants reveal that the heat load was phenomenal, causing the temperature in the vitiated air chamber to reach temperatures of between 89°F (~31°C) and 123°F (~50°C).[82] Meeson voiced his concern that Gurney's system could also affect the temperature inside the House.[83] To allay such fears, Gurney commissioned the engineers Thomas Mather and James Hann to take temperature measurements inside the debating chamber. Thermometers were hung below the glazed panels of the ceiling, but they failed to detect any measurable increase in temperature, even when the lights were on for longer periods.[84]

Meeson remained sceptical of their findings and in his second report he claimed that the lights caused the temperature in the galleries to rise to 74°F (~23°C).[85] To address this, he reactivated the air supply through the side panels, allowing cool air to be delivered directly into the galleries.[86] During Meeson's tenure as superintending engineer, not only was there a lack of commitment to Reid's original plans for the ventilation system but important features of it, such as the intricate supply and extraction network embedded within the floor, were abandoned. Indeed, large parts of the perforated floor were sealed to protect MPs from rising currents.[87]

A final assessment

Despite the alterations that Meeson made to the ventilation system, MPs continued to be dissatisfied, which they voiced during debates held between May 1853 and March 1854.[88] John Bright, MP for Manchester, and Sir Jephson-Norreys, both complained about uneven temperatures across the House,[89] which Richard Spooner, MP for North Warwickshire, described as 'scolding in one part, and freezing in another'.[90] Dust was also being carried up by currents rising through the floor, which reportedly caused irritation of the eyes and lungs, making it difficult for MPs to speak without drinking water. Spooner instigated several debates and lobbied for the adoption of Gurney's 1852 proposal to use a purely stack-driven system.

He was opposed by Sir William Molesworth, First Commissioner of Works, who argued that Meeson needed more time to get Reid's system working optimally.

On 10 March 1854, Spooner successfully tabled a motion for the creation of a Select Committee charged with identifying ways of improving the ventilation system.[91] As a member of the Select Committee, Spooner continued to advocate and lobby for Gurney's system and indeed it was the only proposal that the Select Committee itself considered. Gurney was commissioned to conduct another examination of Reid's ventilation system. Gurney presented a proposal two weeks later, which involved substantially remodeling Reid's system.[92] On 31 March, the Select Committee's preliminary report claimed that Reid's system was 'condemned by common consent' as unsatisfactory and advocated Gurney's scheme, which they recommended should be tested after the Easter recess.[93] On 6 April Gurney met Sir Molesworth to discuss his proposal and, on the following day, he submitted a report on the alterations that could be completed over the Easter recess. The Office of Works commissioned Gurney to test his system on the understanding that if it was unsuccessful Reid's system would be reverted to.[94] In two letters Gurney stresses that more substantial modifications would be needed to fully implement his system.[95] On 10 April Spooner quoted sections of the report to the House and tabled a motion calling for the adoption of Gurney's system,[96] which was approved by a vote.[97] Sir Molesworth was unhappy with the way the process was handled as he felt that Spooner's nine months of advocacy and lobbying – which would ultimately lead to Reid's system being decommissioned[98] – had precipitated the vote in the House before the Select Committee had conducted a full inquiry, in particular as its recommendations were based solely on evidence given by Meeson and Gurney.[99]

Testing Gurney's alternative ventilation system: April 1854

The alterations required to test Gurney's system were made between 14 and 26 April 1854 and its performance was reviewed by two Select Committees, one in the House of Commons and one in the House of Lords (which was appointed to review Barry's ventilation system). MPs were interviewed before and after the alterations were made to determine how far Gurney's system had improved their thermal comfort.

The Elizabeth Tower was converted for the tests into the main up-cast shaft, which was used to extract vitiated air through the centre of the main floor and the central ceiling panels. This air was conveyed to the tower through the former fresh air passages located in the basement. Reid's original up-cast shaft and roof-level inlets were retained, but the floor-level supply was remodelled to allow fresh air to be driven into the House solely through the pull of the two stacks. Instead of introducing air through remote inlets and long passages, which required mechanical assistance, fresh air was admitted directly from the Star Chamber and Commons Courts.[100]

Gurney also replaced Reid's hot-water apparatus with a steam-heating system to allow the temperature to be more rapidly adjusted in response to changes in the House's attendance.[101] Much care was taken to maintain a narrower range of temperatures and more optimal humidity than Reid's system.[102] Attendants were ordered to maintain temperatures between 63°F and 64°F (~17°C and 18°C).[103] Gurney also examined the internal currents of the House by suspending natural down feathers from strings across the chamber.[104] Although the logbooks did not include quantitative data for this period, interviews with MPs between May and July 1854 suggest that the internal climate had significantly improved. The Sergeant-at-Arms reported that the temperature was more tightly controlled, draughts were markedly reduced, and the atmosphere felt much fresher even after long debates.[105] Similarly, Robert Smith, MP for Northampton, testified that the atmosphere felt fresher and no longer became excessively hot.[106] Edward Cayley, an MP for North Riding of Yorkshire, noted that draughts only occurred occasionally,[107] whilst Edward Bouverie, MP for Kilmarnock Burghs, found that adjustments to the temperature by the attendants were felt more quickly than before.[108]

In its second report, dated 26 May 1854, the Select Committee agreed that Gurney's interventions had been successful in improving MPs' thermal comfort and thus formally recommended decommissioning Reid's system and permanently adopting Gurney's alternative. Indeed, MPs perceived the atmosphere as 'sensibly sweeter, fresher and purer' and believed the temperature had been brought under tighter control than before.[109]

Conclusion: Reid's plan remains unrealised

The final verdict of the Select Committee shows how central the lived experiences of MPs were in the evaluation of Reid's system. Over the course of two years, its performance was scrutinised by scientists, Parliamentary committees, and individual MPs. Both Reid and Meeson were confronted with the challenge of improving MPs' experiences. In a way not dissimilar to modern post-occupancy evaluations, several studies at the time were conducted that combined physical measurements and experiments with qualitative evidence from the House's occupants.[110] However, evaluating the performance of the ventilation system became a political process as the Select Committee inquiries were the outcome of initiatives led by a relatively small number of outspoken critics. The post-occupancy history of the House of Commons also shows that some of the difficulties encountered were also the result of political, rather than technological, factors: important features, such as the integrated lighting system and ceiling supply, were never realised due to objections from Barry. Over the short period in which Reid was in charge of running the ventilation system, Reid was also unable to complete, let alone optimise, the design and operational procedures of his system. Meeson, who succeeded Reid as superintending engineer, did not continue Reid's efforts to realise a sophisticated control strategy and when Gurney conducted his last test of Reid's system in 1854, most of the perforated floor, originally intended for the ventilation of the House, had already been sealed. It was therefore never proven whether Reid's strategy could have been made to operate effectively. In the next chapter it will be seen that despite Gurney's alterations, pressure to improve the ventilation of the House continued.

Notes

1 Oral statement by J. Leslie, 27 April 1852, in in HC 402 1852, Second report from the Select Committee on Ventilation and Lighting of the House; together with the proceedings of the committee, minutes of evidence Q3137–73.
2 Coleman S. & Robinson J. Introducing the qualitative performance gap: stories about a sustainable building, *BRI*, 46(5), 485–500 (485).
3 Oral statement by Reid, 26 March 1852, in HC 402, 1852, *Second report, Q331–540.*
4 Letter from Gore to HM Treasury, 13 July 1844, in HC 402, 1852, *Second report*, pp. 578–9.
5 Letter from D. B. Reid to H. M. Commissioners, 3 July 1851 (National Archives: Work 11/14 no. 643).
6 Letter from H. M. Commissioners to D. B. Reid, 15 July 1851 (National Archives: Work 11/14 no. 646).
7 Letter from D. B. Reid to H. M. Commissioners, 25 July 1851 (National Archives: Work 11/14 no. 653).
8 Letter from H. M. Commissioners to D. B. Reid, 18 August 1851 (National Archives: Work 11/14 no. 655).
9 Oral Statement by D. B. Reid, 26 March 1852, in HC 402, 1852, *Second report*, Q471.
10 Letter from C. Barry to Philipps, 12 January 1852 (National Archives: Work 11/14 no. 673).
11 Letter from D. B. Reid to Office of Woods, 3 December 1851, (National Archives: Work 11/16 no. 61); Letter from D. B. Reid to Office of Woods, 15 July 1851 (National Archives: Work 11/14, no. 647); Letter from H. M. Commissioners to Reid, 22 July 1851, (National Archives: Work 11/14 no. 649).
12 'Ventilation of the House of Commons, Dr Reid's reply', *Builder*, 10-475, 13 March 1852, p. 167.
13 HC 1851–52 I.VII 'Estimates for civil services', p. 13; HC 1852–53 I.VII 'Estimates for civil services', p. 12.
14 HC 1851–52 I.VII 'Estimates for civil services', p. 6; HC 1852–53 I.VII 'Estimates for civil services', p. 6.
15 Plan of Principal Floor, 1852 (Strategic Estates Archives, Westminster); Plan and Section of Ventilation Office in South-West angle of House Lobby, November 1852, (National Archives: Work 29/3110).
16 'Plan and Section of offices for ventilation', 22 September 1851 (National Archives: Work 29/3106); 'Plan of ground floor', 1852 (Strategic Estates Archives, Westminster); 'Plans and sections of offices for ventilation', 22 September 1851 (National Archives: Work 29/3106).
17 Alfred Meeson, First Report on the State of the warming, and ventilation and lighting of the Houses of Parliament, 8 January 1853 (National Archives: Work 11/14, no. 768–81).

18 Statement explanatory of the arrangements for warming and ventilating the New House of Commons, 5 April 1852, in HC 402, 1852, *Second report*, pp. 545-48; Oral statement given by D. B. Reid, 20 April 1852, in HC 402, 1852, *Second report*, Q597–8.
19 Oral statement by Goldsworthy Gurney, 26 April 1852, in HC 402, 1852, *Second report*, Q3002–3.
20 D. B. Reid, *Temperature at the House of Commons, taken by the messenger of the Sergeant-at-Arms, 22 March - 4 May 1852*, in HC 402, 1852, *Second report*, pp. 580–5.
21 '*Registers of temperature control and ventilation for the House of Commons 1853-1854*' (Parliamentary Archives: OOW/5).
22 Oral statement by C. Russel, 26 March 1852, in HC 402, 1852, *Second report*, Q255–56.
23 Notes by D. B. Reid, dated 6 November 1852, on drawing 'Plans of Offices for ventilation', 22 September 1851 (National Archives: Work 29/3106).
24 'Imperial Parliament', *Daily News*, 5 February 1852, p. 3.
25 Ventilation of the House, House of Commons, *Hansard's Parliamentary Debates: The Official Report* (6 February 1852 vol 119 cols. 231–4).
26 'The science of ventilation has at length been '*Times*, 9 February 1852, p. 4.
27 'Editorial', *Builder*, 14 February 1852, p. 97.
28 D. B. Reid, *Memorandum*, 7 February 1852 (National Archives: Work 11/14, no. 678).
29 Letter from Office of Works to Barry C, 14 February 1852 (National Archives: Work 11/14, no. 679).
30 D. B. Reid, Letter to Office of Woods, 3 December 1851 (National Archives: Work 11/16, no. 61); Letter from C. Barry to HM Commissioners, 3 July 1851 (National Archives: Work 11/14, 643).
31 D. B. Reid, *Memorandum*, 7 February 1852 (National Archives: Work 11/14, no. 678).
32 Oral statement by Reid, 20 April 1852, in HC 402, 1852, *Second report*, Q1996–7.
33 *Ventilation of the House, House of Commons, Hansard's Parliamentary Debates: The Official Report (11 February 1852. vol. 119 cols. 400–16).*
34 D. B. Reid, Report of measures required for the health and comfort of the House of Commons, 10 March 1852 (National Archives: Work 11/14, no. 711–14).
35 Commissioners of Works, Letter to Barry, 14 February 1852 (National Archives: Work 11/14, 679; Commissioners of Works, Letter to Reid, 16 February 1852 (National Archives: Work 11/14, no. 701); Letter from Office of Works to M. Faraday, 21 February 1852 (National Archives: Work 11/14, no. 702); Letter from C. Barry to Office of Works, 20 February 1852 (National Archives: Work 11/14, 704–5)
36 Ventilation of the House, House of Commons, Hansard's Parliamentary Debates: The Official Report (16 March 1852, vol. 119 cols. 1147–50)
37 Oral statement by D. Jephson-Norreys, 25 March 1852, in HC 402, 1852, *Second report*, Q153–59.
38 Note: Modern standards for air-conditioned environments recommend temperatures of 20–23°C in winter and 22–26°C in summer (ASRAE, 2004), but if historic clothing levels are taken into consideration these lower temperature come very close current standards. 19th century paintings suggest that clothing levels were approximately 1.5 clo. According to *ASHREA Standard 55*, at clothing levels of 1.5 clo a temperature of 65°F is optimal for thermal comfort (Bradshaw V 2006, 15–16).
39 Oral statement by G. Gurney, 8 May 1854, in HL 384 1854, Report from the Select Committee of the House of Lords, appointed to inquire into the

possibility of improving the ventilation and the lighting of the House, and the contiguous chambers, galleries and passages, Q675.

40 Oral statement by G. Gurney, 23 May 1854, in HC 270 1854, Second report from the Select Committee on the ventilation of the House, Q211.

41 Oral statement by J. Leslie, 27 April 1852, in HC 402, 1852, *Second report*, Q3137–43.

42 Oral statement by T. Thornton, 25 March 1852, in HC 402, 1852, *Second report*, Q28–29.

43 Oral statement by the Clerk of the House of Commons, 25 March 1852, in HC 402, 1852, *Second report*, Q124–38).

44 Oral statement by the Speaker of the House of Commons, 26 March 185, in HC 402, 1852, *Second report*, Q316).

45 Oral statement given by S. Daukes, 5 April 1852, in HC 402, 1852, *Second report*, Q1729.

46 S. Daukes and H. Price, *Report on the warming and ventilation of the Houses of Parliament'*, in HC 402 1852, *Second report*, pp. 564–8.

47 G. Gurney, First Report on the ventilation of the new House of Commons, 6 April 1852, in HC 402 1852, *Second report*, p. 586.

48 G. Gurney, Second report on the ventilation, lighting and warming of the new House of Commons, 13 April 1852, in HC 402 1852, *Second report*, pp. 591–2.

49 Oral statement by D. B. Reid, 26 July 1842, in HC 536 1842 Committee on Ventilation, Q4–10.

50 Oral statement by G. Gurney, 1 April 1852, in HC 402, 1852, *Second report*, Q821–8.

51 The New House of Commons from the Bar, Illustrated London News, 7 February 1852, p. 121.

52 D. B. Reid, Statement accompanying Dr Reid's Letter to the Commissioners of Woods, 10 March 1848 (National Archives: Work 11/13, no. 387)

53 D. B. Reid, Nine drawings outlining lighting scheme, 10 March 1848 (National Archives: Work 29/2820-28).

54 Letter from A. Milne to Reid, 22 March 1848 (National Archives: Work 11/13 no. 410).

55 Letter from D. B. Reid to HM Commissioners, 28 October 1848 (National Archives: Work, 11/13 no. 519-20); Letter C. Barry to A. Milnes, 22 March 1848 (National Archives: Work 11/13, 409); Letter from C. Barry to H. Cole, 26 October 1848 (National Archives: Work 11/13 no. 518).

56 Letter from D. B. Reid to HM Commissioners, 23 July 1848 (National Archives: Work 11/13 no. 485).

57 Ibid.

58 Letter from D. B. Reid to Office of Woods, 18 November 1848 (National Archives: Work 11/13, no. 535); Resolution of HM Commissioners, 30 October 1848 (National Archives: Work 11/13, no. 521).

59 Letter from D. B. Reid to H. Cole, 26 April 1848 (National Archives: Work 11/13, 441-2); Letter from D. B. Reid to HM Commissioners, 10 November 1848 (National Archives: Work 11/13, no. 527).

60 *Ventilation of the House, House of Commons, Hansard's Parliamentary Debates: The Official Report (11 February 1852. vol. 119 cols. 400–16).*

61 Letter from D. B. Reid to Office of Woods, 10 March 1848 (National Archives: Work 11/13 no. 391); D. B, Statement accompanying Dr Reid's Letter to the Commissioners of Woods, 10 March 1848 (National Archives: Work 11/13 no. 387).

62 Oral Statement by D. B. Reid, 20 April 1852, in HC 402 1852, *Second report*, Q1991-2008.

63 'Dr Reid System of Lighting the House of Commons,' *Illustrated London News*, 24 April 1852, p. 317; Oral statement by G. Gurney, 26 April 1852, in HC 402 1852, *Second report*, Q2980–2.

64 Oral statement given by Goldsworthy Gurney (26 April 1852, HC 402 Ev 288-92 Q2063–92).

65 Registers of temperature control and ventilation for the House of Commons 1853–1854, (Parliamentary Archives: OOW/5).

66 D.B. Reid, *Supplementary Report*, 29 June 1852 (National Archives: Work 11/14, no. 735).

67 Manners J, Letter to Reid DB, 9 September 1852 (National Archives: Work 11/16 no. 747); Letter from D. B. Reid to J. Manners, 27 August 1852 (National Archives: Work 11/14, no. 741); Letter from Office of Works to the Chancellor of the Exchequer, 22 July 1852 (National Archives: Work 11/16 no. 68); J. Phipps, *Notes on the ventilation work still to be completed*, 3 September 1852 (National Archives: Work 11/14, no. 744–6).

68 Letter from C. Barry to J. Manners, 17 June 1852 (National Archives: Work 11/14, no. 733); Letter from C. Barry to J. Manners, 30 June 1852 (National Archives: Work 11/14, no. 737).

69 Note: In 1853 Reid was paid £3250 in compensation following an arbitration. Letter from J. Gardiner to J. Phipps, 9 May 1853 (National Archives: Work 11/16, no. 140); HC 1852-53 498, Submission to arbitration between Her Majesty's government and Dr Reid.

70 J. Locke and R. Stephenson, Report to Commissioners of Works, 3 August 1852 (National Archives: Work 11/16 no. 73).

71 Letter from Office of Works to D. B. Reid, 21 September 1852 (National Archives: Work 11/16 no. 81).

72 A. Meeson, *First Report on the State of the warming, and ventilation and lighting of the Houses of Parliament*, 8 January 1853 (National Archives: Work 11/14 no. 768–81).

73 Ibid.

74 Letter from A. Meeson to Office of Works, 8 January 1853 (National Archives: Works 11/14, no. 767).

75 Oral Statement by A. Meeson, 30 March 1854, in HC 149 1854, *First report from the Select Committee on the Ventilation of the House; together with the proceedings of the committee*, Q171–4.

76 Oral statement by the Speaker of the House of Commons, 26 March 1852, in HC 402 1852, *Second report*, Q318–28.

77 Oral statement by the Sergeant-at-Arms, 26 March 1852, in HC 402 1852, *Second report*, Q257–69.

78 HC 570 1852-53, *Report of the Standing Committee on the Ventilating and Lighting the House of Commons*; HC 911 1852-53, *Copy of the report of Mr Goldsworthy Gurney to the Commissioners of Works, respecting the lighting of the House of Commons.*

79 Letter from A. Meeson to Office of Works, 11 March 1853 (National Archives: 11/14, no. 829).

80 Oral statement by G. Gurney, 12 June 1854, in HL 384 1854, *Select Committee on improving*, Q931–39.

81 Letter from A. Meeson to Office of Works, 18 April 1853 (National Archives: Work 11/14, no. 788); Letter from A. Meeson to Office of Works, 14 March 1853 (National Archives: Work 11/14, no. 786).

82 Letter from G. Gurney to Office of Works, 12 March 1853 (National Archives: Work 11/14, no. 784).

83 G. Gurney, *Register of thermometer and pressure gauge*, 8 August 1853, in HC 1852-53 911, *report of Mr Goldsworthy Gurney*, p. 4.

84 T. Mather and J. Hann, *Report*, 30 June 1853, in HC 1852–53 911, *report of Mr Goldsworthy Gurney*, pp. 2–3.
85 Oral Statement by A. Meeson, 30 March 1854, in HC 149 1854, *First report*, Q127-34; A. Meeson, *Second Report on the Warming, ventilating and lighting arrangements*, August 1853 (National Archives: Work 11/14, no. 791–812).
86 Oral Statement by A. Meeson, 30 March 1854, in HC 149 1854, *First report*, Q127–34).
87 Oral Statement by G. Gurney, 30 March 1854, in HC 149 1854, *First report*, Q 11–25).
88 Lighting and Ventilation of the House, House of Commons, *Hansard's Parliamentary Debates: The Official Report (*17 February 1853 vol 124 cols. 180-1); Supply – Miscellaneous estimate, House of Commons, *Hansard's Parliamentary Debates: The Official Report (*19 May 1853 vol 127 cols. 388–422); Supply – Miscellaneous estimate, House of Commons, *Hansard's Parliamentary Debates: The Official Report (*4 August 1853 vol. 129, cols. 1297–3180).
89 'Ventilation of the House', *The Times*, 11 March 1854, p. 7; 'Imperial Parliament,' *Daily News*, 20 May 1853, p. 3.
90 'Ventilation of the House', *Daily News*, 1 March 1854, p. 2.
91 'Ventilation of the House', *The Times*, 11 March 1854, p. 7.
92 G. Gurney, *Report*, 30 March 1854, House of Commons, HC 149 1854, *First report, pp.* 2–3.
93 HC 149 1854, *First report*, pp. iii–iv.
94 Letter from Office of Works to G. Gurney, 7 April 1854 (National Archives: Work 11/14, no. 850).
95 Letter from G. Gurney to Office of Works, 7 April 1854, (National Archives: Work 11/14, no. 842); Letter from G. Gurney to Office of Works, 10 April 1854 (National Archives: Work 11/14 no. 847).
96 'Imperial Parliament,' *Illustrated London News*, 15 April 1854, p. 342.
97 Letter from Office of Works to G. Gurney, 12 April 1854 (National Archives: Work 11/14 no. 854).
98 'Imperial Parliament,' *Daily News*, 20 May 1853, p. 3; 'Parliamentary Notices', *The Times*, 5 May 1853, p. 4; 'Lighting and Ventilation of the House,' *Morning Chronicle*, 6 May 1853, p. 2; 'Imperial Parliament', *Daily News*, 5 August 1853, p. 4.
99 'Ventilation of the House,' *Daily News*, 11 April 1854, p. 3.
100 Gurney provided several and detailed oral accounts of his scheme to two Select Committees: Oral statement by G. Gurney, 30 March 1854, in HC 149 1854, *First report*, Q1–126; Oral statement by G. Gurney, 23 May 1854, in HC 270 1854, *Second report*, Q189–232; Oral statement by G. Gurney, 21 July 1854, in HC 1854 (403), *Third report from the Select Committee on Ventilation of the House*, Q233–73, and Oral statement by G. Gurney, 8 May 1854, in HL 384 1854, *Select Committee on improving*, Q 617–734.
101 Oral statement by G. Gurney, 30 March 1854, in HC 149 1854, *First report*, Q1–101.
102 Oral statement given by Gurney, 8 May 1854, in HL 384 1854, *Select Committee on improving*, Q694.
103 Second report, 26 May 1854, HC 270 1854, pp. iii–iv.
104 Oral statement given by G. Gurney, 23 May 1854, in HC 270 1854, *Second report*, Q206-9; Oral statement given by Gurney, 8 May 1854, in HL 384 1854, *Select Committee on improving*, Q618-81, 'House of Commons -27 April', *The Times*, 28 April 1854, p. 3; 'The New House of Commons,' *Morning Chronicle*, 29 April 1854, p. 5.

105 Oral statement by the Sergeant-at-Arms, 15 May 1854, in HL 384 1854, *Select Committee on improving*, Q839–51.
106 Oral statement by Robert Vernon Smith MP, 3 July 1854, in HL 384 1854, *Select Committee on improving*, Q963–73).
107 Oral statement given by Edward Cayley MP, 3 July 1854, in HL 384 1854, *Select Committee on improving*, Q973–94).
108 Oral statement by Edward Bouverie MP, 12 May 1854, in HL 384 1854, *Select Committee on improving*, Q753–9).
109 Second report, 26 May 1854, HC 270 1854, pp. iii–iv.
110 Adrian Leaman and Bill Bordass, 'Assessing building performance in use: the Probe occupant surveys and their implications,' *Building Research and Information*, 29 2001, pp. 129–143.

9 After Reid

judging from the history of the many attempts which were made to improve it, it did not appear to give any satisfaction [...] Old House of Commons never gave satisfaction in spite of the galaxy of talent which was employed in the last hundred years on its installation and improvement, and possibly it may be concluded that some degree of courage (or is the right word foolhardiness?) was required in accepting the responsibility of designing the ventilation of the New House.[1]

Oscar Faber, May 1949

Introduction: stages of experimentation

In a talk given at the Institution of Structural Engineers on 24 February 1949, Dr Oscar Faber – a consulting engineer commissioned to design the air-conditioning system for the new debating chamber of the House of Commons in 1944 – noted that MPs' arguments about the debating chamber 'might have continued indefinitely had not Hitler provided the opportunity for a radical solution'.[2] Faber criticised the original 19th-century system, arguing that the ventilation system did not work and that attempts to address its fundamental problems of thermal comfort were unsuccessful. In his view, the constantly changing approach to its procurement, involving a continual process of adaptation, was unsuccessful and that a break with the past was necessary to resolve these problems.

The post-occupancy history of the Houses of Parliament illustrates that Faber's critique was not unjustified. Although his statement on 24 February was clearly rhetorical, rather than a cool-headed technical analysis, it highlights the fact that technical problems – and the ensuing criticisms from MPs and Lords – persisted even after the installation of Goldsworthy Gurney's new system in the mid-1850s. Both David Boswell Reid's system in the House of Commons and Charles Barry's system in the House of Lords were decommissioned in 1854 because they both failed to satisfy the expectations and needs of their occupants: MPs and Lords. This marked the end of the first generation of building services deployed in the Houses of Parliament. Although both the 1854 House of Commons and House of Lords Select Committees concluded that Gurney's interventions would overcome the defects of Reid's and Barry's systems, this was – as will be seen in this chapter – not to be the case. Throughout its operational life the systems in both Houses continued to be subject to criticism from MPs and Lords, formal inquiries and further alterations.

Gurney's and later alterations were largely adaptations of the original physical infrastructure that Reid and Barry had established. Most of this original network of air shafts, chambers and towers remained in active use until the introduction of modern building services in the mid-20th century. In the House of Lords, which survived World War II comparatively unscathed, the infrastructure was utilised for 111 years and only became redundant in 1966 when it was finally replaced with modern ducts, air handling units and a set of air-conditioning plants. In the House of Commons, the original infrastructure was used for 87 years until the chamber itself was destroyed by fire as a result of the Blitz bombing in 1941.

Reid's legacy can only be fully appreciated if it is treated as a contribution to a continually-evolving system: the 19th-century systems were not static but underwent multiple phases of adaptation and re-evaluations. The introduction of Gurney's system, which will be examined below, was only one amongst several interventions. The historic building services therefore should be understood as a palimpsest of several generations of physical adaptations, which were underpinned by insights gleaned from its day-to-day operation by its operational staff – who continued to observe the sophisticated monitoring and control regimes implemented for Reid's and Barry's systems[3] – and more formal scientific inquiries involving the Office of Works, Select Committees and external scientists and engineers.

Whilst the problems of thermal comfort continued to be a major driver of further investigations and physical adaptations, it should be noted that the ventilation system was also adapted in response to various external environmental factors which had not been fully anticipated in the original design, including cooling the fresh air during the summer, increasing levels of atmospheric pollution from 1868 to the 1890s and the spread of viruses during the cholera (1854–55) and influenza (1889–90) pandemics. Rather than addressing the full history of these adaptations here, two case studies focusing on attempts to solve the persisting problems of thermal comfort will be considered.

The first case study critically examines the use of ventilation in the House of Lords for cooling purposes over two periods, from 1869 to 1878 and from 1911 to 1947; the second case study explores the investigations from 1913 to 1938 into whether the ventilation system of the House of Commons could be remodelled to account for the physical factors believed to affect thermal comfort (such as air velocity), and their impact on the concept behind Faber's air-conditioning system. Although Faber's system was technologically very different from what had come before, conceptually its design was strongly influenced by the inquiries and conclusions of the previous inquiries into the internal climate of the old debating chamber.

Gurney's two systems: 1854–55

Gurney's ventilation systems in both the debating chambers of the Houses of Parliament were driven entirely by stack ventilation (upward flow of air created by thermal buoyancy). By contrast, Reid's and Barry's systems were hybrids that used mechanical fans and to supplement the stack effect. According to Walter Yates each system was a 'very feeble plenum coupled with a heat exhaust'.[4] To achieve this Gurney converted the existing inlet shafts inside the Elizabeth and Victoria Towers into large up-cast shafts that were taller than the Central Tower itself. The shafts inside the Elizabeth Tower, which were connected to the House of Commons, were approximately 70 m (230ft) high, whilst the two shafts in the Victoria Tower, which were connected to the Houses of Lords, were almost 100 m(328ft) high.[5] These shafts were connected to their respective Houses by the former fresh air passages in the basement; each of them was equipped with a coke furnace, also located in the basement, to artificially boost the convection currents in the Houses of Parliament.[6]

Figure 9.1 shows the network that connected the House of Lords to the Victoria Tower. The local up-cast shafts of each chamber, which were separate from the Victoria and Clock Towers, were retained as secondary outlets and were also equipped with separate coke furnaces. In the House of Commons, Gurney installed new gas lights above the painted glass panels in the centre of the ceiling and used a network of iron flues to convey the waste heat and fumes produced by them to the secondary shaft, which enhanced its stack effect following the principles that he had tested during the inquiries of 1854.[7] Fresh air for the debating chambers was now sourced from the courtyards straddling each House using large doors, which acted as air intakes.[8] Those for the House of Commons were in the Commons and Star Chamber Courts, whilst those for the House of Lords were in the Peers' and State Officers Courts.[9]

The fresh air was 'conditioned' in chambers located below each House using similar techniques.[10] In the House of Lords the fresh air was sourced through eight large doors, four of which faced the Peers' Court and four of which faced the State Officer's Court. These doors had louvers, which could be manually adjusted to regulate the air supply. Air chambers behind the intakes were equipped with canvas screens for air filtration, steam batteries for heating and vaporisers for humidification (see Figures 9.1, 9.2).[11] An evaporative cooling system, composed of an array of fine water jets, was placed in the passage outside of the intakes.[12] The ceiling of the chamber had 12 rectangular valves, which directed the 'conditioned' air into an upper air 'equalising' chamber and ensured that the air entered the House at a uniform temperature and velocity.

A similar arrangement was installed below the House of Commons. The conditioned air entered the House through the original perforated cast iron floors, which Gurney had had unsealed. These perforated plates,

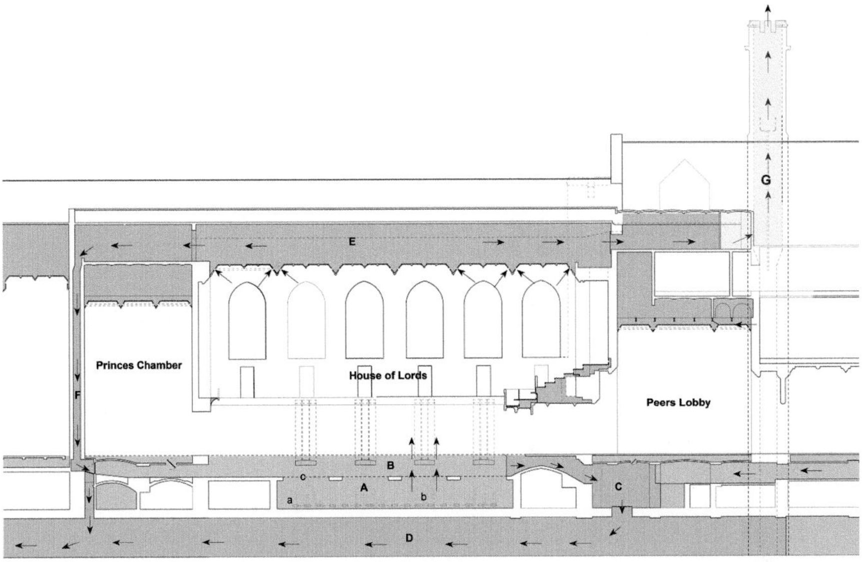

Figure 9.1 Longitudinal section showing Gurney's ventilation and air-conditioning system in the House of Lords (author's own drawing). Key: A. 'Air-conditioning' chamber, B. 'Equalising chamber', a. Steam batteries (heating), b. Inlets facing Peers' Court/State Officers Courts, c. Valves between lower air chamber and equalising chamber, C. Down-pull shaft 1: downward extract through floor at the Bar end of the House, D. Vitiated air channel inside basement that communicates with shafts inside Victoria Tower, E. Vitiated air chamber above ceiling, F. Down-pull shaft 2: conveys vitiated air from ceiling to basement passage, G. Peers' Shaft.

which extended across the entire floor of the two debating chambers and Division Lobbies, were covered with roughly-woven, air-permeable carpets. In the House of Lords cord sisal matting was deployed.[13] On the basis of his trials conducted in the House of Commons in 1854, Gurney argued that the use of perforated floors allowed the introduction of at least 7,000 ft^3/min (198 m^3/min) of air without producing any 'sensible motion' around the body.[14] Air movement was closely regulated through a system of manual controls, which were operated by attendants working under Gurney's supervision. These controls comprised an array of shutters and louvres inside the air chambers with one set affecting the air supply and two sets affecting the quantity of hot air that entered each ventilation shaft (see Figures 9.1 and 9.2). The climate control system in the House of Lords remained largely unchanged throughout the 19th century, whilst that of the House of Commons was gradually extended creating a more complex system.

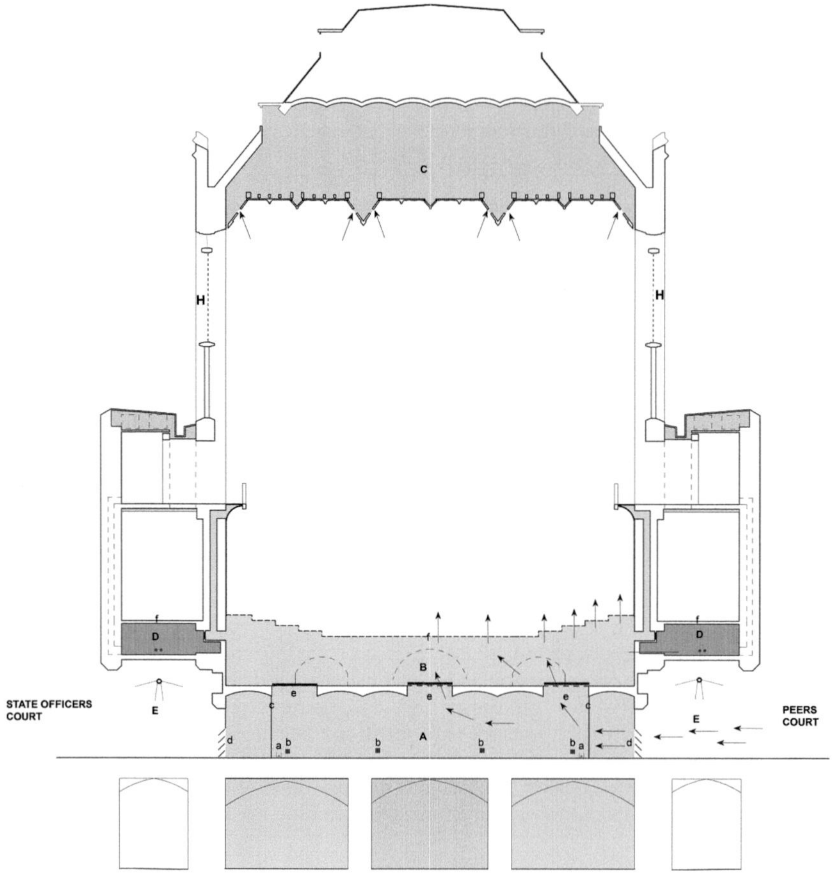

Figure 9.2 Cross-section showing Gurney's ventilation and air-conditioning system in the House of Lords (author's own drawing). *Key: A. 'Air-conditioning' chamber, B. Equalising chamber, C. Vitiated air chamber above timber ceiling, D. Fresh-air chamber under Division Lobby containing steam heating pipes, E. Water jets in front of the air intakes (evaporative cooling), a. Humidification channels, b. Steam batteries (heating), c. Canvas screens (filter), d. Louvers of air intakes, e. Valves between 'air-conditioning' chamber and ‚equalising chamber, f. Perforated cast-iron floors, H. Openable window sections.*

First interventions in the House of Commons: 1866–92

Goldsworthy Gurney was superintendent of the ventilation systems of the Palace of Westminster from 1855 1863, overseeing its operation and maintenance. During this period, the original ventilation arrangements remained largely unchanged. It was only after his successor, Dr John Percy, had taken office that a critical re-examination of Gurney's system was conducted.[15] Between 1866 and 1892 Percy and his successor, Jacob William Prim, remodelled the air-supply arrangements in several stages with three aims. First, providing more effective facilities for cooling during the summer; second, protecting the internal atmosphere from smog; and third, to enhance the ventilation during summer.

Percy's first intervention was to introduce a second air intake to supplement the fresh air from the courtyards when there was smog or during exceptionally hot weather.[16] Located on the Terrace by the River Front, and linked to the House of Commons at basement level through a long passage, this second intake enabled Percy's attendants to switch the supply to accommodate the weather and atmospheric pollution conditions.[17] During periods of smog, the courtyard inlets were closed, the supply was switched to the terrace, and the air was directed through filters. During hot weather, the passage was used for cooling the supply air through evaporation by passing the incoming air through water jets and racks filled with blocks of ice.[18]

An air pump was introduced as a back-up system for periods – typically during hot and windless summer days – when the convection in the Elizabeth Tower shaft, even with the aid of the heat produced by the coke furnaces, was not able to draw in sufficient quantities of fresh air.[19] During the first ten years from 1867 to 1877, the filtration system used two filters: one filter for when there was smog, composed of a canvas screen sprayed with water; and one filter for ordinary days when there was no smog, composed of a dry canvas screen.[20] The original smog filter proved inadequate in the face of the increasingly severe smog experienced in London in the 1870s, with the annual number of smog days in London rising from 40 in 1850 to 70 by 1890.[21] It was complemented in 1878 by a much larger 6 in-thick wool screen, located in the basement below the House of Commons.[22] However, its introduction presented new technical challenges. From 1878 to 1888, the new filter was used in combination with an air pump but the latter proved insufficient to force the required quantities of air to ventilate the House through the thick wool. The air pump was replaced in 1891 with a stronger, more powerful, steam-driven fan.[23] Unfortunately, this fan also proved insufficient to supply enough air when the chamber was crowded. Furthermore, when the smog filters were deployed, the fan was only able to supply enough air to ventilate the House occupied by a maximum of 180 MPs.[24]

The lack of sufficient fresh air to ventilate the House was readdressed at the beginning of the 20th century when more powerful electric fans were introduced. These enabled the Resident Engineer, who took over the role of superintending engineer for the ventilation in the 1890s, to abandon the original courtyard inlets, introduce fresh air exclusively from the Terrace, and deploy the fans continually during sittings.[25]

A partial departure from the hermetically-sealed environment

At the core of Barry and Reid's scheme was the idea of the debating chambers as hermetically-sealed environments.[26] The original glazing in both chamber's windows had been permanently sealed to isolate their interiors from the external atmosphere (i.e. air pollution) and achieve a more tightly-controlled climate. Alfred Meeson, who was superintending engineer of the ventilation during this period, reported that the windows in the House of Lords comprised two layers of fixed glazing: an inner layer of stained glass and an external layer of clear 'plate glass'. Their purpose was 'to preserve an equable temperature in the House in very cold weather, by providing space or stratum of air between the two glazings, which would prevent the cooling action of the external atmosphere upon that of the House'.[27] In addition its purpose was to keep the interior cooler in summer.

However, following the recommendations of the two 1854 Select Committee, Gurney introduced openable windows in both Houses. This was thus a departure from the idea of the Houses as hermetically-sealed environments. In the House of Commons, the narrow band of clerestory windows were made operable. In the House of Lords, where the windows were much larger, two openable sections were introduced in the upper part of its 12 windows (see Figure 9.3). These were equipped with a crank mechanism that enabled attendants outside the House to operate them remotely. Remains of these mechanisms can still be seen on the exterior. Roller blinds were fixed externally on the east and west elevations to protect the Lords from low-angled sunlight in the mornings and afternoons.[28] During the winter months, the windows of the House were temporarily resealed by covering the operable sections with a second layer of exterior glazing, thus providing some insulation and improving airtightness.[29]

The introduction of operable windows may suggest that Gurney was completely discarding Reid's and Barry's original idea of permanent, hermetically-sealed debating chambers. However, such a grand statement was not his original intention. Gurney asserted that the operable windows were merely introduced for the purpose of 'freshening the House' before and after sittings,[30] and advised that they should only be used during sittings when the wind was not too strong or when the outdoor temperature was neither too hot nor too cold, to prevent discomfort.[31] Moreover, he recommended that the windows were kept closed when there were strong winds or when the outdoor temperature was too hot or too cold.[32]

Despite these recommendations, Gurney did not have any fundamental objection to the operable windows in either House being used more extensively than he himself envisioned, noting that it would be at the discretion of each chamber's occupants – MPs and Lords – to determine when the windows were opened. He also believed that it could potentially help improve their thermal comfort during warm weather through increased air movement. As shall be seen in the next section, the windows in the House

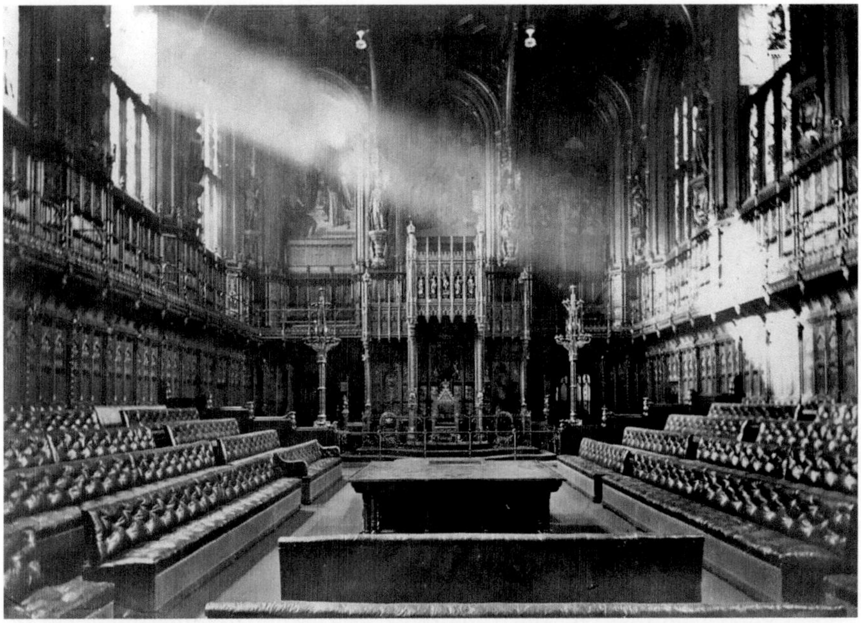

Figure 9.3 Historical photographs (1868) showing the interior of the House of Lords with open windows admitting sunlight (British Library).

of Lords were indeed frequently opened during summer sittings to enhance the Peers' thermal comfort.

Case Study 1: cooling practices in the House of Lords (1869-1937).

Although the Office of Works conducted inquiries into introducing modern mechanical ventilation into the House of Lords in 1911–12, and air conditioning in 1935–38, neither of these schemes were ever realised.[33] The use of coke fires to supplement the stack effect was only discontinued (and replaced with electric fans) in 1923,[34] whilst a mechanical air-conditioning system was first installed in 1966 (a full 16 years after one was installed in the House of Commons).[35]

Thus, unlike the House of Commons, the stack ventilation system and early air-conditioning methods introduced by Gurney in 1855 remained in continual use for over 70 years and 111 years respectively. Nevertheless the system was subject to several post-occupancy studies during this time, which can provide some insight into the practical challenges of implementing Gurney's approach to ventilating and cooling the House.

One of the earliest inquiries was conducted in the summer of 1869 in response to complaints from the Lords. During a sitting on 16 July 1869, several Lords voiced their unhappiness with the state of the ventilation and cooling of the House. Two main concerns were raised: the chill produced by the fresh air floor supply; and the very idea of being placed inside a sealed room.

The Earl of Carnarvon described Gurney's system as an 'extremely complicated and highly scientific arrangement' and asserted that complaints had been made recently 'on all sides, sometimes of the frigid and some of the torrid zone'. Earl Stanhope reported that 'whilst hot air prevailed above, the feet of the noble Lord were exposed to currents of cold air, which came in from below'. The Duke of Montrose also complained that 'it seemed almost impossible to admit fresh air without causing strong draughts, and cold air all came underneath the benches'.

More generally, the Earl of Kimberley and the Marquess of Salisbury both opposed the very idea of a sealed environment and advocated the adoption of a more 'natural' solution using operable windows. Describing the system as 'artificial', the Marquess opined that Peers lacked sufficient personal control over their own environment; the Earl complained that they were at the mercy of the engineers and that 'horrors' were 'endured in that House during the last few days from proceedings of the ventilation philosophers in whose power they were'.

John Percy, who had by this time replaced Gurney as superintendent of the ventilation, was summoned 11 days after the sitting on 27 July 1869 to give oral evidence to a House of Lords Select Committee who were charged with reviewing some of the Lords' complaints. Percy referenced the practical challenge of preventing overheating and draughts, indicating that he was well aware of the problems that the Lords had been experiencing. Although he claimed that the ventilation and cooling system operated as intended when it was cold and relatively uncrowded, he admitted that it was challenging to counteract overheating in the House when it was crowded and the weather was hot. Percy testified that this was because injecting cooled air or boosting the ventilation rate to counteract overheating could affect the Lords' thermal comfort as both methods could lead to cold draughts, depending on the velocity and temperature of the supply air. These issues had come to the fore the previous month on 18 June 1869 when the Lords were engaged in an important debate about the Irish Church Bill. On that day, the debating chamber was unusually crowded and – in an attempt to counteract overheating – Percy's staff increased the ventilation rate and lowered the supply air temperature to 59°F (15°C). However, a large number of complaints were received. Percy testified that 'Peers came down to me in the air chamber under the House, and complained of cold draughts to their feet'.[36] Although he noted that the atmosphere was 'agreeable to the largest number of persons' when it was between 62°F and 64°F (~17 °C and 18°C) and the velocity of fresh air did not exceed 1.5 ft/sec (i.e. 90 ft/min), Percy agreed that 'it is exceedingly objectionable to allow jets of air to impinge upon the legs'.

Percy's account shows that his attendants had difficulties reconciling the control of the three different, yet closely-intertwined, environmental functions of ventilation. First, to renew the atmosphere; second, to reduce the indoor air temperature; and, third, to harness the cooling effect of air currents. These functions had to be carefully balanced by the attendants to avoid untoward consequences as the ventilation rates required to reduce the air temperature or maintain a good air quality could result in an increase in velocity, which was not desirable for the Lords' thermal comfort.

Similar issues to those raised in 1869 were raised nine years later in the summer of 1878. During a sitting on 22 July Lord Granville complained about the excessive heat in the chamber– claiming that the indoor temperatures that month had reached 75°F (~24°C) – and asked the Lord Chancellor if the House could be provided with an ice-cooling system similar to the one used in the House of Commons. In a written reply to Lord Granville, dated 19 July 1878, John Percy claims that the problem was caused by opening the windows as this weakened the fresh air floor supply, resulting in the temperature in the lower part of the chamber to rise.

Percy claimed that the stack system could achieve increased ventilation rates and could maintain a lower air temperature if the chamber were sealed and supplied with cooled air through the floor. During a sitting a few days earlier on 18 July, Percy's staff had closely monitored how the state of the windows affected the internal temperature and observed that from 12pm till 5pm, when the House was sealed, the temperature was 71°F (~22°C), however, it rose to 73°F (~23°C) after five of the 12 operable window units were opened; when all but one of the windows were open at 6.30pm, the temperature reached 75°F (~24°C).[37]

Over the course of their operational lifetime, the windows in the House of Lords were opened more frequently during sittings than Gurney had originally envisaged. Individual Lords frequently requested windows to be opened, claiming that it made the interior feel more comfortable.[38] By the 1870s, natural ventilation was deployed routinely during sittings in the summer, and a formal control regime, that regulated when the 12 windows were opened and closed, was introduced. The aim of this regime was to limit the admission of direct sunlight, which could cause glare and a rise in temperature.[39]

The Lord Chancellor, who was responsible among other things for supervising the operation of the windows during sittings, reported that he ordered them to be opened at 5pm each evening, and would at times give instructions for ad hoc adjustments to be made. Furthermore, solar gain– that is, the increase in the House's thermal energy due to the absorption of incident solar radiation – was a significant issue during sunny weather. To attenuate this problem, permission was only granted for 'windows to be opened on the shady side of the House' (i.e. the west side facing the State Officers' Court) during a sunny day, whilst blinds on the exterior of the House were drawn across the east side facing the Peers' Court to protect the interior from the sunlight.[40] The changes to Gurney's ventilation and cooling system implemented by Percy were significant as they illuminated its technical limitations. Reid's original idea of creating a sealed and climate-controlled space of the House of Lords was abandoned during hot weather. A new operational regime, integrating direct natural ventilation via the openable windows Gurney had installed, was introduced to overcome these limitations. Despite his concerns that this would lead to an increase in air temperatures, Percy recognised that open windows could enhance thermal comfort during hot weather because they caused air movement. After all, 'if air moves at a high velocity, you can cool yourself even with warm air, upon the principle of fanning'.

Two types of ventilative cooling emerged in response to feedback from the Lords. The first was mechanically-induced when the chamber was sealed. In this mode, fresh air was introduced exclusively through the floor due to the stack effect supplemented by mechanical fans. The second was window-induced when the chamber was unsealed typically during hot weather. In this mode, outside air flowed into the chamber increasing air movement within it in order to attenuate the unintended effects of the first type of cooling on the Lords' thermal comfort.

Throughout the lifetime of Gurney's system Lords' complaints, however, suggest that the adaptations to the system did not always prevent thermal discomfort during warm weather, though the logbooks show that overheating issues occurred for relatively short periods in the summer. For example, monitoring data recorded by attendants in the logbook during sittings in the summer of 1869 show that the temperature ranged from 63 to 75°F (~17°C) to 75°F (~24°C) in July of that year.

During the operational life of Gurney's ventilation and cooling system, the Lords became increasingly dissatisfied with the variable conditions of the debating chamber. The pressure to improve the Lords' thermal comfort became the driving force behind the inquiries into Gurney's system in the early 20th century when the Lords experienced longer periods of overheating in the House. The first inquiry of the 20th century conducted between 1911 and 1912 began with an examination of whether it was feasible to remodel the existing ventilation system.

A re-appraisal of the House of Lords: 1911–37

In July 1911 the Secretary of the Commissioner of Works received a note from Earl Spencer, the Lord Chamberlin of the Household, claiming that the 'ventilation is not nearly sufficient in hot weather and the atmosphere is at times very oppressive', and requesting that more windows be made operable.[41] In the same month and the next, overheating issues during several sittings led to an increase in the number of complaints from Peers. The engineers in the Office of Works subsequently reviewed the issue with the chief engineer concluding that the existing system was not capable of providing the higher ventilation rates required to prevent overheating in the summer, especially because debates in the House had become longer and better attended than previously in the past.[42]

On 27 October 1911 the chief engineer of the Office of Works wrote to the Lord Chamberlain that there was 'a great deal of complaint – quite justifiable I thought – of the heat during the debates last August' and that something would need to be done to prevent it from re-occurring. He subsequently requested that the Resident Engineer of the Houses of Parliament, Arthur Patey, develop proposals for improving the existing system. However, Patey argued that the problem could not be simply resolved by opening the windows, and that more substantial and expensive interventions were required.[43] Similar to Percy's claims in the 1860s, Patey claimed that opening the windows in the House would exacerbate, rather than attenuate, the overheating problems.

In April 1912, Patey presented two schemes for replacing Gurney's system with a mechanical ventilation system driven by electric fans. In one scheme, he proposes to reverse the direction of the ventilation by moving the main air supply to the ceiling of the chamber (just as Reid had proposed in 1846). This involved the introduction of two large channels that linked the existing ceiling to new air intakes at roof level.[44] In his final report Patey writes that a ceiling supply would eliminate the existing problems with currents entering near the body, stressing that the 'feet are very sensitive to ascending air currents'.[45]

Minor changes were made to the ventilation system in the House of Lords in the 1920s, replacing the coke furnaces with a mechanical extract fan to enhance the ventilation. Inquiries into more substantial alterations were not conducted until ten years later. The Marquess of Linlithgow tabled a motion in the House of Lords on 24 July 1935 demanding the appointment of a Select Committee to inquire into the feasibility of introducing 'an up-date air-conditioning' system, arguing that this new technology would allow great control over the climates. The motion was withdrawn after an agreement was struck with the Office of Works that its engineers would lead an inquiry into this matter on behalf of the Lords. As a result, a large study that focused on the climate inside the debating chamber was conducted between 1935 and 1936, involving a collaboration between J.A. Macintyre, chief engineer at the Office of Works, and Dr Thomas Bedford from the Medical Research Council's (MRC) Industrial Health Research Board, a physiologist who was renowned for his work on thermal comfort.[46]

The aim of this study was to examine the causes of thermal discomfort, taking into account the effects of temperature, humidity and air movement, and to determine if air-conditioning could overcome it.[47] From October 1935 to September 1936 the temperature and relative humidity of the atmosphere in the chamber was monitored using self-recording (automatic) thermometers and hygrometers, yielding data for a whole year.[48] Bedford and Macintyre also conducted two in-depth studies to examine the velocities of the internal air currents, with one focusing on their behaviour in the winter, and the other on their behaviour in the summer.[49] In addition, Bedford wanted to complement these monitoring studies with qualitative studies that examined the Lords' experience of the internal climate. Bedford proposed to conduct user surveys and interviews, and also to ask selected Peers to keep personal logs of their experience.[50] This part of the study was developed and discussed with the Office of Works between December 1935 and February 1936, but was ultimately abandoned because the Lords were reluctant to participate.[51]

The study that examined the behaviour of the internal air currents in the summer was conducted between June and October 1936.[52] Measurements were taken in nine locations, both at head height and floor level. In addition to temperature, the velocity, variability and cooling rate of air movement were also measured using hot-wire anemometers and kata thermometers. In his final report to the Office of Works, dated 18 November 1936, Bedford concluded that the high indoor temperatures and insufficient air movement in the chamber were the main causes of thermal discomfort. The original data show that high indoor temperatures were common during the summer. Between November 1936 and March 1937 the indoor temperature ranged from 56°F to 70°F (~13–19°C) and on most days stayed within 62°F to 65°F (~17–18°C). However, when it was warmer between May and August, it reached 70°F (~21°C) or above on 33 days. The highest temperatures were recorded in June 1937 when they ranged from 64°F to 77°F (~18°C–25°C), reaching 70°F or above on 14 days in the month. Referencing thermal comfort standards for factory workers, Bedford argued that the temperature in the chamber during the summer typically exceeded that of 64–68°F (~18–20°C), which was perceived as 'comfortable by the majority of persons when sitting still'. Bedford's report did not include data on the occupancy of the debating chamber during the study period nor on the outdoor temperature. However, an analysis using historical weather data from the Kew Observatory suggests that when the internal temperatures in the chamber reached 70°F (~21°C) or higher, it was always marginally cooler indoors than outdoors. Unfortunately, the 1–7°F (~0.5–4°C) difference was not sufficient to mitigate overheating.

The study also found that the velocities of fresh air entering the chamber through the floor were much lower in the summer months than during the heating season, thus further aggravating the sensation of warmth. When heating was required, the natural convection produced by the heated air typically led to air velocities of ~40 ft/min (12m/min) (three times as high as during the winter months), but could reach up to 80 ft/min. Bedford concluded that there was insufficient air movement in the debating chamber. None of the velocities recorded by Bedford reached 100 ft/min, which today is considered the threshold at which cooling effect of the currents become noticeable. He argued that the use of a large perforated floor through which the fresh air was supplied had succeeded in meeting Gurney's objective of keeping velocities very low, even when the ventilation was operating at its maximum rate of eight air changes per hour. According to measurements taken between June and October 1936 the average air velocity was 8–14 ft/min. Bedford concluded that the much maligned chill around the Lords' feet was likely caused by the low temperature of the chamber rather than excessively high velocities. Indeed, the 'impression of draughts near bar may be due to the lower temperature and greater variability of the air currents'.[53]

Another set of measurements was taken between 24 June and 17 August 1936 to determine how the electric fan (introduced in 1912), operable windows and heating system affected the internal air currents. Bedford compared the internal air velocities when the chamber was in a sealed mechanically-induced cooling mode with those when it was in a window-induced cooling mode (i.e. when the windows were open). The data showed that although opening the windows only led to a marginal increase in the indoor temperature, it noticeably improved people's thermal comfort. It not only caused a significant increase in air movement but also resulted in a greater variation of speed, both of which, he argued, were important for comfort.[54] Regarding this, Bedford writes that the 'rather greater velocities which prevailed when the windows were open had a noticeable influence on our comfort', and the air also felt fresher, but when closed 'there was a regular feeling of deadness'.

Bedford's observations were important as they provided empirical evidence, for the first time, that open windows improved thermal comfort. However, the effectiveness of opening windows to naturally ventilate the chamber and ameliorate thermal comfort could be unreliable as it was wholly dependent on the external wind conditions. For example, on 10 July 1936 there was a 'slight breeze outside' and opening the windows significantly increased air movement: when the chamber was sealed, the observed air velocities ranged from 5 to 20 ft/min; when the windows were opened, they increased to between 10 ft/min and 65 ft/min.[55] However, on 17 August 1936 when the weather was 'calm', the recorded velocities were no higher than 35 ft/min.[56]

Despite the problems with maintaining thermal comfort in the summer, Bedford concluded that the benefit of introducing a fully-fledged air-conditioning system would be limited. Referring to a relative humidity of 35–65% and temperatures of 65–70°F (~18–21°C) as adequate for thermal comfort, he argued that the indoor climate of the debating chamber was rarely in a state that would require full air-conditioning.[57] Instead Bedford proposed to replace the evaporative cooling system with one using mechanical refrigeration and to increase air movement using mechanical ventilation. Arguing that 'increased turbulence' was needed to make the air more 'refreshing' during hot weather, he proposed – similar to Hill's scheme for the House of Commons – to move the outlets from the floor to the gallery level and to inject air horizontally above head height. Air was to be introduced at high velocities from both sides of the chamber simultaneously, which would produce eddies in its centre and 'prevent stagnation at the level of the benches'.

In two reports, dated 7 and 17 December 1936, Macintyre, the chief engineer of the Office of Works, reviewed Bedford's findings and made several recommendations. He concluded that the main problem with the ventilation system was its inability to adequately cool the air during the summer, and that the evaporative cooling system that used water sprays and ice was ineffective. In its place, he recommended introducing mechanical refrigeration to cool the air. As this would take some time to implement, Macintyre proposed a simple initial intervention: agitating the atmosphere with the aid of open-ceiling fans.[58]

On 16 February 1937 a committee – composed of scientists from the Medical Research Council (MRC) and Department of Scientific and Industrial Research (DSRI) – reviewed both Bedford's and Macintyre's reports and recommended that night-time ventilation should be used. This would allow the architectural fabric to cool down, and minimise the admission of outside air during the day by keeping the chamber sealed. They also recommended that air movement be increased mechanically using ceiling fans that could be operated at varying speeds, ranging from 40–60 ft/min.[59]

None of these recommendations were implemented, and the idea of enhancing comfort through this new approach to ventilative cooling was discarded after World War II. Further investigations in 1950 and 1963 were conducted but these were limited to feasibility studies that examined the possibility of air conditioning.[60] The Lords continued to complain about the conditions in the debating chamber and an air-conditioning system was piloted in 1965 and fully implemented between 1966 and 1967.[61]

Case study 2: a radical new concept for the House of Commons (1913–38)

The various inquiries into the ventilation system of the House of Commons overlapped those into the system of the House of Commons. Although MPs were concerned about their own thermal comfort, the inquiries– unlike those in the House of Lords – resulted in a radical proposal for remodelling the existing ventilation and climate-control system. This proposal not only utilised modern technologies but also built on contemporary work by two researchers from the MRC, Leonard Hill and Thomas Bedford. Although this proposal was not ultimately implemented in the debating chamber, it strongly influenced the post-war design for the new debating chamber.

The first series of inquiries about the ventilation system of the House of Commons was initiated, as were those of the mid-19th century, by MPs and was conducted between 1913 and 1924.[62] On 13 June 1913 Frederick Banbury, MP for the City of London, and John Rees, MP for Nottingham, called for an inquiry into the possibility of introducing fresh air via the ceiling rather than the floor. They argued in a familiar vein that the existing arrangements resulted in their feet and legs being exposed to cold air whilst their heads were immersed in a hot atmosphere. Banbury said that it should enter 'higher in the Chamber, so that you would get the cool air about your heads, which is the place you want it, and the warmer air about the feet'.[63]

On 10 July, the House of Commons appointed an ad hoc Select Committee to undertake a formal inquiry into this issue.[64] The Committee engaged the physiologist Dr Leonard Hill, Director of the Department of Applied Physiology at the MRC, to reappraise the system.[65] Hill was a leading researcher in the field of thermal comfort, and his work in the 1920s became intimately connected with the development of standardised climates for air-conditioned environments in the United States.[66] Hill confirmed Rees and Banbury's diagnosis of the causes of their cold feet and also endorsed their proposal. The system, he wrote 'caused a draught which had a cooling effect on the feet and legs of the members whereas there was not sufficient movement of air round their heads and shoulders'.[67]

To address this problem Hill proposed replacing the perforated floor with new outlets at gallery level.[68] Located above the head, this would enable fresh air to be injected horizontally into the space without exposing MPs to direct currents. Between December 1913 and January 1914 Hill also collaborated with Arthur Barker – a building services engineer from the Department for Heating and Ventilation at University College London – to conduct a series of demonstrations using full-scale mock ups.[69] In his final report, completed after the trials, Hill and Barker proposed to use two sets of outlets to ventilate the House:[70] In addition to outlets located at gallery level (as proposed by Banbury and Rees), Hill proposed to install another set of outlets into the back of the benches. Hill also observed that the drowsiness reported by MPs was caused by breathing in warmed air and thus suggested that the indoor climate be changed to obviate this problem. Hill thus recommended that the atmospheric temperature of the House be reduced and that MPs be warmed by radiant heaters that could be installed between the benches.[71] Hill also considered the climate to be too uniform and advised introducing more physiologically-stimulating conditions such as gentle variations in temperature and air movement. He suggested that this latter could be achieved by alternating the direction of the fresh air currents.

In February 1914, the Select Committee reviewed Hill and Barker's findings.[72] Hill also conducted measurements during actual sittings, which confirmed that the cooling rate around the legs was twice as high as at the head. Hill argued that these currents not only produced cooled feet but were also responsible for 'a feeling of heaviness' and 'headaches'.[73] In its final report, published in August 1914, the Select Committee endorsed Hill's and Barker's proposal but recommended conducting a full-scale test inside the debating chamber before it was fully adopted.[74] Hill and Barker suggested that nurses and soldiers could be asked to volunteer for this test,[75] but due to the outbreak of World War I the inquiries were intermitted for six years.[76]

The possibility of conducting a full-scale test was raised in the House of Commons by Sir Harry Brittain, MP for Acton, on 24 February 1920 and again on 27 April.[77] Sir Frank Baines, architect of the Office of Works, was subsequently commissioned by the National Physical Laboratory (NPL) to conduct it.[78] The NPL from 1921 to 24 conducted a series of air flow simulations using scale models to empirically evaluate and refine Hill and Barker's proposal.[79] The scale models, which were erected at the NPL's research laboratory in Teddington, were used to compare the behaviour of the internal air currents under the existing arrangements with that under Hill and Barker's proposed new configuration. Air currents were traced visually using fine powder and their velocity was measured using hot wire (or thermal) anenometers. A final test was conducted using a full-scale mock-up installed within the actual debating chamber in 1923. Temporary ducts and nozzles, which were made from cardboard and fitted along the front of the galleries, were installed and the chamber was filled with smoke to trace air movement.[80]

The engagement of the NPL to test Hill and Barker's proposal was the beginning of the long-term involvement of government-funded research institutions in the environmental design of the Houses of Parliament. This culminated in the 1940s with a series of studies that underpinned the design for the modern air-conditioning system installed in the rebuilt House of Commons, which had been destroyed by fire in the 1941 Blitz. In the early 1930s the Office of Works established a formal partnership with the MRC and the DSIR, providing them with access to a wide range of specialists.[81]

Although both the Office of Works and the NPL agreed that the tests had been successful, Hill's and Barker's proposal was not permanently adopted due to its high cost. The temporary installations within the chamber were removed after the short trial, and the Office of Works did not revisit the scheme for another six years, again, following pressure from MPs. In 1931 the Office of Works was receiving an increasing number of complaints, culminating in a petition sent to the First Commissioner of Works in July signed by a total of 225 MPs. The petition's signatories complained about the 'devitalising effect of the atmosphere' and asked for independent experts to be appointed to examine their working conditions.[82] The issue was raised again during a debate on 29 July 1931 and the Office of Works finally acquiesced proposing to conduct further investigations.[83]

Over the following seven years, the Office of Works conducted several trials of Hill and Barker's proposal and, following problems with excessive temperature and humidity in the chamber, a feasibility study on the use of air conditioning in the House. The possibility of installing air conditioning in the Houses of Parliament had initially been raised in the House of Lords in July 1935 and was subsequently debated between May and July 1936 in the House of Commons.[84] In a White Paper, dated 16 July 1936, the Office of Works outlined a scheme that combined mechanical air-conditioning with Hill and Barker's dual outlet scheme.[85] It proposed to implement this scheme in stages with the air conditioning – estimated to be the most expensive feature of the scheme– to be installed in the final stage. However, concerns about the costs of the air conditioning led to delays in its implementation.[86]

Between October 1936 and July 1937, the Office of Works conducted additional trials to further develop the Hill's and Barker's dual outlet scheme. These focused on integrating the air supply and heating into the benches, an aspect of their scheme that the NPL had not examined during the studies conducted in the 1920s. Two alternative proposals were tested.[87] For these tests three rows of benches inside the chamber were remodelled and a committee – composed of engineers and scientists from the MRC, Building Research Station, NPL and the Office of Works – was appointed to review the results. The first mock-up trial was conducted in October 1936 and lasted for three months.[88] During the trial, the internal air temperature was reduced to 58°F (~14°C) and electric heating panels were fixed to the backs of the test benches.[89] The backs had switches that enabled MPs to control the local heat. The committee rejected this version of the scheme on the grounds that it was ineffective as the radiant heat was not sufficient to warm a person or counteract the cooling effect of air currents.

The second mock-up trial, which tested an alternative arrangement of Hill's and Barker's proposal, was conducted in January 1937. In this trial, radiant warmth was produced by underfloor heating located between the benches. The perforated floor was closed and replaced with new inlets located along the back rest of the benches. The committee agreed that this configuration was more successful than the first as 'there was no draughts to the feet and feet were warm'.[90] This arrangement remained in situ for several months to enable a sufficient amount of feedback from MPs to be collected. In July 1937 the chief engineer of the Office of Works reported that no complaints had been received.[91]

Calls for the introduction of an air-conditioning system continued. From June until November 1937 the Office of Works received further complaints. Alfred Bossom, MP for Maidstone, raised the topic during several sittings in June 1937 and in two open letters published during November in *The Times*.[92] In a letter dated 15 November, Bossom pleads for the introduction of mechanical air conditioning to achieve an 'atmospherically modern chamber'.[93] Unfortunately for Bossom and his fellow MPs, the Office of Works judged the costs to be prohibitive and instead argued that Hill and Barker's two outlet scheme alone would be sufficient to address the causes of their thermal discomfort.[94]

Neither scheme was ever implemented. In 1938, the Office of Works postponed the plans as resources were reallocated to rearm the British Military. The outbreak of World War II refocused priorities and the destruction of the Commons in May 1941 brought the inquiries into the ventilation system to an abrupt end. Members of Parliament were initially forced to initially decant to Church House for sittings, and later used the House of Lords debating chamber for sittings. Only a month after the destruction of the House of Commons, in June 1941, MPs had already begun debating how the lost chamber should be rebuilt. The concerns underlying the pre-War investigations still continued to influence the discourse.

Between continuity and invention: the rebuilding of the House of Commons (1944–50)

Despite the destruction of the Houses of Commons in 1941, MPs still continued to voice their concerns about both the past and future ventilation of the House.[95] In February 1942, the Prime Minister, Winston Churchill, successfully argued that the scale and architectural qualities of the original 19th-century debating chamber should be retained.[96] On 28 October 1943 he successfully tabled a motion to appoint a Select Committee to coordinate the development of the new chamber's plans.[97] On 4 November 1943 William Davison, MP for South Kensington, expressed his wish that the new chamber 'should not be ventilated by air driven through Honorable Members' feet'.[98] Churchill's motion on 28 October was followed by an impassioned debate during the same sitting, in which the need to overcome the defects of the old ventilation system was raised by several MPs. In the debate, Arthur Duckworth, MP for Shrewsbury, argued that the system was 'antiquated and calculated to give everyone cold feet and a hot head'; George Buchanan, MP for Glasgow Gorbals, noted that it had a 'terrible atmosphere'; Captain Cunningham-Reid, MP for St Marylebone, claimed that 'it is absolutely essential' that the 'ventilation should be considerably improved'; Somerset de Chair, MP for South West Norfolk, argued that there had been insufficient ventilation inside the old House of Commons and claimed that he had 'seen admirals come into the House red in the face from the quarterdeck and within a week they were as white as lily from the lack of ventilation'; and, Arthur Greenwood, MP for Wakefield, called for improvements to the ventilation, which he had 'always heard defended, though ... never met any Member of the House who thought it was any good'.

The Select Committee subsequently consulted individual MPs, including the Speaker of the House of Commons, about their experience of the old debating chamber, and collaborated with the Ministry of Works (which had been formed in 1940 to requisition property for use during wartime), the consulting engineer Oscar Faber and the architect Giles Gilbert Scott, to develop the design of the new chamber. The Committee insisted that an engineer should be involved in the design of the ventilation and air-conditioning systems from the outset. In July 1944, barely nine months after the formation of the Select Committee, Faber provided a complete scheme that was fully-integrated with Scott's architectural scheme.[99] In his report Faber noted that his objective was 'to conform to the best modern practice in air-conditioning' as well as to overcome the 'defects of old House

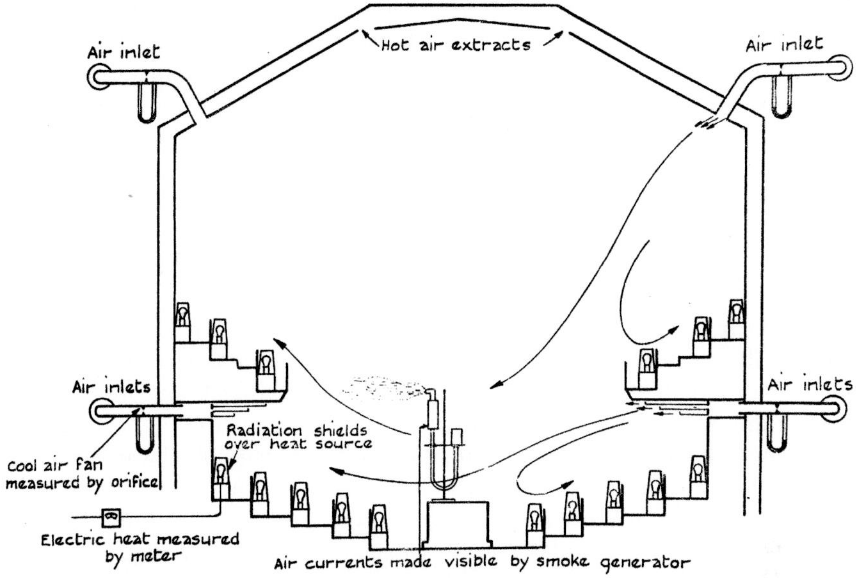

Figure 9.4. Cross-section of scale model showing position of fresh air outlets in the ceiling and below the gallery (NPL, Report on the model investigations, November 1948, National Archives: Work 11/418).

of Commons'.[100] He had reviewed the findings of the pre-war studies, and proposed to introduce fresh air into the debating chamber exclusively at a relatively high level (see Figure 9.4).[101] Faber proposed to introduce one set of outlets under the galleries, which similar to Hill and Barker's earlier proposal, would allow fresh air to be injected horizontally into the body of the House; another set of inlets, to be placed along the edge of the ceiling, would allow air to be introduced downwards.[102] He was to subsequently collaborate with many of the scientific institutions that had been involved in the studies conducted in 1936–37 to refine his scheme using the results of experiments. For example, the NPL assisted Faber in determining the optimal internal configuration for the outlets using air flow simulations conducted inside both a scale model and a full-scale mock-up of a single bay of the chamber;[103] Thomas Bedford from the MRC's Industrial Health Research Board also conducted the first post-occupancy evaluation of Faber's new system, focusing on the behaviour of air currents and their physiological effect on the chamber's occupants.[104]

Conclusion: continuity of practice

The various inquiries that underpinned the rebuilding of the House of Commons were important as they provided substantive insight into how its occupants experienced the environment of the debating chamber. Although technologically the new ventilation system was a radical break with the past, conceptually its design was deeply rooted in longstanding inquiries into the fundamental physiological factors affecting environmental design. Whilst Faber's scheme specifically built on the studies of the modern physiologists Hill and Bedford, solving the problems of thermal comfort had been at the heart of the designs proffered by Reid, Barry and Gurney in the mid-19th century. Thermal comfort – or the lack thereof – was also the subject of several post-occupancy evaluations throughout the debating chamber's operational history. These studies afforded insight into how the Houses of Parliament as an organisation was acquiring, recording and using knowledge of their own building's performance in use. In both Houses understanding the ventilation system's performance in use was largely based on practical experience of its operational staff, user feedback and monitoring data accumulated in the context of its day-to-day operation.

However, in the early 20th-century, there was a strong shift towards involving specialist researchers to study the longstanding issues with the internal climate of the Houses more systematically using a modern scientific method. Whilst earlier inquiries had focused primarily on gradually improving the existing system, the final series of studies in the House of Commons before its destruction in 1941 yielded proposals for more radical interventions.

Notes

1 Oscar Faber, The Construction of the New House of Commons, *Engineering*, May 1949, p. 244.
2 The New House of Commons, *Engineering*, 4 March 1949, p. 208.
3 The environmental and monitoring procedures inside the House of Commons between 1854 and 1941 have been examined in detail in: Henrik Schoenefeldt (2019) The House of Commons: a precedent for post-occupancy evaluation, Building Research & Information, 47:6, pp. 635–665.
4 Walter Yates, The Ventilation of the House of Commons, *Institute of Heating and Ventilating Engineers*, 1907, p.36–62.
5 Cross-section and floor plans of Clock Tower, showing historic ventilation shaft and hidden service floor, undated, circa 1930, Historic England Archive: uncatalogued drawing, Chest 25 'Houses of Parliament'.
6 Report on the ventilation, warming and lighting of the Houses of Parliament, by Dr. Percy, F.R.S. (HC 1866, 98), pp. 4–6.
7 As it was assisting the ventilation, the gaslights were in active use until 1912, before the use of electric fans permitted the introduction of electric light: House of Commons (Lighting) Edward Treacher-Collins, Report on the electric lighting of the debating chamber of the House of Commons (HC 122 1912); Lighting the House of Commons – Report on proposed change, *The Times*, 30 May 1912, p. 5; House of Commons, House of Commons, *Hansard's Parliamentary Debates: The Official Report* (14 October 1912 vol 42 cols. 763–4).
8 Gurney's system is shown in historic surveys drawings of House of Commons and Lords held in the National Archives and Strategic Estates: Arrangement for ventilation of the House of Commons, undated [National Archives, Work

29/2994]; Plan showing arrangements for discharge of foul air from House of Commons, 1867, [National Archives, Work 29/2982]; Plan of mezzanine floor, House of Commons, 1867, [National Archives, Work 29/2981]; Longitudinal section, House of Commons, 1867, [National Archives, Work 29/2983]; Cross-section through House of Lords, 1867, [National Archives, Work 29/2987]; Plan of ground floor, Office of Works, November 1937, [Strategic Estates Archives, nr. 000381/s].

9 Select Committee of the House of Lords, appointed to inquire into the possibility of improving the ventilation and lighting of the House (HL 1854, 384); Cross-section of the House of Lords, 1860s (National Archives, Work29/2985–7).

10 Report 'General outline of the present ventilating arrangemenets', 18 April 1921 [National Archives: Work 11/202]; The plant is shown in 'Houses of Parliament – Plan of ground floor, Office of Works, 1902 [Strategic Estates Archives: uncatalogued drawing).

11 A description of Gurney's can be found in John Billings, The principles of ventilation and heating, and their practical application, (New York: The Sanitary Engineer,1884), pp. 116–18.

12 DSRI Advisory Council, 'Ventilation of the House of Lords. Minutes of a meeting', September 1935 (National Archives: DSIR 3/20)

13 H. Denbon, Report on House of Lords heating and ventilation system, 30 March 1960 (National Archives, Work 11/588, nt. 21–22)

14 Oral statement given by G. Gurney, 8 May 1854, in (HL 1854 (384)), Q643.

15 Dr John Percy, Report on the ventilation, warming and lighting of the Houses of Parliament (HC 1866 98)

16 Report from the Select Committee on House of Commons ventilation, (HC 371 1890–91) Q152.

17 Oral Statement by W. Prim, 1 May 1902, First Report of Select Committee on House of Commons, (HC 227 1902), Q66–86. These new arrangements are shown in the original floor plans (Strategic Estates Archives) and in the drawing enclosed inside the 'Gordon Report of 1906.

18 Oral Statement by W. Prim, 1 May 1902, in HC 227 1902, *First report*, Q170.

19 Ibid, Q66–7.

20 House of Commons (ventilation). Memorandum by Sir Frank Baines, K.C.V.O., C.B.E, 18 November, 1924, regarding the ventilation of the chamber of the House of Commons (1930–31 Cmd. 3871), pp. 12–15.

21 Peter Brimblecombe, The Big Smoke. A history of air pollution in London since medieval times (London: Methuen, 1987)

22 Oral Statement by W. Prim, 1 May 1902, Oral Statement by W. Prim, 1 May 1902, in HC 227 1902, *First report*, Q117–31.

23 Ibid, Q85–6.

24 Oral Statement by G. Rivers, 1 May 1902, in HC 227 1902, *First report*, Q295.

25 Select Committee for House of Commons ventilation (HC 290 1913), Q96; Plan of Ground floor, November 1937, Office of Works (Strategic Estates Archive, number 000381/s).

26 Oral statement given by G. Gurney, 24 March 1854, in (HL 1854 (384)), Q436–37.

27 Oral Statement given by A. Meeson, 3 July 1854, in (HL 1854 (384)), p 104.

28 'State Officer's Court', Photograph, circa 1930 [Historic England Archive: AL 1016].

29 Minute to Secretary of Office of Works, 25 October 1911 (National Archives: WORK 11/124).

30 Letter from G. Gurney to F. Stone, 17 June 1854, in (HL 1854 (384)), pp. 118.

31 Oral Statement given by G. Gurney, 8 May 1854, in (HL 1854 (384)), p. 68.

32 Oral statement given by G. Gurney, 30 March 1854, in (HC 1854 (149) *First report from the Select Committee on the Ventilation of the House,* Q 121.
33 Possibility of applying Faber's air conditioning system to Lords chamber discussed in 1945: Report by the Joint Select Committee of the House of Lords and of the House of Commons appointed to enquire into the accommodation in the Palace of Westminster (HC 64–1 1944–45).
34 T. A. Hoyland, House of Lords Chamber ventilation and heating, 13 August 1959 (NA: Work 11/588, nr. 17).
35 House of Lords Offices, House of Lords, *Hansard's Parliamentary Debates: The Official Report (*9 December 1965 vol 271 cols. 399–402 399); Air-conditioning of House of Lords chamber, House of Lords, *Hansard's Parliamentary Debates: The Official Report (*21 November 1966 vol 278 cols. 7–8).
36 Oral Statement given by J. Percy, 27 July 1869, Select Committee on the Office of the Clerk of Parliaments (HL 1869), Q91–101 (Parliamentary Archives: GB–061).
37 Letter from Percy to Earl Granville, 19 July 1878, in 'Question, observation', *Hansard*, HL Deb 22 July 1878, vol 241 cc2018–21 2018.
38 E.g. 'Construction of the House, Resolution, House of Lords, *Hansard's Parliamentary Debates: The Official Report (*16 July 1869 vol 198 cols. 4–13 4).
39 Letter from Perott to Mr Fletcher, 20 April 1943 (National Archives: Work 11/403).
40 Question Observation, House of Lords, *Hansard's Parliamentary Debates: The Official Report (*22 July 1878 vol 241 cols. 2018–21) A sophisticated operational regime regulating the opening and closing of windows was also adopted in the House of Commons, which has been discussed in detail in, Henrik Schoenefeldt, The House of Commons: a precedent for post-occupancy evaluation, Building Research & Information, 47 6 (2018), pp. 635–665.
41 Letter from Secretary of the Office of Works to the Lord Chamberlain, 6 July 1911 (National Archives: Work 11/124).
42 Letter from Arthur Patey to Chief engineer, 19 October 1911 (National Archives: Work 11/124).
43 Minute to Secretary of Office of Works , 25 October 1911 (National Archives: Work 11/124).
44 Patey, A, Proposed plenum warming and ventilation of the House of Peers, 15 April 1912 (National Archives; Work 11/124).
45 Patey, A. House of Lords Improved Ventilation, 29 April 1912 (National Archives: Work 11/124, unnumbered item).
46 Thomas Bedford,*The Warmth factor in comfort at work* (Industrial Health Research Board: Report no. 76, 1936).
47 Ventilation of the House of Lords, Minutes of a meeting DSRI Advisory Council, September 1935 (National Archives: DSIR 3/20); J. A. Macintyre, Report: 'Houses of Parliament – Lords Chamber', 18 December 1935 (National Archives: Work 11 357).
48 Two sets of charts of climate data: 'Daily maximum and minimum Temperature and humidity, House of Lords, near Bar, 1 October 1835 to 30 September 1936 (National Archives: AE. 2670/4); Daily maximum and minimum Temperature and humidity, House of Lords, near Throne', 1 October 1835 to 30 September 1936 (National Archives: AE. 2670/40).
49 Report House of Lords – Ventilation, Bedford, January 1936. (NA: Work 11 358).
50 Letter from Bedford to Chief Engineer, 10 January 1936 (National Archives: Work 11/358).

51 Letter from Henry Radeley to Patrick Duff, 19 February 1936 [National Archives: Work 11/358]; Letter from Patrick Duff to Henry Radeley, 31 January 1936 [National Archives: Work 11/358]; Letter from Henry Radeley to Patrick Duff, 3 February 1936 (National Archives: Work 11/358).

52 Letter from Bedford to Chief Engineer, 18 November 1936 (National Archives: Work 11/358).

53 Ibid.

54 Sample record of hot-wire anemometer observations, 17 August 1936. (National Archives: Work 11/358); Recordings of velocities, 10 July 1936, enclosed in letter from Thomas Bedford to Chief Engineer of Office of Works, 18 November 1936. [National Archives: Work 11/358].

55 Letter from Thomas Bedford to Chief Engineer of Office of Works, 18 November 1936. (National Archives: Work 11/358).

56 Sample record of hot-wire anemometer observations, 17 August 1936. (National Archives: Work 11/358).

Note: It is to be noted, however, that these were still very low rates from a physiological perspective. According to modern standards velocities of 50 feet per minute (15 metres) or less are unnoticeable, only when above 50 and up to 100 feet per minute (30 metres) produce noticeable yet pleasant currents. Above 100 they become constantly noticeable, but only above 200 feet (60 metres) do they become a nuisance.

57 Bedford T., Notes on questions relating to humidity with special reference to the ventilation of the Houses of Parliament, January 1938 (National Archives: DSIR 3/20).

58 House of Lords ventilation – Report by Board's chief engineer (Mr Macintyre), 7 December 1936 (National Archives: Work 11/358).

59 Committee on Heating and Ventilation problems, Minutes of meeting, 16 February 1937 (National Archives: DSIR 3/20).

60 Report from N. Sizer, 11 July 1950 (National Archives: Work 11/525).

'Chamber of the House of Lords: Air conditioning', 13 December 1955 (National Archives: Work 11/588, nr. 3); H.T. Denbon, Report on House of Lords heating and ventilation system, 30 March 1960 (National Archives: Work 11/588, nr. 21–22).

61 Report on House of Lords proposed Air conditioning, 23 April 1965 (National Archives: Work 11/588, nr. 144).

62 Ventilation, House of Commons, *Hansard's Parliamentary Debates: The Official Report (*3 April 1913 vol 51 cols. 536).

63 Houses of Parliament Buildings, House of Commons, *Hansard's Parliamentary Debates: The Official Report (*13 June 1913 vol 53 cols. 1975–83).

64 House of Commons Ventilation, House of Commons, *Hansard's Parliamentary Debates: The Official Report (*10 July 1913 vol 55 cols. 749)

65 Report from the Select Committee on House of Commons Ventilation (HC 1913, 290).

66 Note: His theories were disseminated internationally through a series of influential publications, which included *The Influence of the atmosphere on our health and comfort in confined and crowded places,* a paper read at the Smithsonian Institution in 1913, and the book *Health and Environment,* published in 1925, which discussed his work in the House of Commons. Leonard Hill, Martin Flack, James McIntosh, R. A. Rowlands, H. B. Walker, *The Influence of the atmosphere on our health and comfort in confined and crowded places* (Washington: Smithsonian Institution, 1913).

67 DSIR Advisory Council,"Report on development of research since 1913," May 1926 (National Archives, DSIR 3/20).

68 "Monotony of air and speech," The Times, 27 March 1914: 12. "Monotony in the commons atmosphere," The Times, 1 May 1914: 5.
69 Ventilation in the Commons, Citizen, 28 February 1921 p. 3; Letter from Baines to LIoyd, 10 April 1923 (National Archives: DSIR 3/20).
70 Frank Baines, Memorandum, 18 November 1924 (HC 1931 Cmd. 3871), p. 24; Monotony in the Commons atmosphere, *The Times*, 1 May 1914, p. 5.
71 Report from the Select Committee on House of Commons (ventilation), together with the proceedings of the committee (HC 1914 445).
72 Ibid.
73 'Monotony of air and speech, *The Times*, 27 March 1914, p. 12.
74 Report from the Select Committee on House of Commons (ventilation), together with the proceedings of the committee (HC 1914 445).
75 'Making a crowd', *The Times*, 25 July 1914, p. 8.
76 DSRI Inter Departmental Conference, Extract from Minutes of Meeting held on 24 November 1925 (NA: DSIR 3/20).
77 House of Commons (Ventilation) House of Commons, *Hansard's Parliamentary Debates: The Official Report (*24 February 1920 vol 125 cols. 1497)
78 House of Commons (Ventilation) House of Commons, *Hansard's Parliamentary Debates: The Official Report (*27 April 1920 vol 128 cols. 1041 1041); T E Stanton, Experiments on wind pressure, National Physical Laboratory Collected Researches, vol 5, 1909, pp. 167–202 (Archive of Institution of Mechanical Engineers, IMS 74/3/3).
79 Air currents in the House of Commons, *Engineering*, 7 July 1922, p. 8; DSRI Report on development of research since 1913, May 1926 (National Archives: DSIR 3/20).
80 Memorandum by Sir Frank Baines, dated 18 November 1924, regarding the ventilation of the chamber of the House of Commons, (London: HMSO, 1931), plate 2] Note: The DSIR was a government body established to maintain strategic oversight over the allocation of research funding, and to promote links between research and industry. It overtook the management of several existing research facilities, such as the National Physical Laboratory, and also established several new laboratories including the Building Research Station (BRS) in 1921.
81 DSIR and MRC – Inter–departmental Committee on terminology in heating and ventilation research, Minutes of special session during the third meeting, 5 November 1931 (National Archives: DSIR 3/20).
82 Letter from to Office of Works to DSRI, 5 August 1931 (National Archives: DSIR 3/20).
83 House of Commons (Ventilation) House of Commons, Hansard's Parliamentary Debates: The Official Report (29 July 1931 vol 255 cols. 2278).
84 House of Commons (ventilation), House of Commons, *Hansard's Parliamentary Debates: The Official Report (*19 July 1937 vol 326 cols. 1771–2); House of Commons (ventilation), House of Commons, *Hansard's Parliamentary Debates: The Official Report (*14 June 1937 vol 325 cols 18–9).
85 'Note by the First Commissioner of Works on the Ventilation of the House of Commons', 16 July 1936 (National Archives: DSIR 3/20).
86 ; Letter from Office of Works to MRC, 8 July 1936; 'Joint committee on Research in Heating and Ventilation Minute for Meeting held on 29 October 1936; Note by the First Commissioner of Works on the Ventilation of the House of Commons, 16 July 1936 (all in National Archives: DSIR 3/20).
87 House of Commons Ventilation – Experiment with some electric panels, *The Times*, 30 July 1936, p. 8.
88 Memoranda 'Houses of Parliament heating, 30 October 1936, National Archives: DSIR 3/20.

89 'Report by Mr. A Bailey and A. Dufton on visits paid to the House of Commons with Mr. Macintyre and Dr. T. Bedford, November 1937' (National Archives: DSIR 3/20).

90 Ibid.

91 Letter from Chief Engineer to Secretary, 15 July 1937 (National Archives: Work 11 357).

92 House of Commons (ventilation), 14 June 1937 Debate; House of Commons (ventilation) 19 July 1937, debate, both in (National Archives: DSIR 3/20); Alfred C. Bossom, 'Changes in the House, Letter to the editor', *The Times*, 9 November 1937.

93 A. Bossom Ventilation in the House, Letter to the editor, *The Times*, 15 November 1937.

94 Letter from Office of Works to DSIR, 19 July 1937 (National Archives: DSIR 3/20).

95 "House of Commons Chamber, House of Commons, *Hansard's Parliamentary Debates: The Official Report (*3 December 1941 vol 376 cols. 1120–1 1120).

96 House of Commons Chamber, House of Commons, *Hansard's Parliamentary Debates: The Official Report (*5 February 1942, Hansard, vol 377 cc1277–8 1278).

97 House of Commons rebuilding, House of Commons, *Hansard's Parliamentary Debates: The Official Report (*28 October 1943 vol 393 cols. 403–73 403).

98 Palace of Westminster, House of Commons, *Hansard's Parliamentary Debates: The Official Report (*4 November 1943 vol 393 cols. 855–855).

99 'Ventilation set' of drawings produced by Oscar Faber in collaboration with Gilbert and Adrian Giles Scott, which is held at the Parliamentary Archives [Ark/Pro/Work11/425]: Key Plan, Oscar Faber, July 1944, DWG NO 4309/1; Plan of Lower ground floor plant room, Oscar Faber, July 1944, DWG NO 4309/2; Plan of ground floor mezzanine, Oscar Faber, Giles Gilbert Scott, Adrian Gilbert Scott, July 1944, DWG NO 4309/5 (Parliamentary Archives); Plan of principal floor, Oscar Faber, July 1944, DWG NO 4309/6; Plan of principal floor mezzanine, Oscar Faber, July 1944, DWG NO 4309/7; Plan of first floor, Oscar Faber, July 1944, DWG NO 4309/8; Plan of second floor, Oscar Faber, July 1944, DWG NO 4309/9; Plan of third floor, Oscar Faber, July 1944, DWG NO 4309/10; Cross-section looking north, Oscar Faber, July 1944, DWG NO 4309/11; Longitudinal Section looking west, Oscar Faber, July 1944, DWG NO 4309/12. Detailed working drawings for the final scheme completed in 1950 are held in the Strategic Estate Archives.

100 Oscar Faber, Abbreviated report on the heating and ventilation of the House of Commons, to the chairman of the Select Committee on House of Commons (Rebuilding), based on full report dated July 1944, pp. 16–17.

101 Oscar Faber, The Construction of the New House of Commons, *Engineering*, May 1949, p. 221–47.

102 Oscar Faber & John Kell, The heating and air-conditioning of the House of Commons, *Journal of the Institution of Heating and Ventilating Engineers*, December 1951, p. 14.

103 Henrik Schoenefeldt (2019) The House of Commons: a precedent for post-occupancy evaluation, Building Research & Information, 47:6, pp. 635–665, DOI: 10.1080/09613218.2019.1547547.

104 Henrik Schoenefeldt (2020) Delivering occupant satisfaction in the House of Commons, 1950–2019, *Building & Cities* (1 – 1), pp. 141–163.

Reflections on the character of the 19th-century environmental tradition

> *along with those means to defense and seclusion which they naturally present, the great and primary object of architecture is to afford the power of sustaining an artificial atmosphere, such as the constitution under each variety of local circumstances may required.*[1]
>
> David Boswell Reid, 1844.

This book has retraced and critically examined the development over 17 years of David Boswell Reid's scheme for the ventilation and climate control system of the Palace of Westminster. This research has afforded insights into the technical, managerial, environmental and political factors that shaped its final design. It has also illuminated the working practices in the new Palace that were used to both develop its design, and evaluate and improve its performance. These practices are important as they show the character of what might be described as a 19th-century tradition of environmental design in architecture. The design of the ventilation system was influenced by two distinct traditions prevalent in the mid-19th century: that of mechanical engineering and that of the natural sciences. Whilst the former exploited steam technology to address the ventilation problems mechanically, the latter followed a simpler, less technological approach using natural ventilation. The Palace of Westminster can therefore be viewed as a precursor of both the modern mechanically-serviced building – with its highly energy-intensive mode of environmental control– and low-energy building designs. Reid was a proponent of the school of natural ventilation, and he drew largely on his knowledge and working practices as a scientist to develop his plans. However, as a specialist with limited experience of architecture and engineering, he had to engage in cross-disciplinary collaborations to realise his objectives. It also relied on defining a relationship between the norms, values and practices (i.e. habitus) of the natural scientist and that of the engineer and architect.

Fluid boundaries in an emerging cross-disciplinary field

Whilst Reid was unable to fulfil the role of a mechanical engineer, he provided Charles Barry's architectural team with the specific skills and knowledge required to address problems of ventilation. In addition to providing an understanding of natural ventilation, Reid also contributed to the non-technical aspects of ventilation, such as indoor air quality, health, thermal comfort and the experience and behaviour of the Houses of Parliament's occupants. This suggests that by the middle of the 19th century 'environmental design' had already begun to be understood as a larger, cross-disciplinary field in which physiological, social and environmental factors were addressed in tandem with the more immediate problems of mechanical engineering. The development of the ventilation systems in the Houses was a complex design problem that could only be resolved through the confluence of multiple disciplines. The skills, knowledge and unique perspectives provided by the engineers and architects of Barry's team complemented, rather than conflicted with, those offered by Reid and other scientists. One of the main challenges, however, was achieving a joint understanding of each discipline's boundaries and commonalities. As building science in the mid-19th century was not yet a formalised discipline, Barry viewed Reid's area of expertise as distinct from those of engineers and architects. The difficulty that both men experienced in collaborating with one another was the result of their joint failure to explicitly recognise that building science and mechanical engineering are distinct, yet interdependent, aspects of environmental design.

A distinctive feature of Reid's investigations was his focus on how the occupants of the new Palace experienced it in use as a way to assess how it was performing. Reid used the temporary Houses of Parliament to test and refine the ventilation, climate control and lighting arrangements over a period of 15 years. The direct involvement of MPs and Lords in this process yielded important insights into the factors that affect thermal comfort in a shared environment. The temporary Houses acted as models in which new technologies and the socio-technical processes Reid established could be tested and improved before they were applied to the design of the very real Palace of Westminster. Along with the physical infrastructure, these processes were Reid's most significant and lasting contribution to the Palace of Westminster: they were observed and refined throughout the operational history of the House of Commons from 1852 until its destruction in 1941, and influenced the operational practices adopted for the new chamber built after World War II.[2]

Reid's and Barry's responsibilities for the performance of the Houses of Parliament continued after the completion of the building and its handover to its occupants when they agreed to supervise and fine-tune their day-to-day operation. The development of the system was characterised by overlapping stages of design, post-occupancy evaluation and facilities management, and as such should be interpreted as an example of evidence-based practice in architecture. Although the concept of this is a modern one, the working methods used in the design of the Palace's ventilation system shared many of the characteristics associated with modern approaches, including evaluating the performance of a building by examining its post-occupancy history. Although there are publications that make reference to the history of modern post-occupancy evaluation (POE) – for example, Bordass and Leaman (2001) and Preiser and Hardy (2015) – they start their accounts in the relatively recent period of the 1960s.[3] The history of the ventilation of the Houses of Parliament shows, to the contrary, that it has been a longstanding practice in architecture. Indeed, the studies conducted in the temporary Houses of Parliament, and later in the new Palace of Westminster, had all the core features of modern POEs, such as environmental monitoring, air flow tests and user satisfaction.

One point of difference between the 19th-century practices observed in the Palace of Westminster and those more generally practised in the 21st century is that the process of assessing and improving the new Palace's performance was ongoing throughout its operational life, whereas modern POEs are typically confined to a period of two to three years. The two mechanisms used to gain knowledge about the performance of the new Palace straddled the boundary between what is currently considered the domain of POEs and the domain of facilities management: first, large formal inquiries led predominantly by either ad hoc Select Committees or the various Offices responsible for overseeing its procurement; and second, continual monitoring and assessment of its performance by operational staff as part of its routine operational and maintenance regimes.

From Reid's time onwards, the Houses of Parliament was a site of environmental experimentation enshrining an ongoing learning process that extended over several generations. Although formal studies were undertaken in different parts of the Palace, the main focus was on the debating chambers, with the House of Commons subject to the largest number of investigations. The participation of MPs, Lords and Parliamentary staff in evaluating the Palace's performance facilitated a wider process of organisational learning that encompassed the perspectives of multiple actors including those who operated, assessed and used the Palace buildings. The 'institutional memory' of the Houses of Parliament– meticulously documented in records kept by the staff of the Office of Works and the two Houses – testify to the continual process of assessing its performance and adaptation throughout its operational life. This was informed by practical experience and significant advances in technology and scientific research. For the House of Commons, this era of gradual refinement and experimentation lasted for 89 years, which itself was preceded by 15 years of tests using the temporary Houses of Commons. This large body of institutional memory was also used by MPs and engineers in the 1940s to inform the design for the House of Commons debating chamber as it is today. The House of Commons should therefore be understand as a physical manifestation of this longstanding history of environmental inquiry.

The engagement of MPs and the Lords in questions of environmental design

The commissioning of the various investigations into the ventilation of the Palace was largely reactive rather than planned. Most inquiries were led by Select Committees appointed by the House of Commons or the House of Lords as a result of their dissatisfaction with the internal environment of the debating chambers. The most prominent problem – which ultimately led both Houses to decommission Reid's and Barry's original systems – was to create an environment that was agreeable to the users of the chambers: the MPs and the Lords themselves. As shown in this book, they – as building users – were the driving force behind the performance evaluations and the alterations ultimately made to the ventilation system. The MPs and Lords, however, were (and still are) unusual occupants: in contrast to the typical user of more ordinary buildings (e.g. commercial office buildings or university faculties), they had an unusual high level of agency with the authority and resources to (re)appraise and change their own working environment based on their own experiences and expectations. Between 1839 and 1854 both Houses had already appointed several ad hoc Select Committees to conduct their own inquiries into the problems of ventilation, climate control and lighting. Although the Office of Works and various specialists were engaged to provide advice and conduct formal studies of the existing systems or trials of alternative technologies, it was the politicians themselves who lead the inquiries and made the important decisions: they determined the focus of the studies, interpreted the findings and made recommendations for any technical changes they believed were needed. Consequently, the process of evaluating and refining the ventilation system was not under the control of the design team or the operational staff. Neither Reid, Barry nor Meeson had any substantive influence over the design or outcome of these inquiries. For example, the Lords decided to limit Reid's involvement in 1846, whilst both Houses sanctioned alteration to Reid and Barry's original system in 1854. In the early 20th century, MPs motivated the inquiries into remodelling the 19th-century system, building on new research into the physiology of thermal comfort and exploiting new mechanical air-conditioning technology. This active engagement with environmental design issues affecting the modern debating chamber also continued after World War II and is examined in more detail elsewhere by the author.[4] The post-war history emphasises the longstanding culture of engagement by MPs and Lords with the design of the Palace of Westminster, one that has been primarily concerned with the environmental rather than architectural aspects. This is important as it shows that environmental control has not been seen as a marginal issue by its occupants.

In the context of the forthcoming Palace of Westminster Restoration and Renewal Programme it will be crucial to revisit the history of engagement with these issues as it will be a new and important phase in the building's history. It will provide both Houses with a rare opportunity to reflect on this past tradition of engagement and how it could offer lessons for a new, more structured and planned framework of participation by MPs and Lords.

Reconciling utilitarian science and architectural historicism

Reid advocated a utilitarian approach to design that was primarily concerned with the environmental function of buildings. He was an outspoken critic of the architectural practice and education of his time, and considered the scientific method as a potential alternative to historicism as the foundation of building design.[5] This was based on two key assumptions: first, that the empirical method can be used to answer questions about architectural form; second, that the functional requirements of ventilation, lighting and acoustics should be the primary determinants of form. The temporary Houses of Parliament, and Reid's earlier experimental structures in Edinburgh, might thus be understood as uncompromising expressions of this largely utilitarian design philosophy. In the design of the new Palace of Westminster, Reid and Barry were faced with the challenge of integrating the (previously divided) architectural and technical approaches.

The use of the historicist and the utilitarian approaches to design, however, were not mutually exclusive. Indeed Reid relied on Barry to create new architectural features to realise his proposals. The new Palace underwent a gradual architectural transformation due to the ever-evolving design of the ventilation system. Throughout the design process, the number, dimensions, form and locations of the architectural towers and their respective functions within the ventilation system were the subject of extensive negotiations amongst stakeholders. Although the form, height and position of the stacks largely followed the requirements of the ventilation system, Barry resorted to traditional Gothic details and tower forms for their architectural treatment. The design of the stacks illuminates the symbiotic relationship between the functional requirements of natural ventilation, mechanical technology and the required Gothic architectural precedents. The stepped cross-section of Gothic spires allowed air intakes and smoke outlets to be accommodated at different altitudes whilst departing from the accepted proportions of their Medieval counterparts. Their extreme height and slenderness was necessary to achieve a sufficiently strong draught and to avoid the emission of smoke and vitiated air at a low altitude.

The relationship seen here between function and architectural form thus exemplified a departure from the purely formalist use of the Gothic style and picturesque principles of composition criticised by the architect Augustus Welby Pugin in his book *The True Principles of Pointed or Christian Architecture*.[6] The famous picturesque composition of the Palace's skyline – with its array of turrets, towers and spires – was not driven by a grand architectural vision but was an accidental outcome resulting from the need to meet the functional requirements of the ventilation system. Concerned about the architectural impact of these new functional features, Barry's practice transformed them into architectural features that harmonised with the overarching Gothic vision that shaped the design of the Palace. This process also showed that the picturesque compositional principles associated with the Gothic style were sufficiently flexible to architecturally assimilate both Reid's original ventilation scheme with its large Central Tower and the array of towers and turrets subsequently added after 1846. By contrast, the shafts of the ventilation system that Reid designed for St George's Hall in Liverpool were partially disguised behind a false façade in order to not interfere with the symmetrical composition of its Neo-classical design.[7]

As argued in this book, in the case of the Palace of Westminster, the close collaboration between architects, scientists and engineers enabled a new relationship between architectural historicism and modern technology to be established that reconciled the compositional principles of the Gothic revival with the rational scientific paradigm followed by Reid. Unfortunately, this was not the result of strategic planning, which led to a long process of intensive and, at times, difficult negotiations amongst stakeholders. Notwithstanding these difficulties, the final design enshrined one approach to reconciling form and technology that was to have a strong influence on the development of 19th-century architecture. It became a distinctive feature of many other public buildings, and of various styles, from the 1860s until the early 20th century. Throughout this period stack ventilation became a widely-adopted method for the ventilation of large public buildings across northern and eastern Europe. Notable 19th-century examples are the Royal Courts of Justice, St Thomas' Hospital, the Natural History Museum and the Victoria and Albert Museum, all in London, and Glasgow University. Examples in the 20th century include the Neo-baroque design of the Reichstag in Berlin, Germany, and the Neo-gothic design of the Hungarian Parliament Building in Budapest.[8] These and many other buildings exemplify the culmination of a wider environmentally-driven transformation of architecture that Reid predicted in his writings. Indeed, in his 1844 book *Illustrations of the Theory and Practice of Ventilation,* Reid wrote that this new architecture would be characterised by roofs with monumental 'towers, spires, pillars and minarets'.[9] The bold landscapes of towers would prove to be the outward expression of the requirements of 19th-century technology. The revival of the principles of natural ventilation over the past 30 years had also led to a modern reincarnation of the ventilation tower as a functional, yet expressive, architectural feature of buildings.

Uncovering forgotten legacies

In hindsight, Reid's long-term reputation as an authority in the field of ventilation was largely founded on the success of his experimental system in the temporary House of Commons. By contrast, his work on the design of the new Palace of Westminster was overshadowed by a narrative of professional failure. The temporary House of Commons continued to feature in key textbooks on building services engineering in Europe and the USA throughout the second half of the 19th century.[10] As a result of its success he received a larger number of commissions, designing the ventilation systems for new courtrooms in the Old Bailey, the Chapel Royal in St James' Palace, the Royal Pavilion in Brighton and for the State Rooms in Buckingham Palace. He also produced schemes for the ventilation of several prisons (e.g. Edinburgh General Prison and General Prison Perth), hospitals (e.g. General Lying-in Hospital in Lambeth and an unrealised proposal for a Military Lunatic Asylum which was to be built in Kent),[11] schools (e.g. Manchester Union School) as well as the ventilation of ships.[12]

Reid finally realised his idea of a fully-integrated ventilation and climate-controlled system in St George's Hall in Liverpool.[13] Although his designs for the Palace of Westminster and the House of Commons received limited, and largely negative, coverage in the 19th-century press and technical literature, it is hoped that this book in examining his important, and until this point largely forgotten, legacy restores some of his reputation as a pioneering 19th-century designer.

Notes

1 Reid, David Boswell, *Illustrations of the theory and practice of ventilation*, (London: Longman, Brown, Green & Longmans, 1844), p. 71.
2 A detailed study on Faber's original social technical practices Henrik Schoenefeldt (2020) Delivering occupant satisfaction in the House of Commons, 1950–2019, *Building & Cities* (1 – 1), pp. 141–163.
3 Leaman A and Bordass B, 'Assessing building performance in use: the Probe occupant surveys and their implications,' *Building Research and Information*, 29 (2001) 129–143; Preiser, Wolfgang & Hardy, Andrea, Historic Review of Building Performance Evaluation, in Architecture Beyond Criticism (London: Routledge, 2015), pp. 146–59.
4 Henrik Schoenefeldt (2020) Schoenefeldt, H. (2020). Delivery of occupant satisfaction in the House of Commons, 1950–2019. Buildings and Cities, 1(1), pp. 141–163.
5 Select Committees on Foundation Schools, Reports from the Select Committees on Foundation Schools and Education in Ireland (HC 1836, 630); Kennedy, David, Dr. D. B. Reid and the Teaching of Chemistry, *Studies: An Irish Quarterly Review*, 31(123), pp. 343–350.
6 Pugin, A. W, *The True Principles of Pointed or Christian Architecture and an Apology for The Revival of Christian Architecture* (London: John Weale, 1841).
7 Mackenzie, William, On the mechanical ventilation and warming of St. George's Hall, Liverpool, *Proceedings of the Institution of Mechanical Engineers*, 1863, pp. 194–208.
8 First Report of the Select Committee on House of Commons, 25 June 1903, HC 227 1903, Q502–558, Cook, Jeffrey; Hinchcliffe, Tanis, Delivering the well-tempered institution of 1873, *Architectural Research Quarterly*, 2 – 1 (September 1996), pp. 66–75; Cook, Jeffrey; Hinchcliffe, Tanis, Designing the well-tempered institution of 1873, *Architectural Research Quarterly*, 1–2 (December 1995), pp 70–78.

9 *Illustrations,*1844, pp. 75–76.
10 John S. Billings, *Ventilation and Heating* (New York: Engineering Record, 1893); Robert Richie, *A treatise on ventilation* (London, 1862), pp. 177–8, 190–1; Robert Tomlinson, *A rudimentary treatise on warming and ventilation* (London, 1858).
11 Statement by D.B. Reid, 24 March 1846, HC 574 1846, *Third report from the Select Committee on Westminster Bridge and new palace*, Q1036–1075; Oral Statement by D.B. Reid, 19 March 1846, HC 574 1846, *Third report*, Q908–10.
12 He developed ventilation system for several ships, which included the HMS Minden, Victoria and Albert, vessels for the 1841 Niger Expedition: Edward J Gillin, Science on the Niger: Ventilation and Tropical Disease during the 1841 Niger Expedition, Social History of Medicine, Volume 31, Issue 3, August 2018, Pages 605–626.
13 Sturrock, N. St George's Hall, Liverpool – A major refurbishment and a new heritage centre for the world's first air-conditioned building in Manfredi, C. (ed.) Architettura e empianti termici (Turin: Umberto Allemandi, 2017), pp. 227–243.

Index

For Product Safety Concerns and Information please contact our EU representative GPSR@taylorandfrancis.com Taylor & Francis Verlag GmbH, Kaufingerstraße 24, 80331 München, Germany

Batch number: 10722122

Printed by Printforce, the Netherlands